Marine Genetics

Developments in Hydrobiology 144

Series editor
H. J. Dumont

Marine Genetics

Edited by

Antonio M. Solé-Cava, Claudia A.M. Russo & John P. Thorpe

Reprinted from Hydrobiologia, volume 420 (2000)

Kluwer Academic Publishers
Dordrecht / Boston / London

Library of Congress Cataloging-in-Publication Data

A C.I.P. Catalogue record for this book is available from the Library of Congress.

ISBN 978-0-7923-6147-3

Published by Kluwer Academic Publishers,
P.O. Box 17, 3300 AA Dordrecht, The Netherlands

Sold and distributed in North, Central and South America
by Kluwer Academic Publishers,
101 Philip Drive, Norwell, MA 02061, U.S.A.

In all other countries, sold and distributed
by Kluwer Academic Publishers Group,
P.O. Box 322, 3300 AH Dordrecht, The Netherlands

Printed on acid-free paper

Contents

Preface vii

Acknowledgements viii

Group photograph ix

Introductory Overview
Stability, equilibrium and molecular aspects of conservation in marine species
J.C. Avise xi–xii

The detection of spatial variation in widespread marine species: methods and bias in the analysis of population structure in the crown of thorns starfish (Echinodermata: Asteroidea)
J.A.H. Benzie 1–14

Molecular systematics of sponges (Porifera)
C. Borchiellini, C. Chombard, B. Lafay, N. Boury-Esnault 15–27

Genetics and marine pollution
D.R. Dixon, J.T. Wilson 29–43

Biotechnology and fish culture
F. Foresti 45–47

Genetic differentiation in Chilean hake *Merluccius gayi gayi* (Pisces: Merlucciidae)
R. Galleguillos, L. Troncoso, C. Oyarzún, M. Astorga, M. Peñaloza 49–54

An overview of marine fish cytogenetics
P.M. Galetti Jr., C.T. Aguilar, W.F. Molina 55–62

Genetics of marine bioinvasions
B.S. Holland 63–71

Molecular genetic analyses of species boundaries in the sea
N. Knowlton 73–90

Gene transfer technology in aquaculture
J.A. Levy, L.F. Marins, A. Sanchez 91–94

Seven *Hox* gene sequences from the asterinid starfish *Patiriella exigua* (Echinodermata: Asteroidea)
S.C. Long, V.B. Morris, M. Byrne 95–98

The karyotype of *Nodipecten nodosus* (Bivalvia: Pectinidae)
E. Pauls, P.R.A.M. Affonso 99–102

Taxonomic status of the '*Patella miniata*' species complex (Mollusca: Gastropoda) in southern Africa
T.M. Ridgway, G.M. Branch, B.A. Stewart, A.N. Hodgson 103–118

Techniques and statistical data analysis in molecular population genetics
E.P. Silva, C.A.M. Russo 119–135

DNA tests of neutral theory: applications in marine genetics
D.O.F. Skibinski 137–152

Molecular genetics and the management and conservation of marine organisms
N.A. Sweijd, R.C.K. Bowie, B.S. Evans, A.L. Lopata 153–164

Exploited marine invertebrates: genetics and fisheries
J.P. Thorpe, A.M. Solé-Cava, P.C. Watts 165–184

Effects of copper on the fitness of the Chilean scallop *Argopecten purpuratus* (Mollusca: Bivalvia)
L. Troncoso, R. Galleguillos, A. Larrain 185–189

Genetics in fisheries management
R.D. Ward 191–201

Genetic population structure of the swimming crab *Callinectes danae* (Crustacea: Decapoda) in southern Brazil
L.I. Weber, J.A. Levy 203–210

Genetic divergence and larval dispersal in two spider crabs (Crustacea: Decapoda)
L.I. Weber, R.G. Hartnoll, J.P. Thorpe 211–219

Preface

Marine genetics is a fascinating subject, which integrates and contributes to various related fields such as oceanography, fisheries biology, aquaculture, conservation biology, molecular, evolutionary and population genetics, ecology, systematics and biotechnology. The field of marine genetics is expanding rapidly. For example in 1981 only 1.7% of the papers cited in the ASFA database used a genetic approach, but by 1998 that number had grown to about 10% of all cited papers. The use of genetic markers for the study of marine organisms has, thus, come of age. Its growth can be seen, for example, in Brazil, which has over 8000 km of coastline, from the large equatorial estuary of the Amazon, to the salt marshes and the rich temperate waters of the south, and has important on-going projects in aquaculture and the preservation of marine biodiversity. From the first published paper, in the early 1980s, Brazilian marine genetics has been growing steadily, and now there are eight laboratories working in that field, with about 15 established academics and many graduate students.

In the first international symposium to deal with the subject, in Venice-1977 ("Marine Organisms: Genetics, Ecology and Evolution"), marine genetics was in its infancy, and the main focus was on the presentation of empirical data on levels of allozyme gene variation and their possible correlation with environmental parameters and taxonomy. The XXIII European Marine Biology Symposium, in Swansea-1988 ("Reproduction, Genetics and Distributions of Marine Organisms") showed the maturing of the field: the articles (mostly using allozymes, but also a few using DNA), were more problem-oriented, and one of the key questions was the relationship between effective gene flow and dispersal capability of benthic organisms. The following symposium, in Bangor-1992 ("Genetics and Evolution of Aquatic Organisms") witnessed the growth of the application of genetics to the study of aquaculture, pollution and fisheries biology, and the increased use of DNA technology, although at this stage mostly for phylogenetic studies in fish.

1998 was the international year of the Oceans, and the idea of a workshop in Rio de Janeiro originated from the desire to bring together scientists working in diverse fields of marine genetics to review the state of the art and to discuss possible trends and the study of marine genetics in the next century. This workshop, held in September 1998, was a short event, three days, and consisted largely of lengthy review papers from invited speakers from many parts of the world, but there were also opportunities for a number of shorter papers to be presented, as well as an extensive number of posters. From the excellent papers presented, and from additional contributions from authors who could not come to the Workshop, but who nonetheless had contributed significantly to the field, we have produced this volume – called simply "Marine Genetics" – which will be the first volume exclusively dedicated to the subject, and which we hope will present a broad overview of our knowledge in the field at the turn of the century.

This volume contains about 20 papers contributed by 40+ authors and reflects the international character of the meeting with a wide range of papers in terms of organisms discussed and geography, as well as in terms of length and the area of marine genetics covered. We have made no attempt to arrange the papers other than alphabetically by first author. This mixture of (mainly invited) reviews and other papers covers so many areas that any effort to divide the book up into sections was always likely to be unsatisfactory. Thus we present a mixture of papers ranging from cytology to theoretical population genetics, from single species studies to reviews of the whole range of marine organisms. All have been extensively reviewed and revised and we regret that space and time limitations meant that some contributions were necessarily excluded. Nonetheless we are very grateful to all who submitted papers and all who attended or otherwise contributed to the workshop.

Rio de Janeiro
June, 1999

ANTONIO M. SOLÉ-CAVA
CLAUDIA A.M. RUSSO
JOHN P. THORPE

Acknowledgements

Thanks are due to very many people; too many to mention most of them by name. We are particularly grateful to Drs/Messrs/Misses C. Aguilar, R.T.N. Araújo, J. Gusmão, C.V.S. Lazoski, R. Schama and E.P. da Silva for their assistance with the organisation and running of the meeting and to Miss M.F.S. Cardoso and Dr B.H. Holland for considerable help with the editing and refereeing of the manuscripts. We acknowledge, with thanks, financial support from CNPq, FAPERJ, the Universidade Federal do Rio de Janeiro, the Universidade Federal Fluminense and SIGMA Chemical Company.

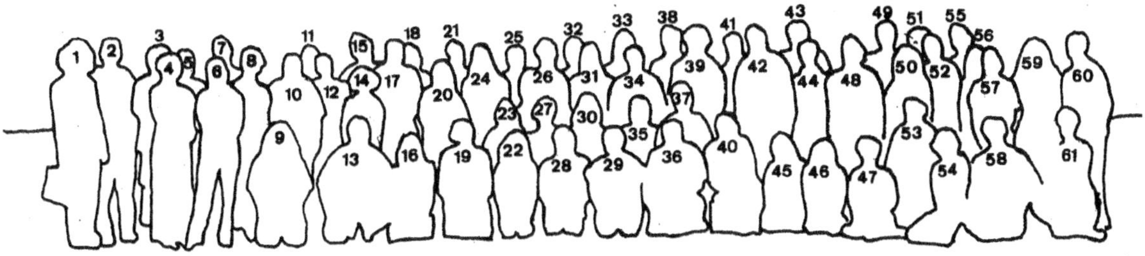

International Workshop on Marine Genetics – Rio 98.

1. Suzzane Williams, Panama
2. David Skibinski, U.K.
3. Edson da Silva, Brazil
4. Rejane Nobrega, Brazil
5. Rodrigo Maggioni, Brazil
6. Claudia Russo, Brazil
7. David Dixon, U.K.
8. Erika Pauls, Brazil
9. Rita Netto, Brazil
10. Jean-Marie Sevigny, Canada
11. Nancy Knowlton, U.S.A.
12. Gisele Lobo-Hajdu, Brazil
13. Brenden Holland, U.S.A.
14. Laura Weber, Brazil
15. João Mansur, Brazil
16. Beatriz Jerez, Argentina
17. Fausto Foresti, Brazil
18. Paulo Affonso, Brazil
19. Eliete Ballestrelos, Brazil
20. Menchie Ablan, Philippines
21. Miguel Couceiro, Brazil
22. Amanda Arnold, U.K.
23. Maria Totaro, Argentina
24. Fernanda Amaral, Brazil
25. Shane Sarver, U.S.A.
26. Wayne Florence, South Africa
27. Elisabeth Neves, Brazil
28. Lilian Troncoso, Chile
29. Nicole Boury-Esnault, France
30. Daniela Cabral, Brazil
31. Tyrone Ridgway, South Africa
32. Gustavo Solha, Brazil
33. Juliana Assumpção, Brazil
34. Pedro Galletti, Brazil
35. Patricia Freitas, Brazil
36. Armand Sanchez, Spain
37. Joanne, Murphy, U.K.
38. Fernando D'Incao, Brazil
39. Antonio Solé-Cava, Brazil
40. José Levy, Brazil
41. Ralf Kersanach, Panama
42. John Thorpe, Isle of Man, U.K.
43. John Avise, U.S.A.
44. Wagner Molina, Brazil
45. Jaqueline Gusmão, Brazil
46. Renata Schamma, Brazil
47. Misaki Takabayashi, Australia
48. Cecilia Aguilar, Brazil
49. Luciano Beheregaray, Australia
50. Margarete Correa, Brazil
51. Rodolfo Pettersen, Brazil
52. Anders Silva, Brazil
53. Terezinha Absher, Brazil
54. Sonia Andrade, Brazil
55. Ciro Oyarzun, Chile
56. Karla Yotoko, Brazil
57. Naiara Pinto, Brazil
58. Luiz Marins, Brazil
59. Suzanne Long, Australia
60. Cristiano Lazoski, Brazil
61. Fatima Cardoso, Brazil

Hydrobiologia **420**: xi–xii, 2000.
A. M. Solé-Cava, C. A. M. Russo & J. P. Thorpe (eds), Marine Genetics.

Introductory overview

Stability, equilibrium and molecular aspects of conservation in marine species

A major challenge in any genetic analysis of population structure is to gain a proper perspective on the relevant spatial and temporal scales that might be involved. For example, with respect to temporal considerations, a species might have restricted dispersal capabilities at most points in ecological time when we observe them, yet they might have broad ranges due to periodic, episodic, or accumulated movements over longer timescales, perhaps following major environmental changes or at other exceptional times.

In the marine environment, many species exhibit a phenomenon known as chaotic patchiness wherein there is great allele frequency variation on microspatial scales at any one time, yet also a temporal kaleidoscopic pattern in which allele frequencies can change quickly at a given locale. Thus, microspatial heterogeneity can be pronounced yet temporally variable such that if a species were assayed multiple times, different pictures of the particulars of population genetic structure might emerge. Many marine species have the potential for explosive reproduction by individuals such that cohorts of larvae, perhaps produced by relatively few breeding individuals, might be the lucky survivors that settle successfully and precipitate a large local change in allele frequency. In a few generations another cohort might come in, and so on. It is this kind of stochastic or chaotic patchiness, demographically driven, that might characterise many marine species. It is even possible that long-term decisions about where stocks exist could be misinformed when based on information gathered at any one slice in time. There is no obvious answer as to what should be done in this sort of situation, except to stress that ecologists and geneticists should at least be aware of the possibilities. It is important to try to come to grips with the spatial and the temporal scales over which population structure might exist. To accommodate temporal and spatial variation in allele frequencies, we need to devote effort to thinking of ways to couple ecological concepts with phenomena that take place episodically or over longer periods of time.

Another point often overlooked in genetic issues is that in most cases the biological world is unlikely to be at equilibrium between opposing evolutionary forces. Many of the traditional equilibrium models of population genetics involve unrealistic assumptions about the way real populations behave in the natural world. One consequence of this is that, despite talk about how the allozyme and DNA revolutions have revealed so much genetic variation in populations, the amount of variation in nearly all species is far less than predicted under neutrality theory if evolutionary effective population sizes approximate current-day population sizes. This statement applies with particular force to currently abundant species. Whether measured by allozymes or at the DNA level, actual levels of genetic variation usually are vastly lower than traditional neutrality expectations. From this perspective, genetic polymorphism is unexpectedly low (rather than high), and this finding is consistent with the possibility that evolutionary effective population sizes in most species are several orders of magnitude below those species' current standing crops. One good way to illustrate this is to consider, for example, some of the mega-abundant populations of particular zooplankton species that have been assayed genetically. These surely consist of many billions of individuals, and yet the coalescent depths in mitochondrial gene genealogies are consistent with evolutionary effective population sizes of the order of perhaps a hundred thousand or less. Similar arguments can be made for other currently abundant species. Historically, almost all species probably have fluctuated dramatically in population size, perhaps in association with speciation events or perhaps at other times in their recent evolutionary history. Species that are abundant today cannot have been abundant consistently through time or, they would have far more genetic variation and deeper coalescent depths than in fact they do.

This is just one line of argument that many natural populations are at non-equilibrium status. It is likely that populations of most species fluctuate dramatically in space and in time at least periodically, and by making field observations today we are merely capturing a snapshot view. Genetic analyses can help to add a temporal dimension to our understanding because the current genetic architecture of a species in large part reflects that species' demographic past as well. Thus, genetic analyses can be informative and influence how we think about the biological world, and in general they indicate that this world is probably in a tremendous state of dynamism. For example, diseases or any of a wide range of environmental impacts can wipe out large fractions of a species and thereby cause dramatic fluctuations in population size.

Overall, marine organisms present some unusual challenges and opportunities for genetic analysis. Marine species generally are less accessible than those on land or in freshwater, so behavioral studies and observations of natural history are more difficult. This opens many opportunities for genetic analyses to reveal otherwise hidden aspects of a species' natural history. Marine organisms offer an exceptionally wide range of dispersal and migratory capabilities, and the geographic ranges of individual species can be vast. Life histories often include potentially enormous fecundity, and the capability for sudden explosive population increases. A high proportion of marine species are also of concern with respect to conservation, with many being harvested commercially on a large scale, in some cases illegally. Thus population management may involve economic, social, jurisdictional or forensic concerns. Molecular biology can play a role in many of these areas, uncovering previously unknown aspects of behavior, natural history, and population demography pertinent to decisions on population management and conservation.

Department of Genetics,
University of Georgia,
Athens, GA 30602, U.S.A.
E-mail: avise@arches.uga.edu

J. C. AVISE

Hydrobiologia **420:** 1–14, 2000.
A. M. Solé-Cava, C. A. M. Russo & J. P. Thorpe (eds), Marine Genetics.

The detection of spatial variation in widespread marine species: methods and bias in the analysis of population structure in the crown of thorns starfish (Echinodermata: Asteroidea)*

J. A. H. Benzie
Australian Institute of Marine Science, PMB No 3, Townsville, M.C., Queensland 4810, Australia
E-mail: j.benzie@aims.gov.au

Key words: population genetics, evolution, allozymes, DNA, marine genetics, *Acanthaster planci*

Abstract

The basic assumptions that widespread marine species should show little spatial variation in genetic structure, given their high potential for dispersal on ocean currents, is being questioned. This has taken some time because there are few studies of widespread marine species over oceanic scales, few data sets that have the high density of sampling required for the detection of fine population structure, and there is little incentive to look further if initial analyses suggest the expected result. The interpretation of the population genetic structure of crown-of-thorns starfish (*Acanthaster planci*) has been found to vary considerably depending on the sample set included in the analyses and on the method of analysis used. Scatter plots of genetic distance or θ, and spatial autocorrelation approaches gave markedly different results ranging from no structure to isolation by distance. Only visual examination of maps of patterns of variation in allele variation first detected that crown-of-thorns starfish occupy large regions with little between population differentiation, but between which there are markedly higher levels of differentiation. These findings highlight the care required in interpreting population structure, particularly where there are few sample points. Many marine species may have population structures where sharp genetic disjunctions, not associated with any obvious environmental boundaries, separate regions of relative genetic homogeneity. Such population structures are very different from those traditionally assumed and are not yet understood. Further advances in understanding the genetic structure of marine species will demand an iterative approach where a greater number of samples are collected over particular regions identified by earlier interpretations.

Introduction

Widespread marine species have attracted attention because they are expected to show little spatial variation in genetic structure given their high potential for dispersal and the perceived lack of barriers to dispersal in the ocean. There are, of course, many marine species with life histories that are not conducive to widespread dispersal, and genetic studies of these have demonstrated restricted dispersal and considerable spatial variation in gene frequencies (Battaglia & Beardmore, 1978; Beaumont, 1994). However, species with a long larval life, and with a wide distribution across one or more oceans, have long been assumed to disperse throughout their range. Early work found little genetic distance among widely separated populations of giant clams (Campbell et al., 1975) and milkfish (Winans, 1980). This finding was consistent with extensive dispersal, and it was assumed that long-distance, contemporaneous dispersal occurred in these species, and that the means by which it was achieved was through dispersal of larvae by ocean currents.

New findings are beginning to question these assumptions. Recent taxonomic research has suggested that many marine species consist of a number of cryptic taxa (Knowlton, 1993). In some cases, each of these cryptic species may have restricted regional distributions, each abutting the other (Foltz et al., 1996). Sharp geographic boundaries have been found

* This is contribution number 964 from the Australian Institute of Marine Science

between populations with markedly different gene frequencies, but which appear to belong to the same species (Avise, 1994). In some cases, several species from diverse taxonomic groups demonstrate rapid shifts in gene frequencies in the same region, suggesting these areas represent major biogeographical boundaries.

The zones of rapid change in gene frequencies do not have any obvious geographical or oceanographic barriers to dispersal, and are often connected by ocean currents. However, the dominant role of ocean currents in influencing the dispersal of widespread marine species was not questioned because the zones of genetic discontinuity were considered special circumstances. The role of ocean currents has only been questioned by the discovery that the patterns of gene flow in giant clams in the Western Pacific are perpendicular to the major surface ocean currents (Benzie & Williams, 1995, 1997). Since then, Palumbi et al. (1997) reported that the patterns of genetic connectedness of sea urchin populations over the Pacific do not reflect ocean circulation patterns. For various reasons, these basic assumptions have remained unquestioned.

The aim of this paper is to illustrate the extent to which interpretation can be biased 1. by the modes of analysis used for even extensive data sets, let alone the limited number of populations that are usually available, and 2. by the expectation that little structure exists. This is done by reviewing the interpretations of population structure of crown-of-thorns starfish, *Acanthaster planci*, made over the last 10 years and by including some reanalysis of this, now relatively extensive, data set.

Methods

The data used in this review were obtained from published work on the crown-of-thorns starfish by Nash et al. (1988), Nishida & Lucas (1988), Benzie & Stoddart (1992a,b) and Benzie (1992, 1999a), with some reanalysis or extensions to the analyses reported in those papers. The combined data for *A. planci* cover much of the Indian and Pacific Oceans and most of the range of this species.

Population genetic statistics and genetic distances were generated using BIOSYS-1 (Swofford & Selander, 1981). *F*-statistic analysis was used to partition genetic variation into within population (F_{IS}) and between populations (θ) components using Weir & Cockerham's (1984) method which explicitly takes account of differences in sample size among populations. The significance of θ values was tested using χ^2, with χ^2 equal to $2N\theta\ (k-1)$ with d.f.=$(k-1)(s-1)$, where N is the total number of individuals sampled, k the number of alleles at the locus and s is the number of populations sampled (see formulae in Waples, 1987).

The average gene flow among populations was assessed from θ by estimating the average number of migrants exchanged between populations (N_em) using the relationship N_em=$[(1/\theta)-1]/4$. Alternative methods of calculating N_em using the frequency of rare alleles (Slatkin, 1985) were not applied to the present data set as most subsets to be analysed had far fewer (i.e. <5) than the recommended number (i.e. 20) of private alleles (Slatkin, 1985). Pairwise θ values, and Nei's (1978) genetic distance were plotted as a function of geographical separation (using the shortest distance between two points) to test whether the genetic structure reflected a pattern of isolation by distance. The significance of the regression between log-transformed θ and log-transformed km, values was tested using Mantel's normalized statistic, Z, (Mantel, 1967), calculated using NTSYS-PC (Rohlf, 1990) and 10 000 random permutations to calculate the significance. Negative θ values were assigned a value of 0.0001 to allow their inclusion in the analyses.

Moran's I and Geary's c were used to test for spatial autocorrelation of allele frequencies among samples, using the software program SAAP (Wartenberg, 1989). Where there is no geographic structure, the expected value of Moran's I is $-1/(N-1)$, where N is the number of localities. Moran's I is positive when the values of a variable at nearby localities are similar, and are negative when the values are dissimilar. Geary's c is 1.0 when values are randomly distributed, less than 1.0 when there is positive spatial autocorrelation and greater than 1.0 when there is negative autocorrelation. Autocorrelation between localities was tested using 5, 10 or 15 distance classes. In each case, the distance classes were defined to keep as even a number of comparisons as possible within each distance class, or the distance classes were equal. Analyses used spherical distances between latitude and longitude coordinates.

Results and discussion

First approaches

A general approach to the analysis of population structure is to first determine whether there is statistically significant heterogeneity among populations. This is an obvious and reasonable approach because, if there is no significant spatial structure in allele or genotype frequencies, there is little more that can be said about such a set of populations. No significant spatial differentiation suggests panmixis (i.e. random mating throughout the populations surveyed), or possibly that sample sizes were inadequate to distinguish any real differences that there might have been. The next step in the analysis is usually to determine whether there is any relationship of genetic differentiation to the geographical separation of populations. Plots of genetic differentiation of population pairs as a function of their geographical separation are often made in exploratory analyses, even if no statistically significant differentiation has been detected. This is because such plots might reveal some consistent trend that, while not significant, might suggest further sampling would be worthwhile, and that the lack of significance of spatial patterns was a result of inadequate sample size or the geographical range of sampling.

Detection of significant differences among populations

F-statistic analyses of *A. planci* allozyme frequencies demonstrated no significant differentiation among populations within most of the reef complexes studied (Table 1). Some differentiation was observed over larger distances. Some population sets were sampled where crown-of-thorns had formed large outbreak populations (Moran, 1986). Crown-of-thorns starfish can suddenly increase greatly in number and cause considerable damage to the corals upon which the starfish feeds. Where this phenomenon occurs in a reef complex, the population growth spreads from reef to reef as the higher population numbers lead to more effective fertilisation and greater production of larvae. Such outbreaks have occurred in many locations throughout the species range and both Japan and the Great Barrier Reef have been considerably affected. Outbreak populations from these regions show no differentiation over several hundred kilometres. This is despite the fact that significant differentiation has been detected among other populations as little as 40 km apart, albeit rarely (in Western Australia). Significant differentiation has also been detected between non-outbreak populations in the Great Barrier Reef sampled over similar scales to the outbreak populations (Nash et al., 1988; Benzie & Stoddart, 1992a; Benzie & Wakeford, 1997).

The results of simple *F*-statistic analysis demonstrated that there are significant differences among *A. planci* populations, despite the long larval life of this species (usually 2–3 weeks, but thought to extend in a teleplanic phase lasting several weeks under certain circumstances; Birkeland & Lucas, 1991). Only under special outbreak conditions, where there is dispersal of large numbers of larvae in short time periods (years to decades) among reefs, does there appear to be sufficient gene flow to homogenise gene frequencies among reefs over larger scales. However, the detection of significant spatial differentiation may either reflect the strong distinction of one or a few outliers, or stochastic variation resulting from the patchy recruitment of larvae from the plankton with no large scale structure, or it may reflect other large-scale patterns such as isolation by distance.

Patterns of spatial variation

Genetic distance as a function of geographical separation

A clear positive relationship between genetic distance and geographical separation of population pairs was evident in the data from Nishida & Lucas (1988) (the Pacific) and an independent data set from Benzie & Stoddart (1992b) (the Western Pacific and eastern Indian Ocean) (Figure 1). The rate of increase of genetic distance with geographical separation was similar in both data sets (Benzie, 1992). All these authors noted that the relationship was made particularly clear by the inclusion of the most distant populations in the data set (California in the case of Nishida & Lucas, 1988; Western Australia in the case of Benzie & Stoddart, 1992b). All noted that particularly isolated populations (Hawaii in the case of Nishida & Lucas, 1988; Lord Howe Island in the case of Benzie & Stoddart, 1992b) were more differentiated than expected given their geographic separation. All also noted that some local populations within reef groups differed significantly in their allele frequencies and that not all the significant spatial variation could be attributed to differences between the more distant population pairs.

The conclusion from these data was that *A. planci* populations were best described by an isolation by distance model, the relationship only clearly discernible when samples were obtained over several thousand kilometres. Some populations showed greater differentiation than that expected on the basis of isolation by distance because they were isolated from others by ocean currents passing perpendicularly between them and their nearest neighbours so blocking routes

Table 1. Differentiation among population sets of crown-of-thorns starfish, tested by determining the significance of θ values from zero. Sources were Nash et al., 1988[a], Benzie & Wakeford, 1997[b], Benzie, 1999a[c], Benzie, 1992[d]

	Number of populations	Population (Nei's D)	Differentiation θ or F_{ST}	Maximum separation (km)
Within reef complexes				
Great Barrier Reef[a]	7	0.009	0.026***	1300
Great Barrier Reef[b]	5	0.013	0.020**	700
Great Barrier Reef[b]	8	0.001	0.007^{NS}	700
Great Barrier Reef[b]	6	0.001	0.001^{NS}	130
Maldives[c]	2	0.000	0.003^{NS}	60
Thailand[c]	3	0.002	0.014^{NS}	20
Japan[c]	2	0.013	0.017^{NS}	30
Gove[c]	3	0.006	0.013^{NS}	20
Western Australia[c]	3	0.005	0.021***	50
Vanuatu[c]	3	0.008	0.008^{NS}	150
Ryukus[d]	4	0.004	0.011^{NS}	400
Between reef complexes				
Pacific Ocean[d]	6	–	0.100***	12 500
Indian Ocean (-West Australia)[c]	5	0.124	0.224***	8160
South-east Asia[c]	3	0.061	0.124***	4303
North-west Pacific[c]	3	0.040	0.051***	2640
Pacific versus Indian Oceans[c]	2	0.319	0.273***	12 300

of larval dispersal. The occasional significant differentiation among local populations was considered the result of stochastic effects of larval recruitment from the plankton.

When Benzie (1999a) plotted genetic distance between population pairs as a function of their geographical separation, using a more extensive data set that included data from through out the Indian Ocean, South-east Asia and the Pacific, and pooling data from local reef populations, the scatter was much greater. There was no obvious relationship between genetic distance and geographical separation. The group of points to the right of that graph might suggest some relationship given that they plot half way up the graph, and none plot near the axis. Benzie's (1999a) data incorporated that from Benzie & Stoddart (1992b), but, because of minor differences in scoring, Nishida & Lucas' (1988) data could not be included. When their data are plotted on the same figure as Benzie's (1999a) data, the lack of any relationship of genetic distance with geographical separation is clear.

θ as a function of geographical separation

An alternative, but similar, approach is to plot the θ calculated between population pairs as a function of geographical separation. Using the data from Benzie (1999a) the plot is similar to that of genetic distance and geographical separation (Figure 2, top graph). Again there was no obvious relationship between genetic distinction and geographical separation, and no statistically significant relationship was detected using the Mantel test (Benzie, 1999a).

Spatial autocorrelation

Plots of the spatial autocorrelation statistics, Moran's I and Geary's c, for all alleles provided no indication of any trend with geographical separation (Figure 3). The pattern resembles that of a 'crazy quilt', whether or not evenly spaced distance categories were used or whether categories with equal numbers of population pairs, and varying distances were used. The crazy quilt pattern was more obvious the greater the number of distance categories used in the analysis. However, when the mean values of Moran's I were displayed on a graph, they declined with increasing distance, implying more geographically separated populations were less likely to have similar allele frequencies, consistent with an isolation by distance model (Figure 4). Mean values of Geary's c also showed that more geographically separated populations were less

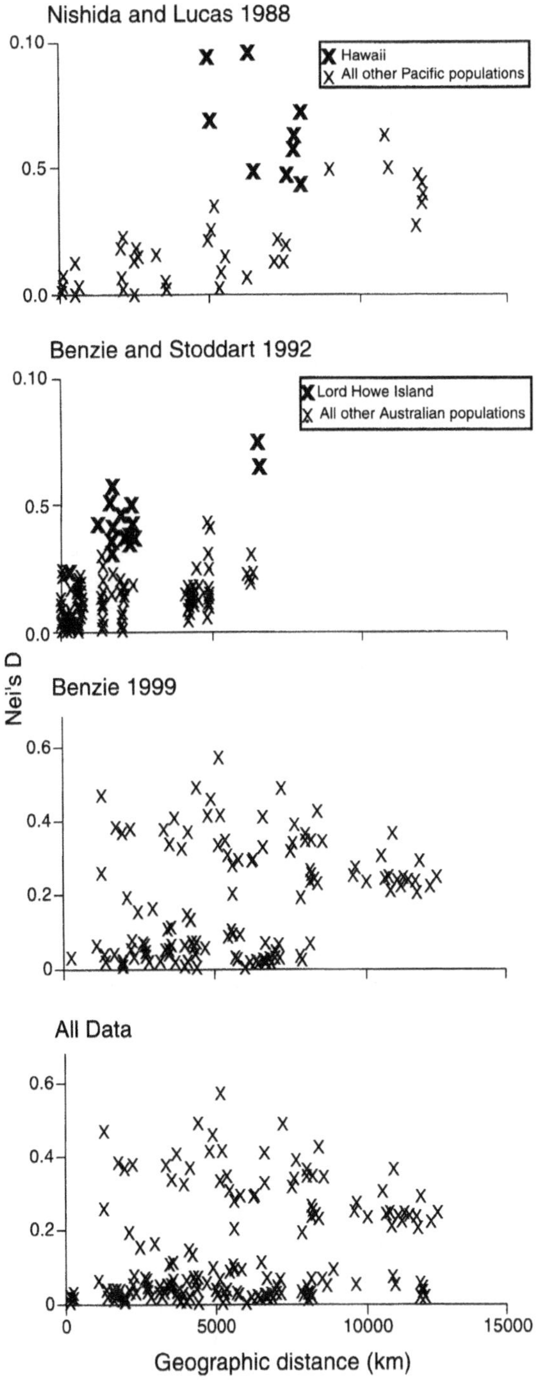

Figure 1. Nei's genetic distance (Nei, 1978) between *Acanthaster planci* populations graphed as a function of the geographical distance between the populations sampled. Data, from top to bottom, are from Nishida & Lucas' (1988) study of Pacific populations, Benzie & Stoddart's (1992b) study of Western Pacific and Western Australian populations, Benzie's (1999a) study of Indian Ocean, South-east Asian and Pacific populations, and a plot including all the data.

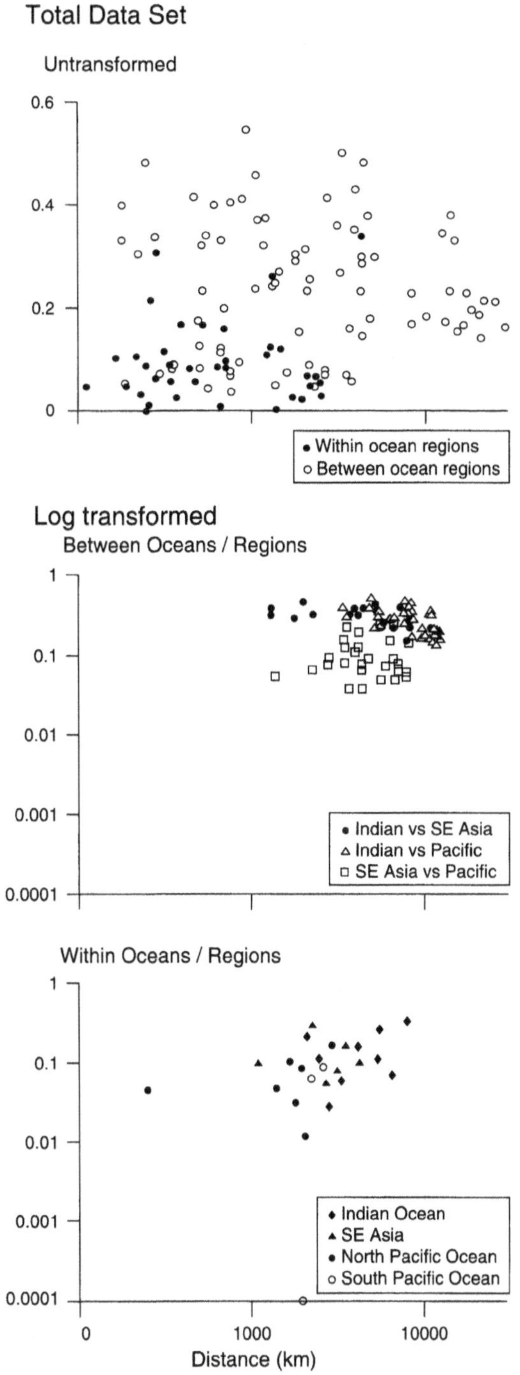

Figure 2. θ between *Acanthaster planci* populations graphed as a function of the geographical distance between the populations sampled. All data are from Benzie's (1999a) study of Indian Ocean, South-east Asian and Pacific populations. From top to bottom, the graphs are untransformed data from pairwise comparisons of all populations, log-transformed data including only comparisons of populations in different oceans or regions, and finally, including only populations within regions or oceans.

likely to have similar allele frequencies, as the value of Geary's c rose with increasing geographical separation. However, the 15-category analyses tend to flatten out suggesting a weaker relationship with distance. This is particularly the case for Geary's c (Figure 4, bottom graphs), especially for the analysis with evenly spaced distance categories, where there appears to be no relationship with geographical separation, and few of the points are significantly different from random (i.e. significantly different from 1).

Further analyses

The results of initial analyses varied depending on the data set and on the method used. The early data sets from the Pacific indicated isolation by distance using plots of genetic distance as a function of geographical separation. One might have expected the more extensive data set that included data from the Indian Ocean and South-east Asia to reveal a stronger relationship of isolation by distance than the others, given the greater geographical spread of the samples. However, plots of both genetic distance and of θ, and graphs of autocorrelation statistics for up to 50 alleles, indicated no relationship between these variables for that data set. In contrast, most of the graphs for average values of the autocorrelation statistics indicated a relationship of increasing differentiation with increasing geographical separation. Even among this one set of analyses though, some indicated no relationship. More detailed analyses of subsets of the data were undertaken to determine the reason for these different results, using the most extensive data set from Benzie (1999a).

Regional variation

θ as a function of geographical separation

Plots of log-transformed θ versus geographical distance, which include only comparisons between populations in different oceans or regions (i.e. the Pacific Ocean, the Indian Ocean and South-east Asia) are given in Figure 2 (middle graph). The points are scattered parallel with the x-axis, and there is no significant relationship between θ and geographical distance. In contrast, when the data include only comparisons between populations within different oceans or regions, the relationship is positive and significant (Figure 2, bottom graph).

Spatial autocorrelation

Excluding the Indian Ocean populations provided sets of graphs that cycled either side of the 0-axis for Moran's I and the 1.0-axis for Geary's c, suggesting no strong pattern of isolation distance (Figure 5). Separate analysis of each ocean and South-east Asia, gave some indication of isolation by distance within the Indian Ocean, although the plots for all alleles still provided an 'crazy quilt' pattern (Figure 6). All other plots cycled around the axis that indicated random distribution.

Once again, the results of the regional analyses depended on the data set and the method of analysis used. Some results suggested isolation by distance, others random distributions. There was, however, an indication in the θ analyses that the genetic structure among populations was different within and between oceans or regions.

A visual assessment of the spatial variation over the Indo-Pacific, illustrated using data from the enolase locus, provides some explanation for these results (Figure 7). There is a marked genetic distinction between populations in the Indian and the Pacific Oceans. Since some of the populations are sampled relatively close to this boundary it is possible to see why there is little relationship between genetic divergence and the spatial distance between these population sets. There is also some genetic distinction between populations in South-east Asia and the Northern Pacific on the one hand, and the southern Pacific and Western Australia on the other. This suggests that there are regions, within which there is relatively little differentiation among populations, but between which levels of differentiation among populations are higher.

Spatial autocorrelation methods do not discriminate well between gradual clines and regions, each of which show no or little variation, and which meet at a sharp boundary (Sokal & Oden, 1978). In the case of *A. planci*, analyses gave a pattern consistent with isolation by distance. However, this simply reflected the fact that the most distant populations (Indian versus Pacific Ocean populations at the extremes of the range sampled) were highly differentiated (Figure 4). Populations that were closer together included a mix of pure Pacific or Indian Ocean types and comparisons that included populations from both Oceans (e.g. Thailand and Malaysia populations). Those very close included only pure Pacific or Indian Ocean types. It is pertinent that the clearest patterns of reducing Moran's I, and increasing Geary's c, with distance are the analyses with fewest distance categories. This effectively led to an averaging of variation across space that most closely approached the patch-size differences between the Indian and Pacific oceans.

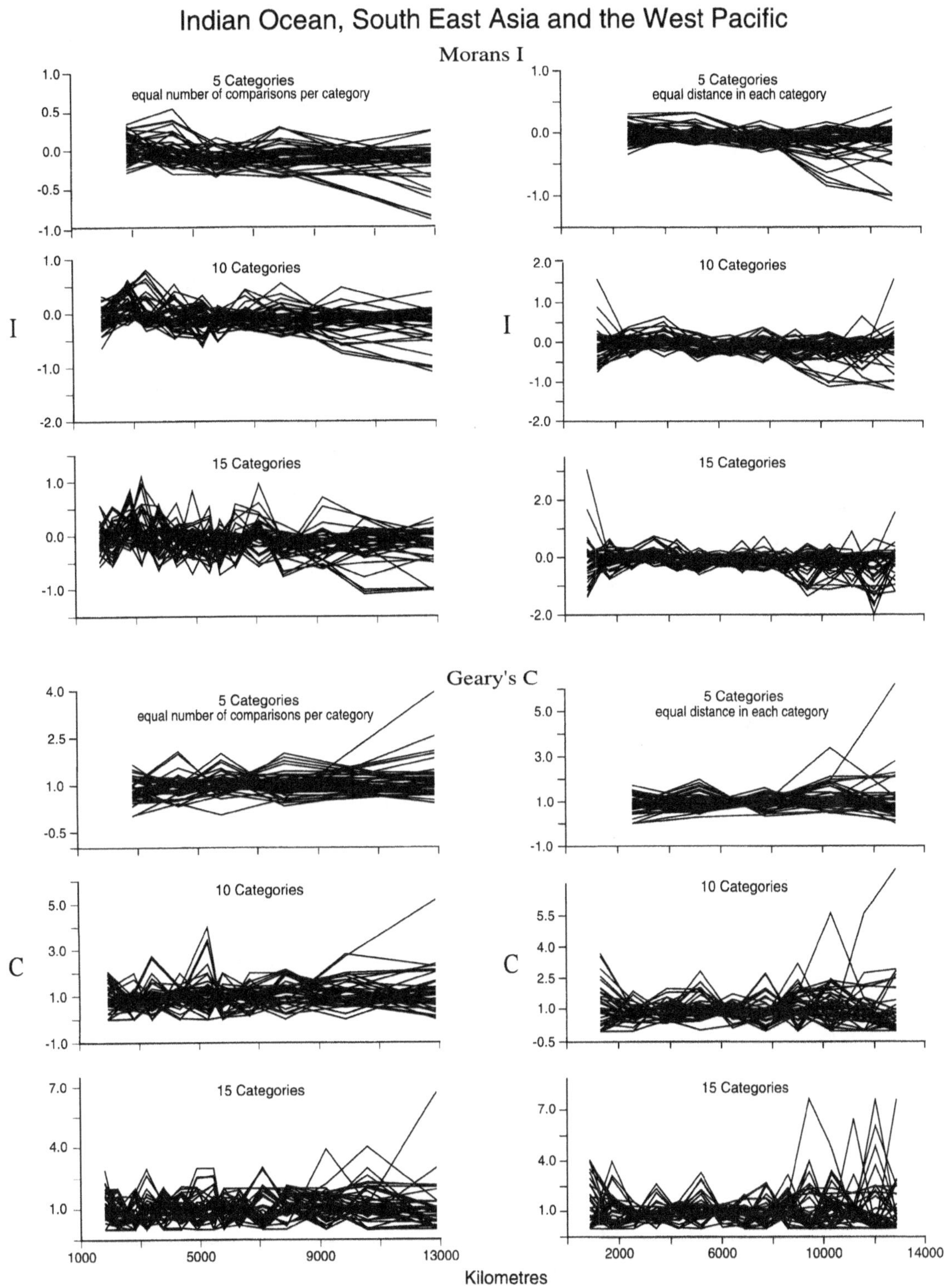

Figure 3. Correllograms illustrating the value of autocorrelation statistics (Moran's I and Geary's c) using 5, 10 or 15 distance classes, for approximately 50 individual alleles among 16 populations sampled from the Indian and Western Pacific Oceans and South-east Asia. The graphs on the left have equal numbers of comparisons in each distance class and variable range of distance. The graphs on the right have equally spaced distance classes and different numbers of comparisons in each class.

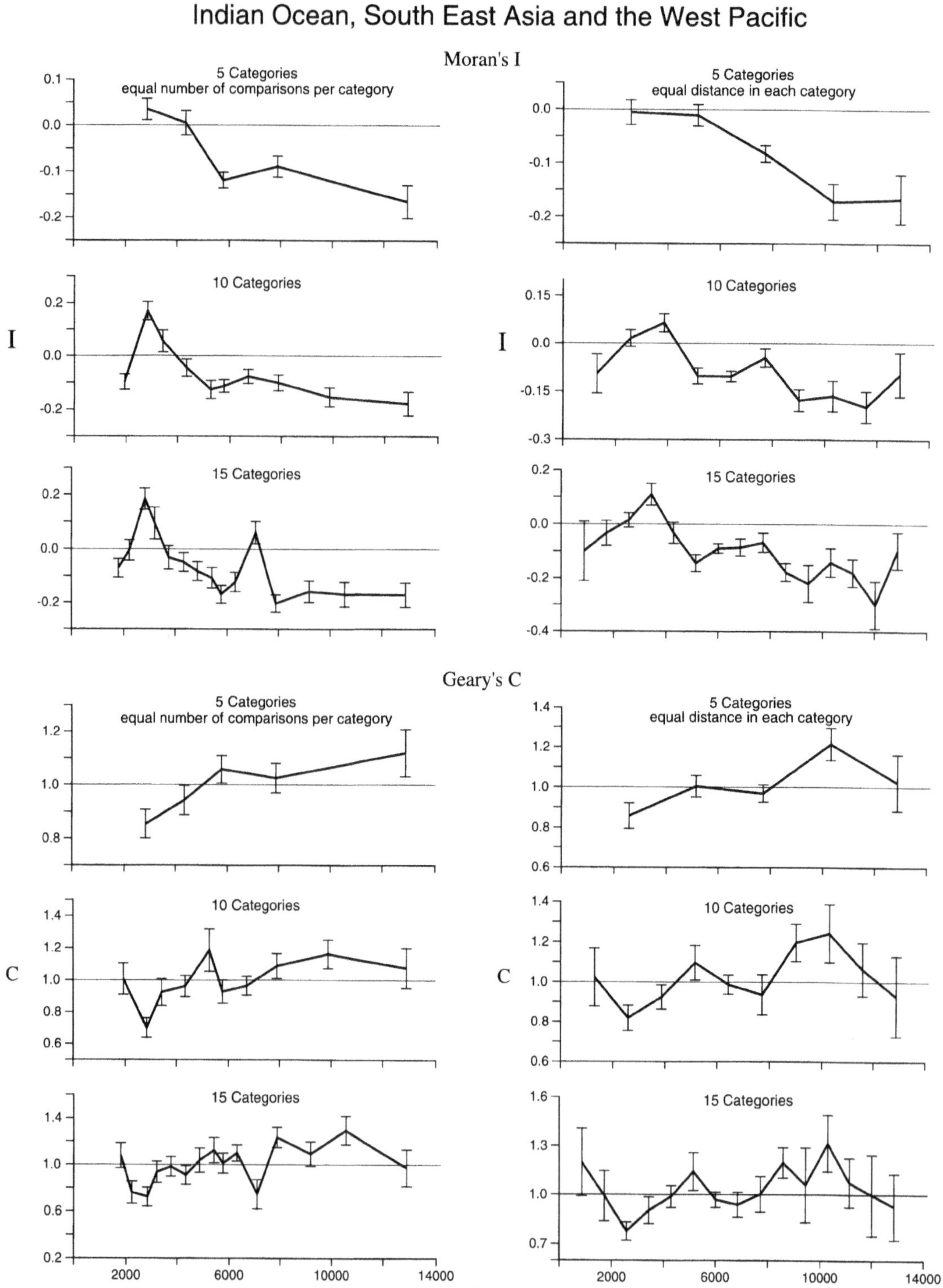

Figure 4. Correllograms illustrating the mean value of autocorrelation statistics (Moran's I and Geary's c) using 5, 10 or 15 distance classes, calculated from approximately 50 individual alleles, among 16 populations sampled from the Indian and Western Pacific Oceans and South-east Asia. The graphs on the left have equal numbers of comparisons in each distance class and variable range of distance. The graphs on the right have equally spaced distance classes and different numbers of comparisons in each class.

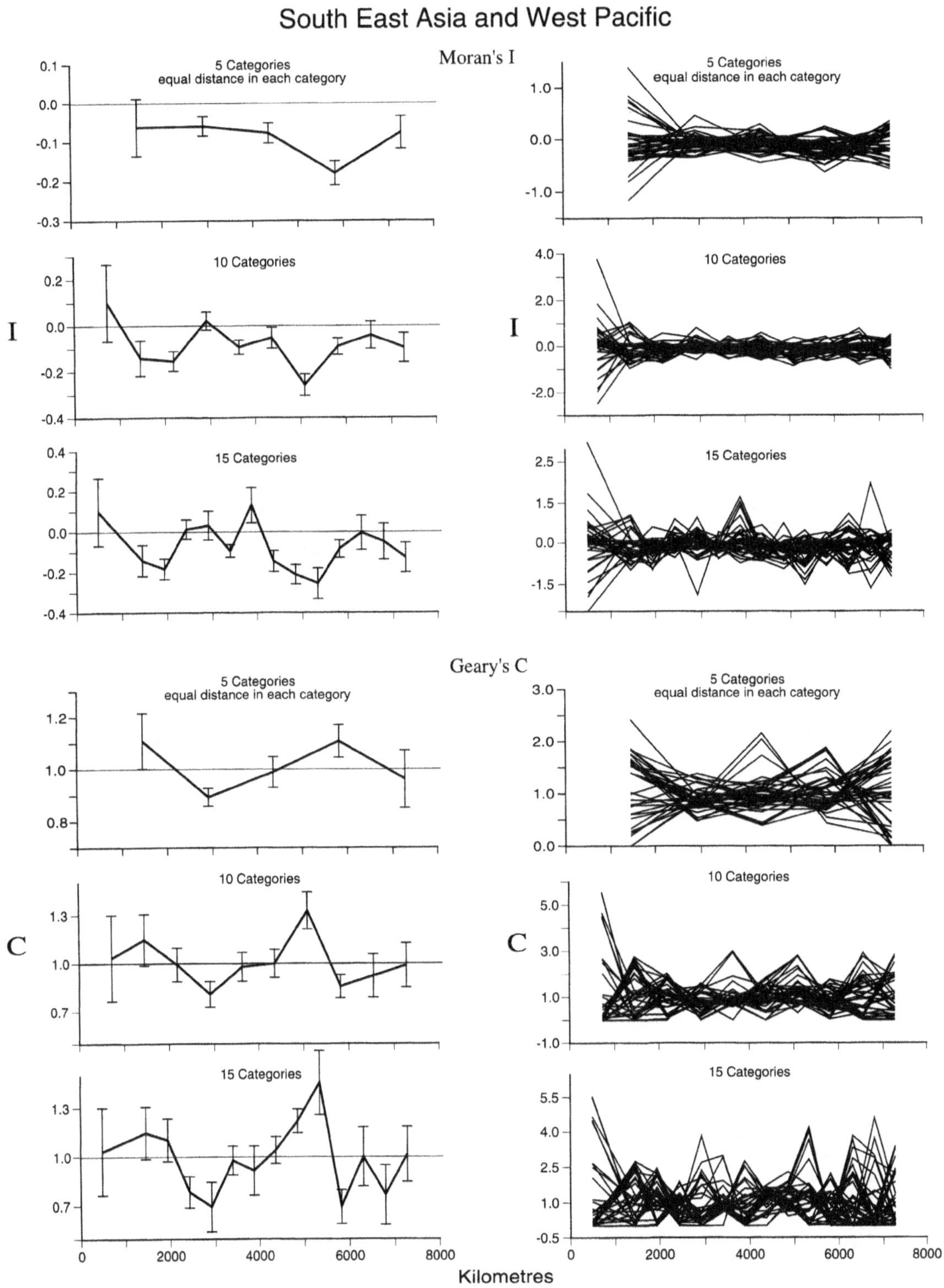

Figure 5. Correllograms illustrating the value of autocorrelation statistics (Moran's I and Geary's c) using 5, 10 or 15 distance classes, for approximately 50 individual alleles (right hand column of graphs), and their mean value (left hand column of graphs), among 11 populations sampled from the Western Pacific Ocean and South-east Asia. All graphs have equally spaced distance classes and different numbers of comparisons in each class.

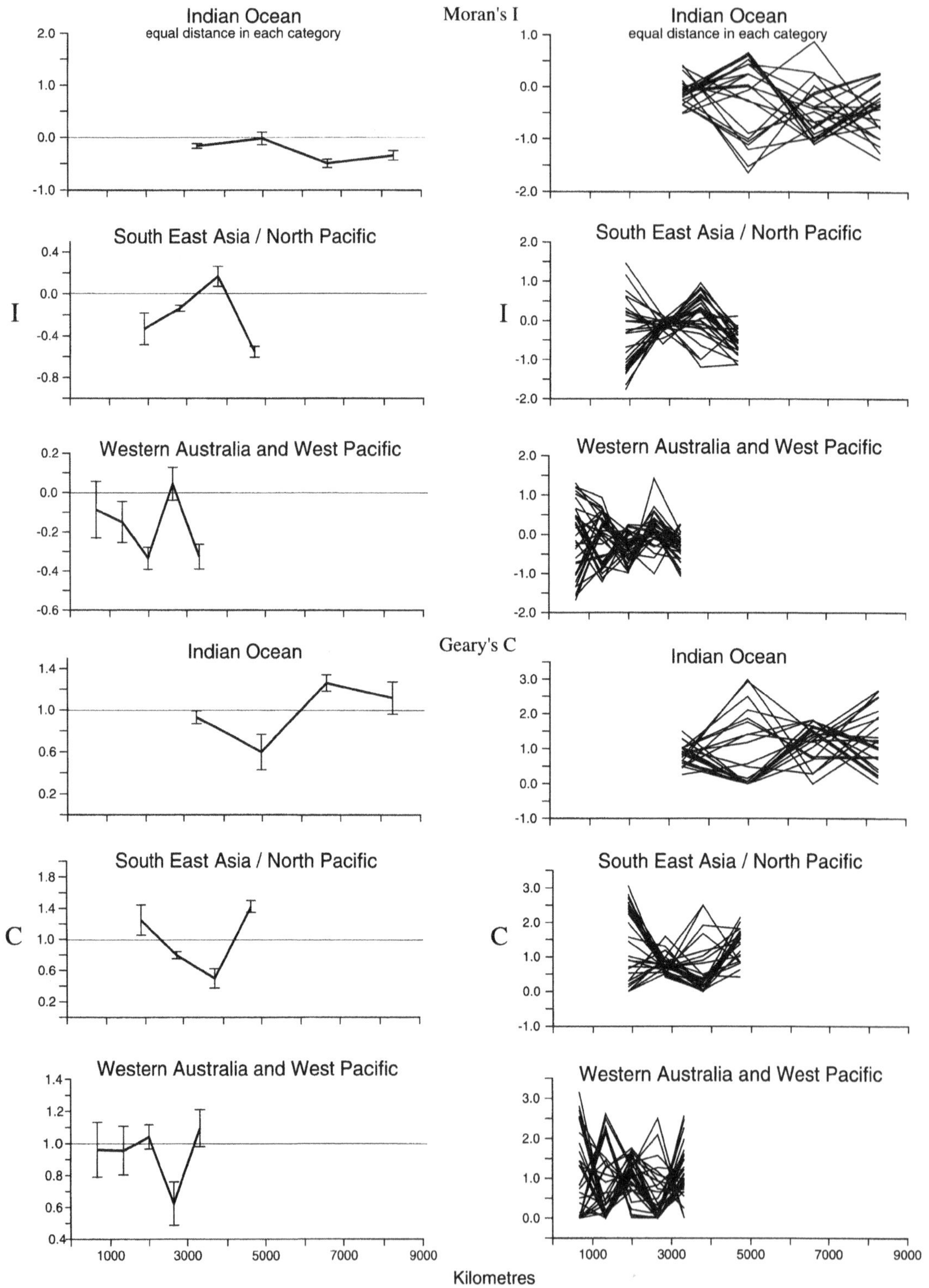

Figure 6. Correllograms illustrating the value of autocorrelation statistics (Moran's I and Geary's c) using 5 distance classes, for approximately 50 individual alleles (right hand column of graphs), and their mean value (left hand column of graphs). The data sets, from the top to bottom, are for 5–6 populations sampled from within either the Indian Ocean, South-east Asia and the Northern Pacific, or Western Australia and the West Pacific. All graphs have equally spaced distance classes and different numbers of comparisons in each class.

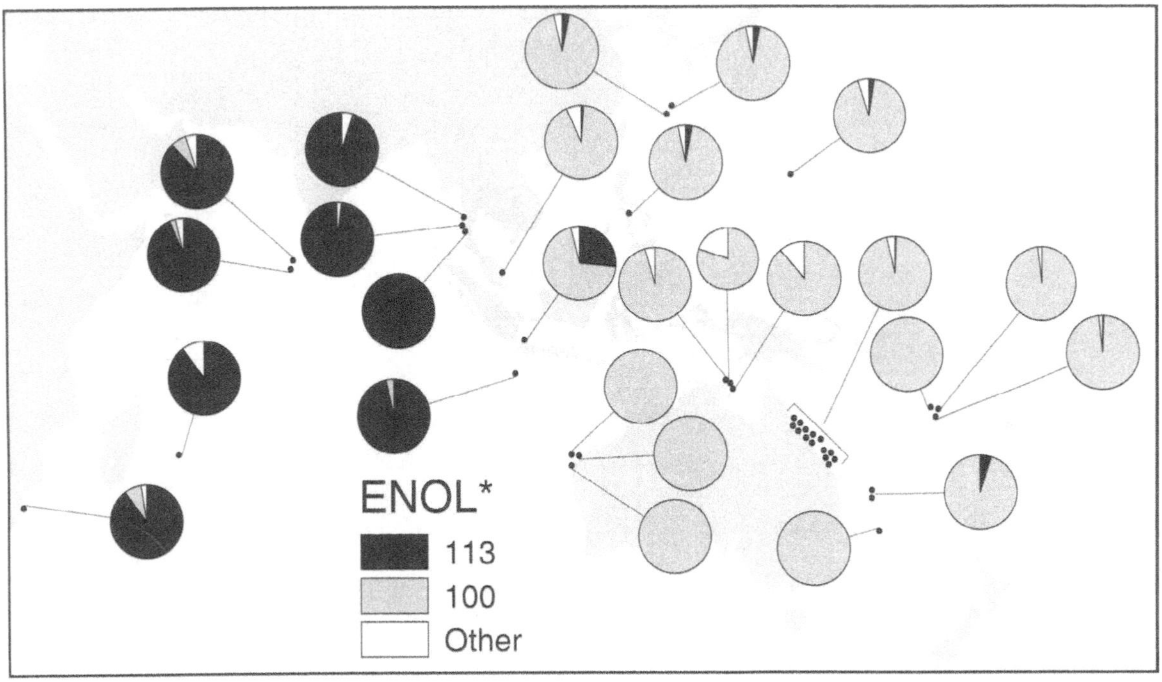

Figure 7. Variation in allele frequencies at *Enol** for 39 populations of *Acanthaster planci* in the Indo-Pacific (after Benzie, 1999a).

Finally, if gene flow is calculated from θ, the patterns of gene flow provide no clear indication of an even and increasing level of gene exchange with increasing distance, or indeed, any relationship with prevailing surface ocean currents (Figure 8).

Conclusions

It is ironic that a geographically extensive data set, in which there is a marked genetic discontinuity between two major sets of populations, one in each ocean, provides a plot of genetic differentiation (either as genetic distance or as θ) that suggests no relationship of genetic distance with geographic separation. In addition, analysis of the same data set using spatial autocorrelation suggests isolation by distance, and provides no hint that there are two regions of markedly different genetic constitution that abut in what is probably a narrow zone of contact.

If one approached the analysis of *A. planci* data simply by plotting θ or genetic distance against geographical separation of populations with an expectation that little structure existed, those expectations would have been met. The result would have been interpreted as indicating significant differences among populations as a result of local stochastic variation in the genetic composition of recruits, but that otherwise gene flow was similar, and high, over the species' range. If spatial autocorrelation of all alleles had been used, a similar conclusion would have emerged. However, spatial autocorrelation analysis based on the average over all alleles would have suggested isolation by distance, or possibly no spatial pattern, depending on the number of distance categories used and whether one had used Moran's I or Geary's c.

In the full data set discussed here, a visual appraisal of the spatial variation displayed in pie diagrams, or observation of tabulated allele frequencies that were very different between oceans, would have provided an impetus for more detailed analysis and interpretation. However, if data sets were more restricted (perhaps limited to five or six locations), such patterns may not have been as easy to observe, particularly if samples had only been obtained from South-east Asia and the Pacific. The interpretations of the early results from the Pacific (Nishida & Lucas, 1988; Benzie & Stoddart, 1992b) quite reasonably inferred isolation by distance over large geographical scales, with some interruption by outliers. The collection of more data within the region suggests more outliers have been obtained, which spread the scatter of points to one approaching a random pattern.

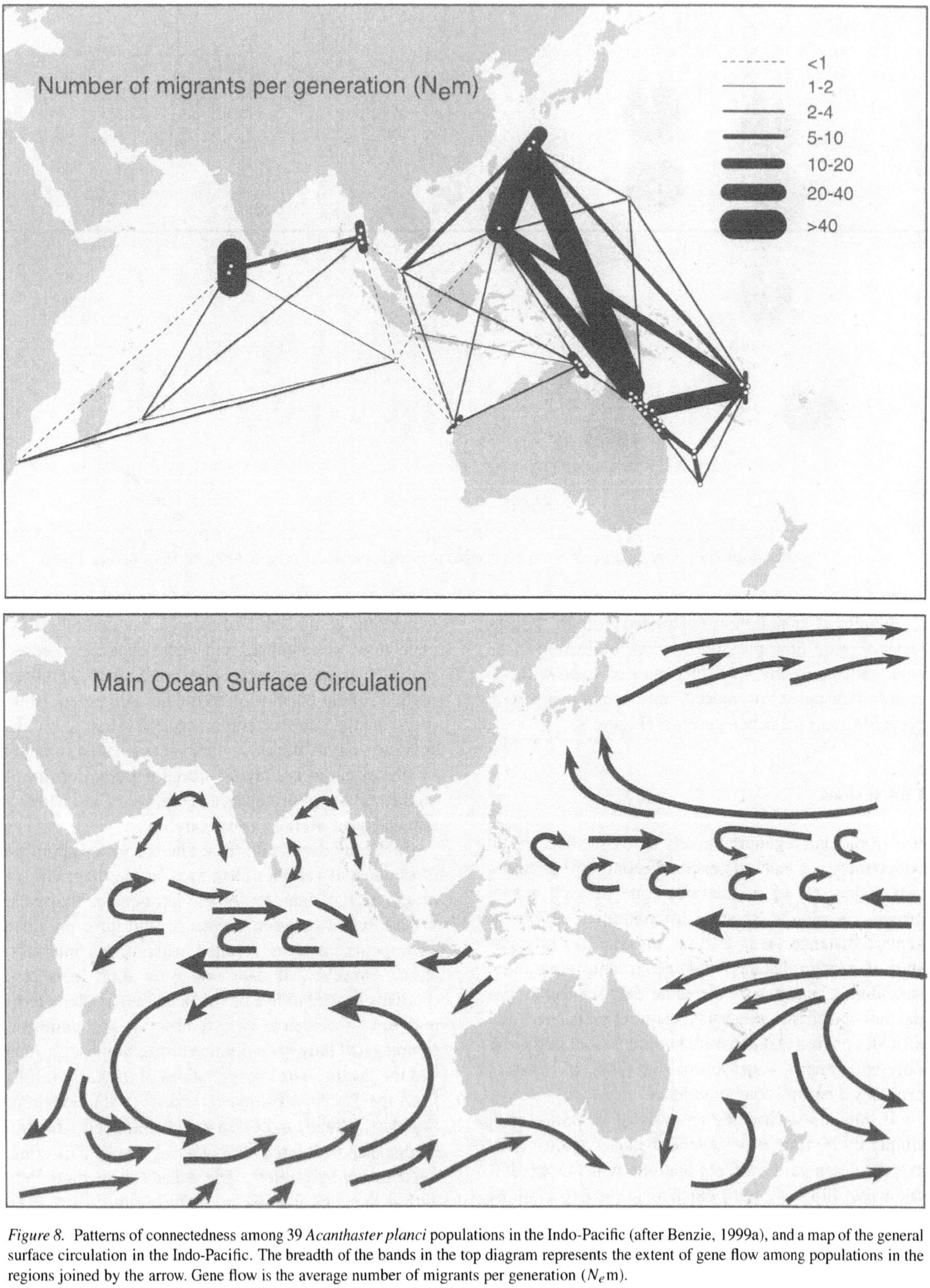

Figure 8. Patterns of connectedness among 39 *Acanthaster planci* populations in the Indo-Pacific (after Benzie, 1999a), and a map of the general surface circulation in the Indo-Pacific. The breadth of the bands in the top diagram represents the extent of gene flow among populations in the regions joined by the arrow. Gene flow is the average number of migrants per generation (N_em).

The determination of genetic structure of widespread marine species

There are a number of reasons why the basic assumptions that widespread marine species should show little spatial variation in genetic structure given their high potential for dispersal on ocean currents have been questioned only recently. Firstly, there have been few studies of widespread marine species over oceanic scales. There have been a number of studies of whales, turtles and pelagic fish, but these species are not constrained by oceanic currents, and in the case of the first two groups, their genetic structure appears to be dominated by social interactions and their return to specific breeding grounds. Recent reviews of genetic work on animals dependent on a larval phase for their dispersal list only 8–10 studies (Benzie, 1998, 1999b).

Secondly, where there is unlikely to be strong structure, any structure that does exist is likely to be detected only with a reasonable array of samples. The logistical and financial difficulties of working over oceanic scales means that there are few data sets that have the high density of sampling required for the detection of fine population structure. Thirdly, where an expectation exists that there is likely to be little structure, there is little incentive to look further if initial analysis suggests none exists, or suggests a simple isolation by distance model or another predictable pattern.

Fourthly, the marine environment is not familiar to humans, and they have limited ability to determine the meaningful boundaries to species in the aquatic habitat. This means that sampling designs may not be adequately distributed within and between regions that are likely to be differentiated. Studies are, therefore, dependent upon the results of analyses of preliminary sample sets taken almost at random, to determine the best sampling regime. However, the likelihood of being able to undertake preliminary work, with more intensive follow-up sampling is reduced because of the first three factors listed above.

Recent work suggests that a number of widespread species with strong dispersal potential have population structures that are characterised by regions between which there is significant differentiation, but within which there is relatively little (e.g. Duke et al., 1998; Williams & Benzie, 1998). It has been suggested that such a structure may reflect that limits of expansion of different refugial populations that existed at times of low sea level (Benzie, 1999b). This interpretation also highlights the restriction in gene flow among regions, and the lack of a dominant influence of present-day ocean circulations in determining gene flow and the genetic structure of these species. This is a very different view from the traditional one, and emphasises our lack of understanding of the principal factors influencing the origin and maintenance of genetic diversity in the marine sphere.

Further advances in understanding the genetic structure of marine species will demand an iterative approach, where a greater number of samples are collected using the results of initial sampling to plan future collections. This will provide information on different spatial scales, but taking into account the particular geographical pattern of genetic variation in the species being investigated. They will also depend on a flexible approach to the analysis of population structure, and the recognition that the true patterns of variation may be obscured in a brief analysis of even extensive data sets. Superficially, these may demonstrate the patterns traditionally expected of widespread marine species.

References

Avise, J. C., 1994. Molecular Markers, Natural History and Evolution. Chapman and Hall, New York. 511 pp.

Battaglia, B. & J. A. Beardmore (eds), 1978. Marine Organisms: Genetics, Ecology and Evolution. Plenum Press, New York. 757 pp.

Beaumont, A. (ed.), 1994. Genetics and Evolution of Aquatic Organisms. Chapman and Hall, London. 539 pp.

Benzie, J. A. H., 1992. Review of the genetics, dispersal and recruitment of crown-of-thorns starfish (*Acanthaster planci*). Aust. J. mar. Freshwat. Res. 43: 597–610.

Benzie, J. A. H., 1998. Genetic structure of marine organisms and SE Asian biogeography. In Hall, R. & J. D. Holloway (eds), Biogeography and Geological Evolution of SE Asia. Backhuys Publishers, Leiden, The Netherlands: 197–209.

Benzie, J. A. H., 1999b. Genetic structure of coral reef organisms – ghosts of dispersal past. Am. Zool. 39: 131–145.

Benzie, J. A. H., 1999a. Major genetic differences between crown-of-thorns starfish (*Acanthaster planci*) populations from the Indian and Pacific Oceans. Evolution 53: 132–145.

Benzie, J. A. H., & J. A. Stoddart, 1992a. Genetic structure of outbreaking and non-outbreaking crown-of-thorns starfish (*Acanthaster planci*) populations on the Great Barrier Reef. Mar. Biol. 112: 119–130.

Benzie, J. A. H., & J. A. Stoddart, 1992b. The genetic structure of crown-of-thorns starfish (*Acanthaster planci*) in Australia. Mar. Biol. 112: 631–639.

Benzie, J. A. H. & M. Wakeford, 1997. The genetic structure of crown-of-thorns starfish (*Acanthaster planci*) outbreak populations occurring ten years apart on the Great Barrier Reef, Australia. Mar. Biol. 129: 149–157.

Benzie, J. A. H. & S. T. Williams, 1995. Gene flow among giant clam (*Tridacna gigas*) populations in the Pacific does not parallel ocean circulation. Mar. Biol. 123: 781–787.

Benzie, J. A. H. & S. T. Williams, 1997. Gene flow among giant clam (*Tridacna maxima*) populations in the west Pacific is not consistent with dispersal by present-day ocean currents. Evolution 51: 768–783.

Birkeland, C. & J. S. Lucas, 1990. *Acanthaster planci*: major management problem of coral reefs. CRC Press, Boca Raton, Florida. 257 pp.

Campbell, C. A., J. W. Valentine & F. J. Ayala, 1975. High genetic variability in a population of *Tridacna maxima* from the Great Barrier Reef. Mar. Biol. 33: 341–345.

Duke, N., J. A H. Benzie, J. A. Goodal & E. R. Ballment, 1998. Genetic structure and evolution of the mangrove genus *Avicennia* (Avicenniaceae) in the Indo-Pacific. Evolution 52: 1612–1626.

Foltz, D. W., J. P. Breaux, E. L. Campagnaro, S. W. Herke, A. E. Himel, A. W. Hrincevich, J. W. Tamplin & W. B. Stickle, 1996. Limited morphological differences between genetically identified cryptic species within the *Leptasterias* species complex (Echinodermata: Asteroidea). Can. J. Zool. 74: 1275–1283.

Knowlton, N., 1993. Sibling species in the sea. Ann. Rev. Ecol. Syst. 24: 189–216.

Mantel, N., 1967. The detection of disease clustering and a generalised regression approach. Cancer Res. 27: 209–220.

Moran, P. J., 1986. The *Acanthaster* phenomenon. Oceanogr. mar. Biol. ann. Rev. 24: 379–480.

Nash, W. J., M. Goddard & J. S. Lucas, 1988. Population genetic studies of the crown-of-thorns starfish *Acanthaster planci* (L.) in the Great Barrier Reef region. Coral Reefs 7: 11–18.

Nei, M., 1978. Estimation of average heterozygosity and genetic distance from a small number of individuals. Genetics 89: 583–590.

Nishida, M. & J. S. Lucas, 1988. Genetic differences between geographic populations of the Crown-of-thorns starfish throughout the Pacific region. Mar. Biol. 98: 359–368.

Palumbi, S. R., G. Grabowsky, T. Duda, L. Geyer & N. Tachino, 1997. Speciation and population genetic structure in tropical Pacific sea urchins. Evolution 51: 1506–1517.

Rohlf, F. J., 1990. NTSYS-PC: Numerical Taxonomy and Multivariate Analysis System. Exeter Software, New York. 37 pp.

Slatkin, M., 1985. Rare alleles as indicators of gene flow. Evolution 39: 53–65.

Sokal R. R. & N. L. Oden, 1978. Spatial autocorrelation in biology 1. Methodology. Biol. J. linn. Soc. 10: 199–228.

Swofford, D. L. & R. B. Selander, 1981. BIOSYS-1: a fortran program for the comprehensive analysis of electrophoretic data in population genetics and systematics. J. Hered. 72: 281–283.

Waples, R. S., 1987. A multispecies approach to the analysis of gene flow in marine shore fishes. Evolution 41: 385–400.

Wartenberg, D., 1989. SAAP: a spatial autocorrelation analysis program. Department of Environmental and Community Medicine, Robert Wood Johnson Medical School, University of Medicine and Dentistry of New Jersey, Piscataway, New Jersey. 31 pp.

Weir, B. S. & C. C. Cockerham, 1984. Estimating F-statistics for the analysis of population structure. Evolution 38: 1358–1370.

Williams, S. T. & J. A. H. Benzie, 1998. Evidence of a biogeographic break between populations of a high-dispersal starfish: congruent regions within the Indo-West Pacific defined by color morphs, mtDNA and allozyme data. Evolution 52: 87–99.

Winans, G., 1980. Geographic variation in the milkfish *Chanos chanos*. I. Biochemical evidence. Evolution 34: 558–574.

Hydrobiologia **420:** 15–27, 2000.
A. M. Solé-Cava, C. A. M. Russo & J. P. Thorpe (eds), Marine Genetics.

Molecular systematics of sponges (Porifera)

C. Borchiellini[1], C. Chombard[2], B. Lafay[1] & N. Boury-Esnault[1]
[1]*Centre d'Océanologie de Marseille, Station marine d'Endoume, UMR-CNRS 6540, Université de la Méditerranée, Rue de la Batterie des Lions, F-13007, Marseille, France*
[2]*MNHN, Service de Systématique Moléculaire CNRS GDR 1005, 43 Rue Cuvier, F-75005, Paris, France*

Key words: phylogeny, evolution, allozymes, rDNA, DNA sequence

Abstract

The first application of molecular systematics to sponges was in the 1980s, using allozyme divergence to discriminate between conspecific and congeneric sponge populations. Since this time, a fairly large database has been accumulated and, although the first findings seemed to indicate that sponge species were genetically more divergent than those of other marine invertebrates, a recent review of the available dataset indicates that levels of interspecific gene identities in most sponges fall within the normal range found between species of other invertebrates. Nevertheless, some sponge genera have species that are extremely divergent from each other, suggesting a possible polyphyly of these genera. In the 1990s, molecular studies comparing sequences of ribosomal RNA have been used to reappraise the phylogenetic relationships among sponge genera, families, orders and classes. Both the 18S small subunit and the 28S large subunit rRNA genes have been sequenced (41 complete or partial and 75 partial sequences, respectively). Sequences of 18S rRNA show good support for Porifera being true Metazoa, but they are not informative for resolving relationships among genera, families or orders. 28S rRNA domains D1 and D2 appear to be more informative for the terminal nodes and provide resolution for internal topologies in sufficiently closely related species, but the deep nodes between orders or classes cannot be resolved using this molecule. Recently, a more conserved gene, Hsp70, has been used to try to resolve the relationships in the deep nodes. Metazoan monophyly is very well supported. Nevertheless, the divergence between the three classes of Porifera, as well as the divergence between Porifera, Cnidaria and Ctenophora, is not resolved. Research is in progress using other genes such as those of the homeodomain, the tyrosine kinase domain, and those coding for the aggregation factor. For the moment the dataset for these genes is too restricted to resolve the phylogenetic relationships of these phyla. However, whichever the genes, the phylogenies obtained suggest that Porifera could be paraphyletic and that the phylogenetic relationships of most of the families and orders of the Demospongiae have to be reassessed. The Calcarea and Hexactinellida are still to be studied at the molecular level.

Introduction

The Porifera are considered the structurally simplest and the oldest metazoan phylum (Delage & Hérouard, 1899; Bergquist, 1978; Li et al., 1998), and have a very controversial phylogenetic history and a highly problematic taxonomy. Sponge classification systems essentially use the characteristic features of the skeleton, and the shape and size of its constitutive elements, the spicules and the spongin fibers. However, classical taxonomy as well as recent cladistic analyses (van Soest, 1990, 1991; Rosell & Uriz, 1997) are hampered by the difficulty of formally establishing homologies within the skeletal elements and the skeletal organization (Fromont & Bergquist, 1990; Chombard, 1998). Thus, there is no phylogenetic consensus among poriferan experts.

The questions of the true metazoan nature of the Porifera, their monophyletic or paraphyletic origin, the phylogenetic relationships among classes, orders and families, and the cosmopolitanism of numerous species with a simple skeleton, are old problems which remain unresolved. In the last thirty years, the development of molecular biology has made available a range of new techniques (enzyme electrophoresis, protein and gene sequence comparisons, RFLPs, etc.)

Table 1. List of species in which 18S rRNA, 28S rRNA and Hsp70 have been sequenced: p = partial sequence; c = complete sequence; in bold, sequences available in sequence databases; underlined, published sequences; in italics, sequences present in Ph.D. theses; plain text, sequences not available. 1 = Lafay et al., 1992; 2 = Alvarez, 1998; 3 = Borchiellini et al., 1998; 4 = Chombard et al., 1997; 5 = Kelly-Borges et al., 1991; 6 = Cavalier-Smith et al., 1996; 7 = Christen et al., 1991; 8 = Itskovich et al., 1999; 9 = Koziol et al., 1998a; 10 = West & Powers, 1993; 11 = Chombard et al., 1998; 12 = Chombard, 1998; 13 = Koziol et al., 1996; 14 = Kelly-Borges & Pomponi, 1994; 15 = Kobayashi et al., 1993; 16 = Dams et al., 1982; 17 = Koziol et al., 1998b; 18 = Wainwright et al., 1993; 19 = Lafay in Chombard, 1998; 20 = Odorico & Miller, 1997; 21 = Collins, 1998

Species/genes	rRNA					Hsp
	18S rRNA	ITS1	5,8S rRNA	ITS2	28S rRNA	-70
CALCAREA						
Calcinea						
Clathrina cerebrum (1, 6)	**c**				**C1-C2**	
Calcaronea						
Leucosolenia sp. (21)	**c**					
Petrobiona massiliana (1, 3)					**C1-C2**	**x**
Sycon (Scypha) ciliatum (18)	**c**					
Sycon calcaravis (15)	**c**					
Sycon raphanus (9)						**x**
DEMOSPONGIAE						
Homosclerophorida						
Corticium candelabrum (19)					*C1-C2*	
Oscarella lobularis (19)					*C1-C2*	
Plakortis sp. (21)	**c**					
Hadromerida						
Aaptos aaptos (12)			*c*	*c*	*C1-D2*	
Aaptos sp. (5)	p					
Acanthochaetetes wellsi (4)			**c**	**c**	**C1-D2**	
Cliona orientalis (12)			*c*	*c*	*C1-D2*	
Cliona viridis (4)			**c**	**c**	**C1-D2**	
Cliona sp, (5)	p					
Diplastrella bistellata (12)			*c*	*c*	*C1-D2*	
Ficulina ficus (12)			*c*	*c*	*C1-D2*	
Paratimea sp . (12)			*c*	*c*	*D1-D2*	
Polymastia fusca (5)	p					
Polymastia mammillaris (12)			*c*	*c*	*C1-D2*	
Suberites domuncula (12, 17)			*c*	*c*	*C1-D2*	**x**
Suberites ficus (21)	**c**					
Spirastrella hartmani (4)			**c**	**c**	**C1-D2**	
Tectitethya crypta (12)			*c*	*c*	*C1-D2*	
Tethya aurantium (12)			*c*	*c*	*C1-D2*	
Tethya taboga (12)			*c*	*c*	*C1-D2*	
Tethya sp1 (12)			*c*	*c*	*C1-D2*	
Tethya sp2 (5)	p					
Timea sp. (12)			*c*	*c*	*C1-D2*	
Axinellida						
Acanthella acuta (2)						*D3*
Acanthella cavernosa (2)						*D3*
Acanthella pulcherrima (2)						*D3*

Continued on p. 17

Table 1. contd.

Species/genes	rRNA					Hsp
	18S rRNA	ITS1	5,8S rRNA	ITS2	28S rRNA	-70
Agelas oroides (1)					**C1-C2**	
Agelas mauritiana (2)						*D3*
Astrosclera willeyana (2, 4)			**c**	**c**	**C1-D2**	*D3*
Axinella aruensis (2)						*D3*
Axinella carteri (2)						*D3*
Axinella damicornis (1, 2, 4)			**c**	**c**	**C1-D2**	*D3*
Axinella polypoides (2, 6)	**c**					*D3*
Cymbastella coralliophila (2)						*D3*
Cymbastella stipitata (2)						*D3*
Cymbastella vespertina (2)						*D3*
Dictyonella incisa (1, 7)					**C1-C2**	
Hemiasterella complicata (12)			*c*	*c*	*C1-D2*	
Phakellia sp, (2)						*D3*
Pseudaxinella australis (2)						*D3*
Pseudaxinella durissima (2)						*D3*
Pseudaxinella reticulata (2)						*D3*
Pseudaxinella sp. (2)						*D3*
Ptilocaulis (2)						*D3*
Reniochalina stalagmitis (2)						*D3*
Halichondrida						
Ciocalypta confossa (2)						*D3*
Epipolasis n.sp. (12)			*c*	*c*	*C1-D2*	
Halichondria japonica (8)	**c**					
Halichondria panicea (2, 4)		**c**	**c**	**c**	**C1-D2**	*D3*
Halichondria phakelloides (2)						*D3*
Hymeniacidon heliophila (20)	**p**	**c**	**c**	**c**	**p**	
Hymeniacidon sanguinea (16)		**c**				
Monanthus ciocalyptoides (5)	p					
Spongosorites genitrix (1)					**C1-C2**	
Scopalina ruetzleri (2)						*D3*
Poecilosclerida						
Anchinoe tenacior (1)					**C1-C2**	
Asbestopluma hypogea (3, 12)			*c*	*c*	*C1-D2*	**x**
Crambe crambe (1)					**C1-C2**	
Hamacantha implicans (12)			*c*	*c*	*C1-D2*	
Microciona prolifera (18)	**c**					
Mycale fibrexilis (21)	**c**					
Dictyoceratida						
Cacospongia scalaris (19)					*C1-C2*	
Spongia agaracina (19)					*C1-C2*	
Chondrosida						
Chondrilla nucula (12)			*c*	*c*	*C1-C2*	
Chondrosia reniformis (3, 12)			*c*	*c*	*C1-D2*	**x**

Continued on p. 18

Table 1. contd.

Species/genes	rRNA					Hsp
	18S rRNA	ITS1	5,8S rRNA	ITS2	28S rRNA	-70
Tetractinellida						
Cinachyrella apion (5)	p					
Cinachyrella cf *schuzei (4)*			**c**	**c**	**C1-D2**	
Corallistes masoni (11)			**c**	**c**	**C1-D2**	
Discodermia polydiscus (4)			**c**	**c**	**C1-D2**	
Discodermia sp, (5)	p					
Erylus euastrum (11)			**c**	**c**	**C1-D2**	
Geodia cydonium (12, 13)			*c*	*c*	*C1-D2*	x
Geodia neptuni (14)	**p**					
Leiodermatium lynceus (5)	p					
Pachastrella sp. (5)	p					
Placospongia sp. (12)			*c*	*c*	*C1-D2*	
Poecillastra compressa (11)			**c**	**c**	**C1-D2**	
Stelleta grubei (12)			*c*	*c*	*C1-D2*	
Stelleta sp. (5)	p					
Stryphnus mucronatus (11)			**c**	**c**	**C1-D2**	
Pachastrissa pathologica (11)			**c**	**c**	**C1-D2**	
Pachymatisma johnstoni (11)			**c**	**c**	**C1-D2**	
Penares helleri (11)			**c**	**c**	**C1-D2**	
Tetilla japonica (15)	**c**					
Theonella atlantica (5)	p					
Scleritoderma sp. (5)	p					
Siphonidium ramosum (5)	p					
Vetulina stalactites (5)	p					
Haplosclerida						
Haliclona oculata (16)		**c**				
Lubomirskia abietina (8)	**p**					
Petrosia ficiformis (1,3)					**C1-C2**	x
Reniera fulva (1)					**C1-C2**	
Reniera mucosa (1)					**C1-C2**	
Spongilla lacustris (8, 12)	**p**		*c*	*c*	*C1-D2*	
Swartschewskia papyracea (8)	**p**					
HEXACTINELLIDA						
Oopsacas minuta (19)					*C1-C2*	
Rhabdocalyptus dawsoni (3, 9, 21)	**c**					x
Farrea occa (10)	p					

that have been used to bring insight into the phylogeny and evolution of this phylum. Cladistic analysis, when based on true homologies has also provided a new approach.

Molecular studies on sponges were first initiated in the late 1980s, using allozyme polymorphism to detect sibling species (Solé-Cava & Thorpe, 1986). In the 1990s, molecular sequence comparisons were used to reappraise phylogenetic relationships among sponges (Kelly-Borges et al., 1991; Lafay et al., 1992; Kelly-Borges & Pomponi, 1994; Chombard et al., 1997, 1998; Alvarez, 1998; Chombard, 1998), and the metazoan nature of the Porifera was also addressed, thus updating a controversy going back to the end of the last century (Delage & Hérouard, 1899; Christen et al., 1991; Lafay et al., 1992; Müller, 1997).

The aims of this paper are, thus, to review the major results obtained using molecular genetics and tentatively evaluate them in the context of a general strategy for the study of the as-yet-unanswered questions in sponge phylogeny and evolution. All data were gathered from literature and gene databases (Genbank and EMBL) (Tables 1 and 2).

Allozyme polymorphism

Among marine sponges, studies investigating allozyme divergence have been used to evaluate gene flow between allopatric populations of various species (Solé-Cava et al., 1991a; Klautau et al., 1994; Boury-Esnault et al., 1999) and to detect sibling species (Solé-Cava & Thorpe, 1986; Solé-Cava et al., 1991b, 1992; Boury-Esnault et al., 1992; Sarà et al., 1993; Muricy et al., 1996) including those within supposedly cosmopolitan species (Solé-Cava et al., 1991a; Boury-Esnault et al., 1992; Klautau et al., 1999). A compilation of available data (Solé-Cava & Boury-Esnault, 1999) indicates that the frequency distribution of intraspecific and intergeneric gene identities for the different genera of marine sponges that have been studied (12 Demospongiae and 1 Calcarea) was similar to that found for other organisms. The mean value of genetic identity I (Nei, 1972) between congeneric sponges species is 0.42 as compared to a value of 0.54 for invertebrates generally (see Thorpe, 1983). However, Solé-Cava & Boury-Esnault (1999) suggested that the data for the distribution of I values may show a degree of bimodality, with a few sponge genera having mainly very low (0.30) I values between species. This may indicate inadequate definition of such genera, possibly resulting from a restricted range of useful morphological characters. Furthermore, some species within these very broad genera have been found (from allozyme studies) to be apparently more closely related to species in other genera than to some congeneric species. Thus polyphyly is probable, a hypothesis corroborated by results from other types of taxonomical studies. For example, the polyphyly of *Cliona* has been suggested recently following a re-evaluation of morphological characters (Rosell & Uriz, 1997), and the monophyly of *Axinella* has been questioned from a chemotaxonomical work (Braekman et al., 1992); this later result was further supported by a recent phylogenetic molecular analysis (Alvarez, 1998).

Sequence comparisons

rRNA molecules

At the dawn of the new era of molecular phylogeny, ribosomal RNAs (rRNAs), with a function conserved from bacteria to vertebrates, appeared to be the molecules of choice for reconstructing molecular phylogenies. The first phylogenetic studies of sponges and other metazoans (molluscs, tunicates, vertebrates, etc.) (Dams et al., 1982; Komiya et al., 1983) compared sequences of two small rRNA molecules (5S and 5.8S). These sequences appeared to be highly conserved and upheld the metazoan nature of the Porifera (Komiya et al., 1983). However, it was later recognized that these rRNAs are too short and the level of saturation too high to provide a signal to unambiguously resolve phylogenetic relationships (Halanych, 1991).

In the 1990s, sequences of longer rRNA molecules, 18S rRNA (Kelly-Borges et al., 1991) and 28S rRNA (Christen et al., 1991), were compared to reappraise the phylogenetic relationships between sponge genera, families, orders and classes (currently, 41 complete or partial sequences of 18S rRNA, and 75 partial sequences of 28S rRNA are available) (Table 1).

Among the 18S rRNA sequences available, 30 are short (about 250–500 bases) with few overlapping zones (Figure 1). These partial sequences have been used in an attempt to resolve phylogenetic relationships among the Hadromerida (Kelly-Borges et al., 1991) and among lithistids (Kelly-Borges & Pomponi, 1994). Low variability of 18S rRNA partial sequences in sponges limits their value in phylogeny (Kelly-Borges & Pomponi, 1994). There are only 11 near complete (about 1500 bases) sequences of 18S rDNA (six for Demospongiae, four for Calcarea and one for Hexactinellida; Figure 1). Another sequence of Hexactinellida was mentioned in West & Powers (1993), but the sequence itself does not appear in the publication or in a sequence database. Both sub-classes of Calcarea are represented, but by only one genus in Calcinea and by two genera in different families in Calcaronea.

In the Demospongiae, the dataset is composed of six species belonging to six different genera and families spread over five orders. Available sequences of

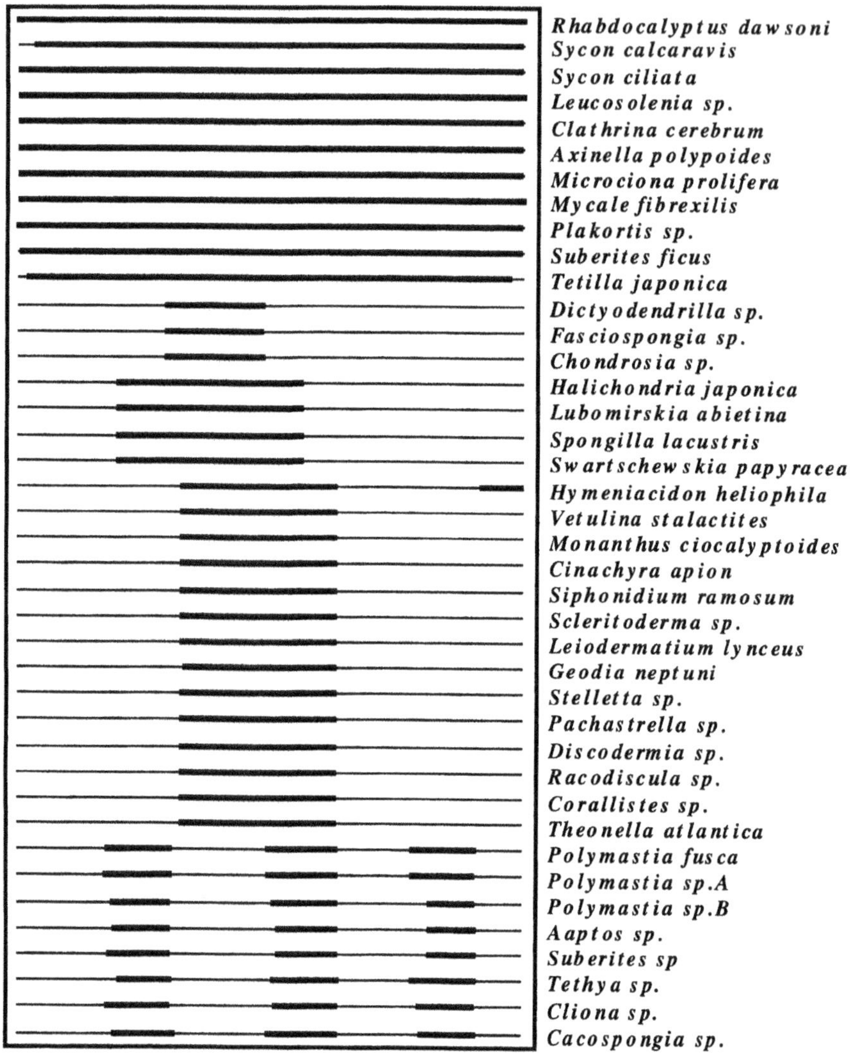

Figure 1. 18S rRNA of 40 Porifera (1 for Hexactinellida, 4 for Calcarea, top of the figure, and 35 for Demospongiae). Thick line corresponds to the part of the molecule for which a sequence is available.

the classes of Porifera are inadequate to produce a phylogenetic tree resolving the relationships between Demospongiae and Calcarea and between Porifera and other diploblasts (Figure 1). These sequences have mainly been used within a very large dataset to investigate the monophyly of metazoans (Kobayashi et al., 1993; Cavalier-Smith et al., 1996) and the origin of the bilaterians (Collins, 1998). In Porifera paraphyly has often been suggested (e.g. Lafay et al., 1992; Cavalier-Smith et al., 1996), although without producing any robust tree and without support from morphological characters. In a recent work (Zrzavy et al., 1998), based on the frequent grouping in molecular trees of Calcarea and Ctenophora, the Porifera have been divided into two distinct phyla: Silicospongea and Calcispongea. However, the methodology used is controversial: the analysis was performed on few sequences – two Calcarea, two Demospongiae, two Ctenophora and two Cnidaria – and the morphological characters used are questionable.

The 28S rRNA molecule is composed of 24 domains, alternately conserved and variable. Depending on the level of variability, the different domains can be used to resolve problems at various levels of taxonomic divergence. The first studies investigating sponge phylogeny (Christen et al., 1991; Lafay

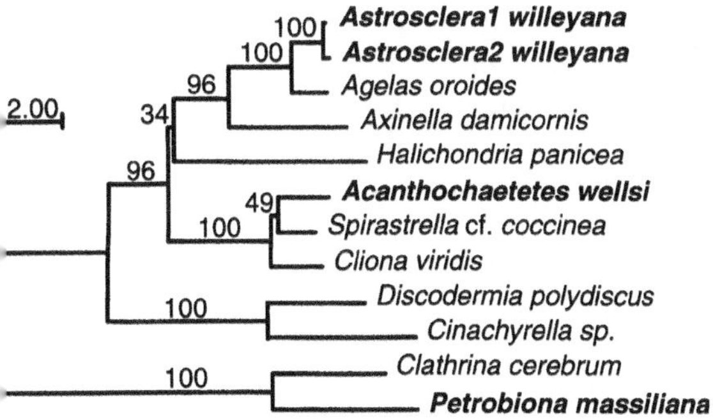

igure 2. Phylogenetic tree obtained with PAUP by exhaustive ana-sis of short-length aligned 28S rRNA sequences (374 bp) for 7 pecies of Demospongiae, 3 species of "sclerosponges", 2 species f Calcarea (one Calcinea, one "pharetronid") using ACCTRAN ptimization. Tree length = 268, CI = 0.795, RI = 0.767, and G1 −1.50. Bootstrap proportions (1000 replicates using a Branch and ound search) are shown above internal branches. Sclerosponges pecies are shown in bold (from Chombard et al., 1997).

t al., 1992) used about 400 nucleotides belonging o the first three domains of the 5' extremity. La-ay et al. (1992) were the first to sequence a large umber of diploblastic species to attempt to resolve heir phylogenetic relationships: 11 marine sponges nine Demospongiae and two Calcarea), three Cten-phora, one Placozoa and 10 Cnidaria. The molecular ree supported the division between the two classes alcarea and Demospongiae, but the rRNA sequence omparison suggested a separation earlier than usually hought. According to Lévi (1956, 1973), the nine spe-ies of Demospongiae belong to six orders grouped nto two subclasses: Ceractinomorpha and Tetractino-norpha. The molecular tree contradicts this generally ccepted view (Bergquist, 1978) and shows an early livergence between Haplosclerida-Petrosida on one ide and Poecilosclerida-Halichondrida-Axinellida-Agelasida on the other. Each of these two groups eems to form a well separated monophyletic unit, ut the latter is composed of orders belonging to both ubclasses defined by Lévi (1956).

The deep branching and the calculated genetic listances between the three different marine sponge groups identified by Lafay et al. (1992), as well as etween each of these groups and the phyla Cnid-ria, Ctenophora and the Placozoa, again suggest he paraphyly of the Porifera. However, the appar-ent phylogenetic relationship of the Calcarea with the Cnidaria and Ctenophora occurs within an area of poor esolution of the tree, the internodes being short and he external branches long, so that the branching or-der of the deep divergences cannot be ascertained with precision. This work demonstrated that 28S rRNA is suitable for the resolution of questions dealing with a relatively low taxonomic level; with the use of longer sequences and a larger set of species it might also be of use at higher taxonomic levels. Indeed, Chombard et al. (1997), analysing about 800 nucleotides from the first four domains of the 5' extremity of 28S, proved the polyphyly of the class Sclerospongiae; their data showed that the two hypercalcified genera, *Astrosclera* and *Acanthochaetetes* are Demospongiae (Figure 2). Within the Demospongiae, *Astrosclera* comprises a clade along with the axinellids, *Axinella* and *Agelas,* and *Acanthochaetetes* with the hadromerids, *Spirastrella* and *Cliona* (as previously suggested by morphological data; Vacelet, 1985).

The combination of morphology and molecular systematics seems to be a promising approach to the understanding of the phylogeny of a group (Smith et al., 1992, 1995; Lafay et al., 1995) and to testing the homology of diagnostic characters. Chombard et al. (1998) tested the monophyly of the order Tetractinellida, as well as the monophyly of different taxa within this order, and the homology of tetractine spicules and of different asterose microscleres. The monophyly of Tetractinellida was supported as well as the homology of the tetractine spicules, which therefore constitute the main synapomorphy of the group. The molecular tree, also supported by a scanning electron microscopy study of microcleres, sustains a primary homology of streptoscleres and sanidasters and leads to the removal of *Stryphnus* from the Euastrophorida and to a re-evaluation of the taxonomical position of this genus as well as those of *Asteropus*, *Holoxea* and all Ancorinidae possessing sanidasters. However, the appearance and secondary loss of spicule types has been indicated in *Penares,* which, according to the molecular tree, should actually be classified as an *Erylus* without aspidasters.

In conclusion, the 5' extremity of the 28S rRNA seems to be appropriate for resolving sponge phylogenetic relationships between genera and families using divergent domains (D2) and at the order or class level using the more conservative domains (C1, D1 and C2). A longer sequence including the domains D3, D6 and D7 would be more informative, as would the secondary structure of this molecule (Kjer, 1995), but the latter requires the entire 28S rRNA sequence. The resolution of the recent 28S rRNA trees (Chombard et al., 1998) is higher than that obtained with 18S rRNA sequences (Kelly-Borges et al., 1994). For example, the genera *Corallistes* and *Discodermia* differ only by one nucleotide out of 535 aligned sites for 18S rRNA, whereas there are 83 variable nucleotides among the 885 aligned sites for 28S rRNA. Nevertheless, complete sequences of 18S rRNA give valuable data on relationships between classes or phyla.

Hsp70 gene

Members of the heat-shock protein (Hsp) family carry out a highly conserved molecular chaperone function and are involved in the intracellular transport of proteins and in the protection of organisms against stress (Lindquist & Craig, 1988; Gething & Sambrook, 1992). Within this family, Hsp70 is a promising candidate as a phylogenetic marker among eukaryotes, and sequence data for this molecule are accumulating rapidly. Indeed, the primary structure of Hsp70 is highly conserved during evolution, and the high degree of sequence conservation, in conjunction with its large size and an ancient conserved function, makes it a particularly useful gene for investigating deep phylogenetic relationships at the class, phylum and kingdom levels (Gupta & Golding, 1993; Koziol et al., 1996, 1998a, b; Borchiellini et al., 1998; Budin & Philippe, 1998).

The first sequence of Hsp70 from marine sponges was obtained for the demosponge *Geodia cydonium* (Koziol et al., 1996). Koziol et al. (1998b) sequenced this molecule in the two other classes of Porifera: Hexactinellida (*Rhabdocalyptus dawsoni*) and Calcarea (*Sycon raphanus*). These sequences were compared to the DnaK sequence of *Escherichia coli*. The resulting tree indicated the clustering of the three sponges, but no further phylogenetic information could be inferred, because of the low number of metazoan sequences included and the use of such a remote outgroup (Koziol et al., 1998b). Recently, Borchiellini et al. (1998) sequenced Hsp70 of four Demospongiae, one Hexactinellida, one Calcarea, two Cnidaria and one Ctenophora to investigate the monophyly of metazoans and that of Porifera. The phylogenetic tree (Figure 3) indicates the monophyly of the Metazoa with strong supporting bootstrap values. The data also support the suggestion that the Fungi are a sister-group to Metazoa. Within the animal clade, the topology is unresolved, but the results suggest that diploblasts and triploblasts diverged very early and that the ancestor of the triploblastic animals may have emerged much earlier than is classically assumed. These results support those obtained with the 28S rRNA tree (Lafay et al., 1992) as well as the old hypothesis of Haeckel (1873). Among the diploblasts the Porifera seem to be monophyletic with a relatively good resolution (the bootstrap value being 88%) in congruence with the traditional view based on morphological characters. However the number of sequences available for the Calcarea and the Hexactinellida is currently too small to allow the resolution of the conflict between rRNA and Hsp70 trees.

Other genes

Many other genes have been sequenced in sponges but for too few species to be of major use in a phylogenetic analysis of the phylum (Table 2). Among these, genes coding for receptor tyrosine kinase (RTK) (Schäcke et al., 1994), integrin (Pancer et al., 1997) lectins (Pfeifer et al., 1993a; Gamulin et al., 1994; Hirabayashi & Kasai, 1998), collagen (Exposito & Garrone, 1990; Boute et al., 1996), ets transcription factors (Degnan et al., 1993), ubiquitin (Pfeifer et al., 1993b; Müller et al., 1994), and elongation factor α (Kobayashi & Satoh, 1998) could be good candidates because of their highly conserved function among animals. Another group of genes which could carry phylogenetic information are the homeobox genes (Finnerty, 1998), which so far have been sequenced in *Ephydatia fluviatilis* (Coutinho et al., 1994; Seimiya et al., 1994, 1997) and *E. mülleri* (Richelle-Maurer et al., 1998). A homeobox gene has also been reported for *Geodia cydonium* (Kruse et al., 1994), but this has been questioned (Seimiya et al., 1998) and may be an integrin sequence. Degnan et al. (1995) identified two homeobox genes in *Tethya* cf. *aurantia* and *Haliclona* sp., but the sequences are very short and the presence of these genes has yet to be confirmed in these species (Seimiya et al., 1998).

Conclusions

There is no miracle molecular marker suitable for all taxonomic levels and with which the entire phylogenetic history of sponges can be reconstructed. However, at each taxonomic level, certain markers have provided valuable insights.

Undoubtedly allozyme electrophoresis is a very powerful method in Porifera for the detection of sibling species, false cosmopolitanism, and polyphyletic genera. This method is most useful coupled with a precise re-evaluation of morphological characters.

At higher taxonomic levels, the initial hope put in rRNA gene sequences to resolve phylogenetic re-

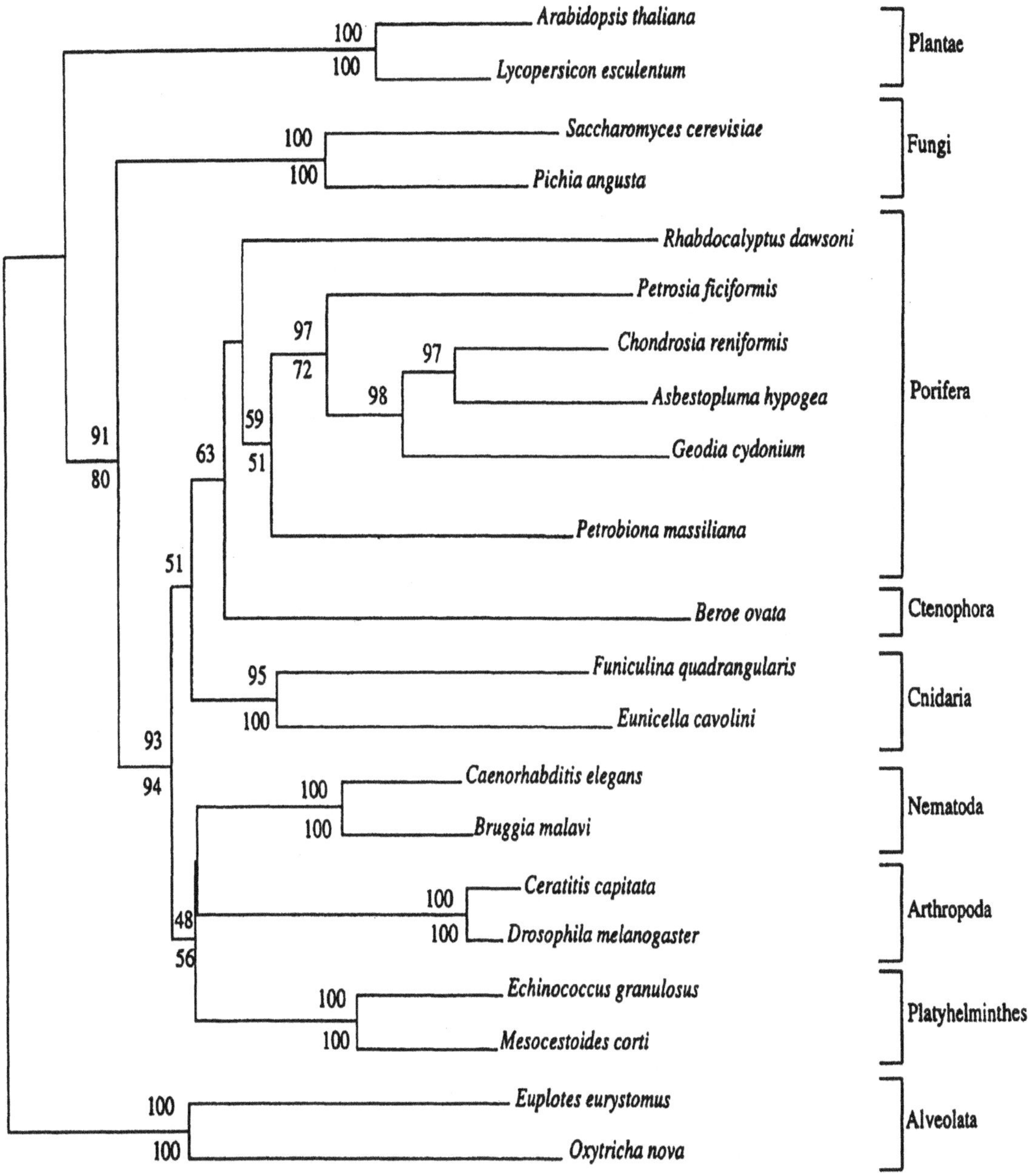

Figure 3. Phylogenetic tree based on a comparison of nucleic acid sequences of cytoplasmic Hsp 70. The tree depicts relationships among Hsp sequences from 21 species (2 Plantae, 2 Fungi, 6 Porifera, 1 Ctenophora, 2 Cnidaria and 6 Triploblasts) using neighbor-joining analysis with the Alveolata lineage as the outgroup. The tree was based on nucleotide sequence at the first two codon positions only. Results of 500 neighbor-joining bootstrap replicates are shown above the branches. The maximum-parsimony tree for the same data set shows a similar topology. The result of a bootstrap (maximum-parsimony) analysis is noted below the branches (from Borchiellini et al., 1998).

Table 2. List of genes which have been completely or partially sequenced in Porifera. ? = sequences that have been considered doubtful by other authors (see text). 1 = Pfeifer et al, 1993a; 2 = Degnan et al., 1993; 3 = Exposito & Garonne, 1990; 4 = Boute et al., 1996; 5 = Coutinho et al., 1994; 6 = Degnan et al., 1995; 7 = Gamulin et al., 1994; 8= Hirabayashi & Kasai, 1998; 9 = Kobayashi & Satoh, 1998; 10 = Kruse et al., 1994; 11 = Müller et al., 1994; 12 = Pancer et al., 1997; 13 = Pfeifer et al., 1993b; 14 = Richelle-Maurer et al., 1998; 15 = Schäcke et al., 1994; 16 = Seimiya et al., 1994; 17 = Seimiya et al., 1998; 18 = Seimiya et al., 1997; 19 = Biesalski et al., 1992; 20 = Gamulin et al., 1997; 21= Kruse et al., 1996; 22 = Krasko et al., 1997a; 23 = Krasko et al., 1997b; 24 = Koziol et al., 1997; 25 = Fernàndez-Busquets & Burger, 1997; 26 = Nikoh et al., 1997; 27 = Kanzawa et al., 1995; 28 = Lorenz et al., 1996; 29 = Davies, 1997

Class	Demospongiae											Calcarea
Order	Tetractinellida	Hadromerida		Poecilosclerida		Haplosclerida				Halichondrida	Homosclero morpha	
Genes/Species	*Geodia cydonium*	*Suberites domuncula*	*Tethya cf aurantia*	*Microciona prolifera*	*Ophlitaspongia tenuis*	*Haliclona sp*	*Ephydatia fluviatilis*	*Ephydatia muelleri*	*Spongilla lacustris*	*Halichondria okadai*	*Pseudocorticium jarrei*	*Sycon raphanus*
Lectins 1,7,8	x											
Ets proto-oncogene 2			x			x						
Retinoic acid 19	x											
Triosephosphate isomerase 26							x					
Aldolase 26							x					
Elongation factor alpha 9							x					
Integrin 12	x				x							
Aggregation factor 25				x								
Homeobox genes 5, 6, 10, 14, 16, 17, 18	?		?			?	x					
Collagen gene 3				x				x			x	
Collagen type IV 4											x	
ß,ρ, crystalline 22	x											
GDP dissociation inhibitor 23	x											
Myosin 27, 28	x									x		
Polyubiquitin 11, 13	x											
Protein Kinase 21	x	x										
Receptor Tyrosine Kinase 15, 21	x	x										x
Enolase 29									x			
Telomerase 24	x	x										

lationships has lately been tempered, and metazoan phylogenies based on rRNA have been questioned by several authors (Rodrigo et al., 1994; Aguinaldo et al., 1997; Abouheif et al., 1998; Budin & Philippe, 1998).

Hsp70 gene is a promising alternative (Budin & Philippe, 1998). However, the data set is currently too small to be evaluated for Porifera, although the first results have been encouraging (Borchiellini et al., 1998); more work is needed to assess the adequacy of Hsp70 sequences for resolving phylogenetic problems in the Porifera.

Among the other genes that have been sequenced, homeobox genes appear to have the greatest potential for the study of animal evolution, since they constitute a link between the knowledge of phylogenetic relationships and the understanding of developmental processes. However, they remain difficult to study, and unfortunately it is unlikely that a sufficient number of sequences, from a large enough sample of taxa to allow comprehensive analyses, will become available very rapidly.

The molecular systematics of Porifera based on comparisons of gene sequences is still in its early stages after a promising start (Kelly-Borges et al., 1991; Lafay et al., 1992). Since then only three papers (Kelly-Borges & Pomponi, 1994; Chombard et al., 1997, 1998) and two Ph.D. theses (Alvarez, 1998; Chombard, 1998) have dealt with molecular phylogeny, versus 18 papers during the same period using allozyme electrophoresis. This paucity of sequence studies could be linked to the absence of specific primers for Porifera (Chombard, 1998). Nevertheless, it is puzzling that this phylum, considered as the most ancient among multicellular animals (Haeckel, 1873; Nielsen, 1995), has not received more attention from zoologists at the molecular level.

Comparison between datasets obtained with the different molecules is for the moment difficult or impossible because different taxa have been analyzed for different genes. In future work it would be interesting to increase the number of taxa common to the different gene sequence datasets, and to integrate the results from molecular studies within a broader framework involving all aspects of classical taxonomy.

Acknowledgements

This work benefited from fruitful discussions with Jean Vacelet, Yannick Le Parco and Michael Manuel and from the useful comments of anonymous referees. This work was supported by Réseau National de Biosystématique (ACC-SV7) and the European Program MAS3-CT97-018. Nicole Boury-Esnault had a grant from CNPq to participate in the International Workshop on Marine Genetics in Rio de Janeiro in September 98, where this work was presented.

References

Abouheif, E., R. Zardoya & A. Meyer, 1998. Limitations of metazoan 18S rRNA sequence data: implications for reconstructing a phylogeny of the animal kingdom and inferring the reality of the Cambrian explosion. J. mol. Evol. 47: 394–405.

Aguinaldo, A. M. A., J. M. Turbeville, L. S. Linford, M. C. Rivera, J. R. Garey, R. A. Raff & J. A. Lake, 1997. Evidence for a clade of nematodes, arthropods and other moulting animals. Nature 387: 489–492.

Alvarez de Glasby, B., 1998. The phylogenetic relationships of the family Axinellidae (Porifera, Demospongiae). Ph.D. Thesis, The Australian National University, Canberra.

Bergquist, P. R., 1978. Sponges. Hutchinson, London.

Biesalski, H. K., G. Doepner, G. Tzimas, V. Gamulin, H. C. Schröder, R. Batel, H. Nau & W. E. G. Müller, 1992. Modulation of myb gene expression in sponges by retinoic acid. Oncogene 7: 1765–1774.

Borchiellini, C., N. Boury-Esnault, J. Vacelet & Y. Le Parco, 1998. Phylogenetic analysis of the Hsp70 sequences reveals the monophyly of Metazoa and specific phylogenetic relationships between animals and fungi. Mol. Biol. Evol. 15: 647–655.

Boury-Esnault, N., M. Klautau, C. Bézac, J. Wulff & A. M. Solé-Cava, 1999. Comparative study of putative conspecific sponge populations from both sides of the Isthmus of Panama. J. mar. biol. Ass. U.K. 79: 39–59.

Boury-Esnault, N., A. M. Solé-Cava & J. P. Thorpe, 1992. Genetic and cytological divergence between colour morphs of the Mediterranean sponge *Oscarella lobularis* Schmidt (Porifera, Demospongiae, Oscarellidae). J. nat. Hist. 26: 271–284.

Boute, N., J. Y. Exposito, N. Boury-Esnault, J. Vacelet, N. Noro, K. Miyazaki, K. Yoshigato & R. Garrone, 1996. Type IV collagen in sponges, the missing link in basement membrane ubiquity. Biol. Cell. 88: 37–44.

Braekman, J.-C., D. Daloze, C. Stoller & R. W. M. van Soest, 1992. Chemotaxonomy of *Agelas* (Porifera: Demospongiae). Biochem. Syst. Ecol. 20: 417–431.

Budin, K. & H. Philippe, 1998. New insights into the phylogeny of Eukaryotes based on ciliate Hsp70 sequences. Mol. Biol. Evol. 15: 943–956.

Cavalier-Smith, T., M. T. E. P. Allsopp, E. E. Chao, N. Boury-Esnault & J. Vacelet, 1996. Sponge phylogeny, animal monophyly, and the origin of the nervous system: 18S rRNA evidence. Can. J. Zool. 74: 2031–2045.

Chombard, C., 1998. Les Demospongiae à asters: essai de phylogénie moléculaire. Homologie du caractère 'aster'. Ph.D. Thesis, Museum, national d'Histoire naturelle Paris.

Chombard, C., N. Boury-Esnault & S. Tillier, 1998. Reassesment of homology of morphological charcters in tetractinellid sponges based on molecular data. Syst. Biol. 47: 351–366.

Chombard, C., A. Tillier, N. Boury-Esnault & J. Vacelet, 1997. Polyphyly of 'sclerosponges' (Porifera, Demospongiae) supported by 28S ribosomal sequences. Biol. Bull. 193: 359–367.

Christen, R., A. Ratto, A. Baroin, R. Perasso, K. G. Grell & A. Adoutte, 1991. An analysis of the origin of metazoans, using comparisons of partial sequences of the 28S rRNA, reveals an early emergence of triploblasts. Embo J. 10: 499–503.

Collins, A. G., 1998. Evaluating multiple alternative hypotheses for the origin of bilateria: An analysis of 18S rRNA molecular evidence. Proc. natn. Acad. Sci. U.S.A. 95: 15458–15463.

Coutinho, C. C., S. Vissers & G. Van de Vyver, 1994. Evidence of homeobox genes in the freshwater sponge *Ephydatia fluviatilis*. In van Soest, R. W. M., T. M. G. van Kempen & J. C. Braekman (eds), Sponge in Time and Space; Biology, Chemistry, Paleontology. A. A. Balkema, Rotterdam: 385–388.

Dams, E., A. Vandenberghe & R. de Wachter, 1982. Nucleotide sequence of three Poriferan 5S ribosomal RNAs. Nucl. Acids Res. 10: 5297–5302.

Davies, R. E., 1997. Surprising diversity and distribution of spliced leader RNAs in flatworms. Mol. Biochem. Parasitol. 87: 29–48.

Degnan, B. M., S. M. Degnan, A. Giusti & D. E. Morse, 1995. A hox/hom homeobox gene in sponges. Gene 155: 175–177.

Degnan, B. M., S. M. Degnan, T. Naganuma & D. E. Morse, 1993. The ETS multigene family is conserved throughout the Metazoa. Nucl. Acids Res. 21: 3479–3484.

Delage, Y. & E. Hérouard, 1899. Traité de zoologie concrète (Mésozoaires-Spongiaires), vol. 2. Schleicher Frères, Paris.

Exposito, J.-Y. & R. Garrone, 1990. Characterization of a fibrillar collagen gene in sponges reveals the early evolutionary appearance of two collagen gene families. Proc. natn. Acad. Sci. U.S.A. 87: 6669–6673.

Fernàndez-Busquets, X. & M. M. Burger, 1997. The main protein of the aggregation factor responsible for species-specific cell adhesion in the marine sponge *Microciona prolifera* is highly polymorphic. J. biol. Chem. 272: 27839–27847.

Finnerty, J. R., 1998. Homeoboxes in sea anemones and other nonbilaterian animals: implications for the evolution of the Hox cluster and the zootype. Curr. Top. dev. Biol. 40: 211–254.

Fromont, J. P. & P. R. Bergquist, 1990. Structural characters and their use in sponge taxonomy; when is a sigma not a sigma? In Rützler, K. (ed.), New Perspectives in Sponge Biology. Smithsonian Institution Press, Washington, D.C: 273–278.

Gamulin, V., K. Pfeifer, H. Bretting, I. Spreitzer & W. E. G. Müller, 1994. Identification and characterization of the first S-type lectins from invertebrates: isolation from the marine sponge *Geodia cydonium*. In Müller, W. E. G. (ed.), Use of Aquatic Invertebrates as Tools for Monitoring of Environmental Hazards. Gustav Fischer Verlag, Stuttgart: 213–223.

Gamulin, V., A. Skorokhod, V. Kavsan, I. M. Müller & W. E. G. Müller, 1997. Experimental indication in favor of the intronslate theory: the receptor tyrosine kinase gene from the sponge *Geodia cydonium*. J. mol. Evol. 44: 242–252.

Gething, M. J. & J. P. Sambrook, 1992. Protein folding in the cell. Nature 355: 33–45.

Gupta, R. S. & G. B. Golding, 1993. Evolution of Hsp70 gene and its implications regarding relationships between Archaebacteria, Eubacteria, and Eukaryotes. J. mol. Evol. 37: 573–582.

Haeckel, E., 1873. On the Calcispongiae, their position in the animal kingdom. Ann. Mag. nat. Hist. 4: 241–262.

Halanych, K. M., 1991. 5S ribosomal RNA sequences inappropriate for phylogenetic reconstruction. Mol. Biol. Evol. 8: 249–253.

Hirabayashi, J. & K.-I. Kasai, 1998. Evolution of animal lectins. In Müller, W. E. G. (ed.), Molecular Evolution: Evidence for Monophyly of Metazoa. Vol. 19. Springer, Berlin: 45–88.

Itskovich, V. B., S. I. Belikov, S. M. Efremova & Y. Masuda, 1999. Phylogenetic relationships between Lubomirskiidae, Spongillidae and some marine sponges according partial sequences of 18S rDNA. Mem. Queensland Mus. 44: 275–280.

Kanzawa, N., H. Takano-Ohmuro & K. Maruyama, 1995. Isolation and characterization of sea sponge myosin. Zool. Sci. 12: 765–769.

Kelly-Borges, M., P. R. Bergquist & P. L. Bergquist, 1991. Phylogenetic relationships within the order Hadromerida (Porifera, Demospongiae, Tetractinomorpha) as indicated by ribosomal RNA sequence comparisons. Biochem. Syst. Ecol. 19: 117–125.

Kelly-Borges, M. & S. A. Pomponi, 1994. Phylogeny and classification of lithistid sponges (Porifera: Demospongiae): A preliminary assessment using ribosomal DNA sequence comparisons. Mol. mar. Biol. Biotechnol. 3: 87–103.

Kjer, K. M., 1995. Use of rRNA secondary structure in phylogenetic studies to identify homologous positions: an example of alignment and data presentation from the frogs. Mol. Phylogenet. Evol. 4: 314–330.

Klautau, M., C. A. M. Russo, C. Lazoski, N. Boury-Esnault, J. P. Thorpe & A. M. Solé-Cava, 1999. Does cosmopolitanism in morphologically simple species result from overconservative systematics? A case study using the marine sponge *Chondrilla nucula*. Evolution 53: 1414–1422.

Klautau, M., A. M. Solé-Cava & R. Borojevic, 1994. Biochemical systematics of sibling sympatric species of *Clathrina* (Porifera: Calcarea). Biochem. Syst. Ecol. 22: 367–375.

Kobayashi, M. & N. Satoh, 1998. Early evolution of the Metazoa: an inference from the elongation factor-1α. In Müller, W. E. G. (ed.), Molecular Evolution: Evidence for Monophyly of Metazoa. Vol. 19. Springer, Berlin: 177–185.

Kobayashi, M., M. Takahashi, H. Wada & N. Satoh, 1993. Molecular phylogeny inferred from sequences of small subunit ribosomal DNA, supports the monophyly of the Metazoa. Zool. Sci. 10: 827–833.

Komiya, H., M. Hasegawa & S. Takemura, 1983. Nucleotide sequences of 5S rRNAs from sponge *Halichondria japonica* and tunicate *Halocynthia roretzi* and their phylogenetic positions. Nucl. Acids Res. 11: 1969–1974.

Koziol, C., R. Borojevic, R. Steffen & W. E. G. Müller, 1997. Sponges (Porifera) model systems to study the shift from immortal to senescent somatic cells: the telomerase activity in somatic cells. Mech. Ageing Dev. 100: 107–120.

Koziol, C., N. Kobayashi, I. M. Müller & W. E. G. Müller, 1998a. Cloning of sponge heat shock proteins: evolutionary relationships between the major kingdoms. J. Zool. Syst. Evol. Res. 36: 101–109.

Koziol, C., S. P. Leys, I. M. Müller & W. E. G. Müller, 1998b. Cloning of Hsp70 genes from the marine sponge *Sycon raphanus* (Calcarea) and *Rhabdocalyptus dawsoni* (Hexactinellida). An approach to solve the phylogeny of sponges. Biol. J. linn. Soc. 62: 581–592.

Koziol, C., C. Wagner-Hülsmann, A. Mikoc, V. Gamulin, M. Kruse, Z. Pancer, H. Schacke & W. E. G. Müller, 1996. Cloning of a heat inducible biomarker, the cDNA encoding the 70-kDa heat shock protein, from the marine sponge *Geodia cydonium*: response to natural stressors. Mar. Ecol. Prog. Ser. 136: 153–161.

Krasko, A., I. M. Müller & W. E. G. Müller, 1997a. Evolutionary relationships of the metazoan $\beta\chi$-crystallins, including that from the marine sponge *Geodia cydonium*. Proc. r. Soc. Lond. B 264: 1077–1084.

Krasko, A., U. Scheffer, C. Koziol, Z. Pancer, R. Batel, F. A. Badria & W. E. G. Müller, 1997b. Diagnosis of sublethal stress in the marine sponge *Geodia cydonium*: application of the 70 kDa heat-shock protein and a novel biomarker, the Rab GDP dissociation inhibitor, as probes. Aquat. Toxicol. 37: 157–168.

Kruse, M., V. Gamulin, H. Cetkovic, Z. Pancer, I. M. Müller & W. E. G. Müller, 1996. Molecular evolution of the Metazoan protein kinase C multigene family. J. mol. Evol. 43: 374–383.

Kruse, P. D., A. Mikoc, H. Cetkovic, V. Gamulin, B. Rinkevich, I. Müller & W. E. G. Müller, 1994. Molecular evidence for the presence of a developmental gene in the lowest animals: identification of a homeobox-like gene in the marine sponge *Geodia cydonium*. Mech. Ageing Dev. 77: 43–54.

Lafay, B., N. Boury-Esnault, J. Vacelet & R. Christen, 1992. An analysis of partial 28S ribosomal RNA sequences suggests early radiations of sponges. Biosystems 28: 139–151.

Lafay, B., A. B. Smith & R. Christen, 1995. A combined morphological and molecular approach to the phylogeny of the Asteroids. Syst. Biol. 44: 190–208.

Lévi, C., 1956. Etude des *Halisarca* de Roscoff. Embryologie et systématique des démosponges. Arch. Zool. exp. gén. 93: 1–184.

Lévi, C., 1973. Systématique de la classe des Demospongiaria (Démosponges). In Grassé, P. P. (ed.), Spongiaires. Vol. 3(1). Masson & Co., Paris: 577–632.

Li, C.-W., J.-Y. Chen & T.-E. Hua, 1998. Precambrian sponges with cellular structures. Science 279: 879–882.

Lindquist, S. & E. A. Craig, 1988. The heat shock proteins. Annu. Rev. Genet. 22: 631–677.

Lorenz, B., R. Bohnensack, V. Gamulin, R. Steffen & W. E. G. Müller, 1996. Regulation of motility of cells from marine sponges by calcium ions. Cell. Signal. 8: 517–524.

Müller, W. E. G., 1997. Molecular phylogeny of Eumetazoa: experimental evidence for monophyly of animals based on genes in sponges (Porifera). Progr. mol. subcell. Biol. 19: 89–132.

Müller, W. E. G., I. Müller & V. Gamulin, 1994. Phylogenetic relationship of ubiquitin repeats in the polyubiquitin gene from the marine sponge *Geodia cydonium*. In Müller, W. E. G. (ed.), Use of Aquatic Invertebrates as Tools for Monitoring of Environmental Hazards. Gustav Fischer Verlag, Stuttgart: 187–200.

Muricy, G., A. M. Solé-Cava, J. P. Thorpe & N. Boury-Esnault, 1996. Genetic evidence for extensive cryptic speciation in the subtidal sponge *Plakina trilopha* (Porifera: Demospongiae: Homoscleromorpha). Mar. Ecol. Prog. Ser. 138: 181–187.

Nielsen, C., 1995. Animal Evolution. Interrelationships of the Living Phyla. Oxford University Press, Oxford.

Nikoh, N., N. Iwabe, K. Kuma, M. Ohno, T. Sugiyama, Y. Watanabe, K. Yasui, Z. Schi-cui, K. Hori, Y. Shimura & T. Miyata, 1997. An estimate of divergence time of parazoa and eumetazoa and that of cephalochordata and vertebrata by aldolase and triose phosphate clocks. J. mol. Evol. 45: 97–106.

Odorico, D. M. & D. J. Miller, 1997. Internal and external relationships of the Cnidaria: implications of primary and predicted secondary structure of the 5'-end of the 23S-like rDNA. Proc. r. Soc. Lond. B 264: 77–82.

Pancer, Z., M. Kruse, I. Müller & W. E. G. Müller, 1997. On the origin of metazoan adhesion receptors: cloning of integrin α subunit from the sponge *Geodia cydonium*. Mol. Biol. Evol. 14: 391–398.

Pfeifer, K., W. Frank, H. C. Schröder, V. Gamulin, B. Rinkevich, R. Batel, I. M. Müller & W. E. G. Müller, 1993b. Cloning of the polyubiquitin cDNA from the marine sponge *Geodia cydonium* and its preferential expression during reaggregation of cells. J. Cell Sci. 106: 545–554.

Pfeifer, K., M. Haasemann, V. Gamulin, H. Bretting, F. Fahrenholz & W. E. G. Müller, 1993a. S-type lectins occur also in invertebrates: high conservation of the carbohydrate recognition domain in the lectin genes from the marine sponge *Geodia cydonium*. Glycobiology 3: 179–184.

Richelle-Maurer, E., G. van de Vyver, S. Vissers & C. C. Coutinho, 1998. Homeobox-containing genes in freshwater sponges: characterization, expression, and phylogeny. In Müller, W. E. G. (ed.), Molecular Evolution: Evidence for Monophyly of Metazoa. Vol. 19. Springer, Berlin: 157–175.

Rodrigo, A. G., P. R. Bergquist, P. L. Bergquist & P. R. Reeves, 1994. Are sponges animals? An investigation into the vagaries of phylogenetic inference. In van Soest, R. W. M., T. M. G. van Kempen & J. C. Braekman (eds), Sponge in Time and Space; Biology, Chemistry, Paleontology. A. A. Balkema, Rotterdam: 47–54.

Rosell, D. & M. J. Uriz, 1997. Phylogenetic relationships within the excavating Hadromerida (Porifera), with a systematic revision. Cladistics 13: 349–366.

Sarà, M., G. Corriero & G. Bavestrello, 1993. *Tethya* (Porifera, Demospongiae) species coexisting in a Maldivian coral reef lagoon – taxonomical, genetic and ecological data. Mar. Ecol. 14: 341–355.

Schäcke, H., H. C. Schröder, V. Gamulin, B. Rinkevich, I. Müller & W. E. G. Müller, 1994. Molecular cloning of a tyrosine kinase gene from the marine sponge *Geodia cydonium*: a new member belonging to the receptor tyrosine kinase class II family. Mol. Membrane Biol. 11: 101–107.

Seimiya, M., H. Ishiguro, K. Miura, Y. Watanabe & Y. Kurosawa, 1994. Homeobox-containing genes in the most primitive metazoa, the sponges. Eur. J. Biochem. 221: 219–225.

Seimiya, M., M. Naito, Y. Watanabe & Y. Kurosawa, 1998. Homeobox genes in the freshwater sponge *Ephydatia fluviatilis*. In Müller, W. E. G. (ed.), Molecular Evolution: Evidence for Monophyly of Metazoa. Vol. 19. Springer, Berlin: 133–155.

Seimiya, M., Y. Watanabe & Y. A. D. Kurosawa, 1997. Identification of POU-class homeobox genes in a freshwater sponge and the specific expression of these genes during differentiation. Eur. J. Biochem. 243: 27–31.

Smith, A. B., B. Lafay & R. Christen, 1992. Comparative variation of morphological and molecular evolution through geologic time: 28S ribosomal RNA versus morphology in echinoids. Phil. Trans. r. Soc. Lond. B 338: 365–382.

Smith, A. B., G. L. Paterson & B. Lafay, 1995. Ophiuroid phylogeny and higher taxonomy: morphological, molecular and paleontological perspectives. Zool. J. linn. Soc. 114: 213–243.

Soest, R. W. M. van, 1990. Toward a phylogenetic classification of sponges. In Rützler, K. (ed.), New Perspectives in Sponge Biology. Smithsonian Institution Press, Washington, D.C: 344–350.

Soest, R. W. M. van, 1991. Demosponge higher taxa classification re-examined. In Reitner, J. & H. Keupp (eds), Fossil and Recent Sponges. Springer-Verlag, Berlin: 54–71.

Solé-Cava, A. M. & N. Boury-Esnault, 1999. Patterns of intra and interspecific divergence in marine sponges. Mem. Queensland Mus. 44: 591–602.

Solé-Cava, A. M., N. Boury-Esnault, J. Vacelet & J. P. Thorpe, 1992. Biochemical genetic divergence and systematics in sponges of the genera *Corticium* and *Oscarella* (Demospongiae: Homoscleromorpha) in the Mediterranean Sea. Mar. Biol. 113: 299–304.

Solé-Cava, A. M., M. Klautau, N. Boury-Esnault, R. Borojevic & J. P. Thorpe, 1991a. Genetic evidence for cryptic speciation in allopatric populations of two cosmopolitan species of the calcareous sponge genus *Clathrina*. Mar. Biol. 111: 381–386.

Solé-Cava, A. M. & J. P. Thorpe, 1986. Genetic differentiation between morphotypes of the marine sponge *Suberites ficus* (Demospongiae: Hadromerida). Mar. Biol. 93: 247–253.

Solé-Cava, A. M., J. P. Thorpe & R. Manconi, 1991b. A new Mediterranean species of *Axinella* detected by biochemical genetic methods. In Reitner, J. & H. Keupp (eds), Fossil and Recent Sponges. Springer-Verlag, Berlin: 313–321.

Thorpe, J. P., 1983. Enzyme variation, genetic distance and evolutionary divergence in relation to levels of taxonomic separation. In Oxford, R. S. & D. Rollison (eds), Protein Polymorphism: Adaptive and Taxonomic Significance. Academic Press, London: 131–152.

Vacelet, J., 1985. Coralline sponges and the evolution of the Porifera. In Conway-Morris, S., J. D. George, R. Gibson & H. M. Platt (eds), The Origin and Relationships of Lower Invertebrates. Clarendon Press, Oxford: 1–13.

Wainwright, P. O., G. Hinckle, M. L. Sogin & S. K. Stickel, 1993. Monophyletic origins of the Metazoa: an evolutionary link with Fungi. Science 260: 340–342.

West, L. & D. Powers, 1993. Molecular phylogenetic position of hexactinellid sponges in relation to the Protista and Demospongiae. Mol. mar. Biol. Biotechnol. 2: 71–75.

Zrzavy, J., S. Mihulka, P. Kepka & A. Bezdek, 1998. Phylogeny of the Metazoa based on morphological and 18S ribosomal DNA evidence. Cladistics 14: 249–285.

Hydrobiologia **420**: 29–43, 2000.
A. M. Solé-Cava, C. A. M. Russo & J. P. Thorpe (eds), Marine Genetics.

Genetics and marine pollution

D. R. Dixon & J. T. Wilson
Southampton Oceanography Centre, Empress Dock, Southampton SO14 3ZH, U.K.

Key words: cytogenetics, mutations, DNA, chromosomes, karyotypes

Abstract

The development of cytogenetic methods applied to cells and tissues of marine invertebrates has been hampered by (1) a lack of *in vitro* cell lines, (2) inadequate karyotypic information (partly as a result of too few workers chasing too many organisms), and (3) the failure of their chromosomes to band satisfactorily. Compared to mammalian cytogenetics, our knowledge of marine invertebrates lags behind by several decades. With the current concern about mutagens in the marine environment, and the recognition that the cells of marine species have sensitivities to DNA-damaging agents similar to those of higher organisms, there is a need for methods which can be used (a) in environmental monitoring and (b) to screen potentially harmful substances in the laboratory. In the absence of *in vitro* cell lines, embryos and larvae have been used to provide a supply of dividing cells for mutation studies, although the advent of molecular methods has now brought with it the means to detect DNA damage without any need for the cells to be in a dividing state. Moreover, the use of FISH (Fluorescence *In Situ* Hybridisation) now makes it possible to study numerical and structural chromosomal aberrations with far greater accuracy than was previously possible. A new marine genotoxicity assay is described, based on the embryos and larvae of a tube-dwelling polychaete worm (*Pomatoceros lamarkii*), suitable for both laboratory studies and field monitoring. This new *Pomatoceros* assay provides, at the same time, a useful model for studying the consequences of adult exposure on the offspring. A novel application of marine cytogenetic research is the study of the evolutionary adaptations of invertebrates living in naturally polluted extreme environments *viz.* deep sea hydrothermal vents, which are typified by high levels of toxic heavy metals and radionuclides, substances known to inflict damage to DNA. Given these new methodological and conceptual advances, it is predicted that our understanding of the role played by mutation in the marine environment, both in an evolutionary and toxicological context, will increase dramatically over the next decade.

Introduction

Pollution can act on the genetic material in two ways. It may exert selective pressure on the genetic structure of populations, by influencing gene frequencies, or inflict unscheduled and largely detrimental changes on DNA, i.e. mutation. It is the latter category of effects which is the subject of this review. For convenience, methods used for the detection of DNA damage can be divided into two categories based on target size: effects visible at the chromosome level (i.e. the macro scale) and effects operating at the level of the gene (the molecular or DNA scale). Of course, this distinction is somewhat artificial since chromosomes are the physical manifestation of replicating DNA, and structural chromosomal abnormalities are, in turn, the result of DNA damage (Savage, 1993). Nevertheless, for practical purposes, this distinction is useful since methods which are used to detect chromosomal aberrations are usually different from those used to detect changes at the molecular level. In a similar way, different terms are used to describe agents which can inflict damage at either the molecular or chromosome level. For chromosome damage the term used is clastogen, whereas for agents that have their effect at the molecular level, the term used is mutagen; it should be noted, however, that many clastogens are capable of acting as mutagens and *vice versa*. Finally, brief mention should be made of those agents which can cause cancer (carcinogens). While cancer and carcinogenicity fall outside the scope of this review, it should be noted that in the case of marine invertebrates, there is little supporting evidence for the existence of chemical-induced tumours in this group of organisms. In the majority of pollution studies where tumours have been reported (Mix, 1986) there remains scope for a viral aetiology, i.e. the contaminant has activated a latent viral genome leading to the onset of a tumour-like disease condition, through a suppressing effect on the immune system of the host organism. Given this lack of evidence for chemically-induced tumours in at least some groups of marine invertebrates (Annelida and Mollusca are the best studied groups) it is tempting to speculate that a molecular mechanism for the generation of tumours, of the multi-tiered type currently recognised for mammals, appears to be lacking. This means that great care must be exercised when predicting the genetic consequences of pollution exposure on individuals and populations of marine organisms, and in particular when transposing mammalian-based findings to lower groups. It is, however, recognised that, at the level of the DNA molecule and the chromosome, marine invertebrates express similar types of chemically-induced damage to that recorded in higher organisms.

Structural chromosomal aberrations

Due to the large body of evidence derived from human and mammalian studies, our understanding of the causes and consequences of structural and numerical chromosomal damage is far greater than for any genetic endpoint. For this reason, chromosomal aberrations (CAs) feature predominantly in mammalian studies both from a medical viewpoint (i.e. screening for birth defects) and in the context of environmental monitoring/mutation testing. The spontaneous chromosome aberration rate in humans has been reported to be 7% of all conceptuses (Hook, 1983). The consequences of chromosomal aberrations are universally detrimental and include congenital abnormalities, reduced fertility and cancer (e.g. Hagmar et al., 1994). Our understanding of the types and consequences of structural and numerical chromosomal abnormalities is greater than for any other genetic endpoint.

Metaphase analysis provides the most comprehensive approach to aberration detection, and it is commonly claimed, albeit incorrectly, that using this method the entire genome is surveyed at once. A major disadvantage with this approach is that metaphase analysis is extremely laborious, thus expensive, to perform since it requires a high degree of expertise on the part of the scorer. Furthermore, the high rates of artefactually-induced chromosome loss, resulting from slide preparation, have stimulated a search fo simpler, albeit less comprehensive, methods. Non etheless, some attempts have been made to appl metaphase analysis to marine invertebrate subjects with some degree of success (e.g. Dixon & Clarke 1982; Jha et al., 1996). The difficulty in transposin CA analysis to marine invertebrates stems from (1 a lack of *in vitro* cell lines; (2) inadequate inform ation relating to the karyotype; and (3) the failur by marine invertebrate chromosomes to band success fully. This general lack of success when applyin chromosome banding (CB) techniques has prevente individual chromosome pairs from being identifie unequivocally. The marine environment contains a di versity of animal groups unmatched anywhere els on earth. However, with a few exceptions, curren knowledge of the chromosomes of most invertebrat taxa is at best rudimentary and in most cases en tirely lacking. Where cytogenetic studies have bee performed this usually amounts to no more than a attempt to describe chromosome number, often wit only a limited degree of success (see discussion i Dixon et al., 1998a). In only two marine inverteb rate taxa, the bivalve molluscs and polychaetes, ca our understanding of their chromosomal features b considered to have reached a stage where it is nov possible to consider conducting CA analysis with an degree of confidence (e.g. *Mytilus* species, Dixon & Flavell, 1986; *Neanthes arenaceodentata*, Pesch an co-workers, 1981a; *Platynereis dumerilii*, Jha et al. 1996). Using a traditional mammalian CA approach it has been possible to identify a range of structura effects, including dicentric chromosomes, fragments and deletions, under laboratory exposure conditions The general lack of success with CB with marin invertebrates probably reflects something fundament ally different about the packaging of the histone an chromatin. Where CB has been demonstrated this ha tended to be restricted to specific chromosome region (Dixon & McFadzen, 1987), or appears to have bee a product of semi-conservative chromatin condensa tion during prometaphase (e.g. Moore et al., 1986) Recently, Martinez-Lage et al. (1994) reported hav ing some success with applying a range of banding methods (G-banding, C-banding and chromomycin A staining) to the chromosomes of *Mytilus galloprovin cialis*, but the utility of this approach for mutatio testing has yet to be demonstrated.

The lack of established cell lines, due to the failure by mammalian mitogens to work on the cells of marine invertebrates, represents a major barrier for CA analysis, where a high mitotic index and synchronised cell divisions are ideal. In marine invertebrates in general (a special case will be described later), there is a naturally low cell turnover linked with poikilothermy (seasonal/temperature dependent growth rate), a problem made worse by the lack of haemopoietic tissues (meristematic centres of cell division). Consequently, the few dividing cells naturally present in the organism at any one time tend to be heavily diluted by large numbers of interphase nuclei during slide making. In an attempt to overcome this problem, a number of studies have focused on larvae or other early life-history stages where the natural rates of cell division tend to be higher than in the adult somatic tissues (e.g. Pesch et al., 1981a; Harrison & Jones, 1982; Jha et al., 1996). Other studies have attempted to promote increased rates of cell division by artificially enhancing growth rate (Dixon & Clarke, 1982). However, in the absence of *in vitro* culture, CA analysis will have only limited application in marine invertebrate studies until such time as new molecular methods, particularly FISH (Fluorescence *in situ* Hybridisation) and chromosome painting, are introduced. The time is not far off when it will become possible to detect with great accuracy clastogen-induced structural changes (i.e. insertions, deletions, translocations, etc.) in chromosomes which are currently intractable to conventional chromosome banding methodology. This will then lead on to the introduction of automated scoring methods, thus removing much of the labour from this type of analysis. In the meantime, efforts in the marine field will tend to focus on simpler, albeit less comprehensive, ways to detect CAs.

Sister chromatid exchange

The sister chromatid exchange (SCE) technique has been demonstrated to be a sensitive cytogenetic endpoint applicable to a range of aquatic invertebrates and fish species (e.g. Zakour et al., 1984; Dixon & Prosser, 1986; Jones & Harrison, 1987; Jha et al., 1996). First described by Taylor et al. (1957), using autoradiography, the method has since been refined and modified so that tritiated thymidine has been replaced by a more sensitive chemical label, bromodeoxyuridine (BrdU) (Korenberg & Freelander, 1974; Perry & Wolff, 1974). Based on the semi-conservative nature of DNA replication, the SCE method usually requires exposing cells to BrdU for two DNA replication cycles (Perry & Wolff, 1974). At the next metaphase, one member of each pair of sister chromatids, which is partially substituted by BrdU, stains darkly with giemsa stain, while the other chromatid, which is completely substituted by BrdU, stains palely due to its reduced affinity with the nuclear stain after UV irradiation. Unscheduled rearrangements of chromosome segments across the two sister chromatids are clearly visible as a characteristic 'harlequin' staining pattern. It is the number of these unscheduled, so-called sister chromatid exchange events which provides a sensitive indicator of mutagen-induced chromosomal disruption. However, a number of practical and conceptual difficulties mean that the SCE method must remain a sensitive adjunct to less equivocal cytogenetic methods for measuring DNA damage e.g. metaphase analysis.

At a practical level, exposing cells *in vivo* to BrdU can present difficulties. Also, a lack of *in vitro* cell lines reduces the value of this method in the laboratory. Conceptually, there are difficulties in interpreting the significance of this genetic endpoint, which stem from the fact that SCEs involve homologous chromosome segments. There is also a lack of specificity in the SCE response which has been shown to be sensitive to a variety of non-mutational influences, including animal condition. In a number of marine invertebrate groups a wide variety of cellular functions are carried out by lysosomal enzymes, with the result that at the time of reproduction lysosomal enzymes, which includes hydrolases, are released into the tissues as part of a natural process of tissue cleansing following gametogenesis. It appears that some of these enzymes enter cell nuclei where they inflict damage to DNA. As a result, SCE levels in wild mussel populations may increase significantly at the time of reproduction and for some time afterwards, making it impossible to distinguish between natural and anthropogenic-induced effects. There is also clear evidence for life-cycle related differences in SCE levels in some species. For example, marine mussel larvae (*Mytilus*) from Tomales Bay, northern California showed significantly higher and variable levels of SCE (Jones & Harrison, 1987) compared with gill cells of adult mussels from both Tomales Bay and the UK (e.g. Dixon & Clarke, 1982). Also, in some invertebrates (e.g. mussel larvae and polychaete worms) the SCE response is limited to a simple doubling of the rate compared to the baseline level, irrespective of toxicant dose (Pesch et al., 1981b; Harrison & Jones, 1982).

Over the past two decades a number of authors have demonstrated the utility of the SCE approach for detecting chemically-induced changes in the chromosomes of a range of marine invertebrate species, although in the absence of *in vitro* cell lines, the majority of these studies have focused on embryos/larvae or other immature life stages (Pesch et al., 1981a,b; Dixon et al., 1985; Dixon & Prosser, 1986; Jha et al., 1996). Given the problems associated with labelling animals with BrdU in the field, Dixon & Pascoe (1994) showed that it was possible to detect an increased level of SCE in the chromosomes of 2-cell stage mussel embryos originating from field-exposed animals. However, the highly synchronised and seasonal reproductive behaviour of the marine mussel, coupled with the failure by the labelled metaphases to remain intact during slide making, still poses problems for the applicability of this method.

Numerical chromosome aberrations

Like structural CAs there is a large mammalian literature which shows that aneuploidy (single chromosome deviations from the normal diploid condition) is an important chromosomal endpoint linked to a wide range of birth defects and, in some cases, certain types of cancer (e.g. Levan & Mittelmann, 1977; Tsutsui et al., 1983). In humans, a high frequency of spontaneous abortuses have been shown to carry numerical aberrations (e.g. Hassold, 1986). Unlike structural CAs, numerical aberrations are usually caused by damage inflicted on the microtubular apparatus leading to individual chromosome loss or gain during cell division. In contrast, in the case of haplo/polyploidy (entire chromosome complement variation from the haploid or diploid condition), there are a number of potential causes including failure by the egg nucleus to complete meiosis, endoreplication (DNA replication without cell division) and failure by the mitotic spindle apparatus to operate during anaphase. While some attempts have been made to detect aneuploidy directly in the cells of selected marine invertebrates (e.g. Dixon, 1982), this direct counting method, like metaphase analysis, is extremely time-consuming and laborious to perform. Also, unless one is careful to select a cell type where the cell membrane remains intact and clearly visible, there is the risk of artefactually-induced variations. For this reason, Dixon (1982) restricted chromosome counts to 2-cell mussel embryos where there is a clearly visible membrane surrounding each cell. However, there are other methods which can be applied in the study of aneuploidy which are less labour-intensive while allowing a large number of cells to be scored. One method involves no special pretreatment apart from staining cells from embryos or larvae with a DNA-specific stain (e.g. aceto-orcein). Anaphase spreads are then analysed under a light microscope for abnormalities such as lagging chromosomes, tri-polar anaphases, and lagging metaphases (Dixon et al., 1999 – described later). As with other cytogenetic endpoints, aneuploidy can be caused by a wide variety of agents, many of which are also mutagenic and clastogenic. In addition to numerical abnormalities, the anaphase aberration assay also allows the identification of structural effects such as acentric fragments and chromosome bridges (dicentric chromosomes). An alternative approach involves the use of DNA probes and FISH for detecting aneuploidy affecting individual chromosomes. An advantage of this method is that it allows aneuploidy to be scored in interphase cells, and, when coupled with the use of fluorescent probes, may lead to automated detection and screening (at least in mammals), where a supply of isolated intact cells can be easily obtained. This approach holds considerable promise in the future for marine subjects once the problem of establishing *in vitro* cell cultures has been overcome. It must be remembered that cells which have yet to undergo division will not express any evidence of chromosome loss or gain stemming from spindle or centromere disruption. In the micronucleus test described below, it has become general practice in mammalian genotoxicology to use cytochalasin-B (CYT-B) to produce binucleate cells based on cytokinesis block (e.g. Fenech & Morley, 1985). However, this only applies to cases where *in vitro* cell lines are available. In marine invertebrates, where these are lacking, it has become practice when working on numerical aberrations to concentrate on embryonic or larval material, where cell division rates are intrinsically higher than for somatic tissue. However, even when the generation time is known to be relatively short, there remains the likelihood that cells which are stressed or damaged will not behave in an exactly similar way to cells which are normal (see Wilson et al., 1998).

Limiting factors

In vitro techniques are a mainstay of mammalian cytogenetics (Table 1). An ability to grow cells in culture ensures the high rates of cell division which are an essential prerequisite for effective metaphase and anaphase analyses. Cell culture also allows cell division to be synchronised using temperature or chemical inhibitors, in order to maximise on a desired divisional stage, e.g. anaphase. Certain chemical manipulations

Table 1. Summary of differences underlying the difficulties associated with cytogenetic studies of marine invertebrates

	Mammals	Marine Invertebrates
Thermoregulation	Homeothermic	Poikilothermic
Cell division rate	High	Low
Haemopoietic tissue	Present	Absent or diffuse
Karyotypes	Extensively studied	Poorly studied
Cell culture	Many established cell lines	No established cell lines
Reproductive output	Low	Very high

are only possible when cells are grown in culture (e.g. cytokinesis block micronucleus test and, to a lesser extent, SCE). Unfortunately, invertebrate cells do not respond to vertebrate mitogens (e.g. Phytohemagglutinin, Concanavalin A, etc.), so little progress has been made in the development of *in vitro* cell lines. Some success has been reported, however, at producing primary cell lines (e.g. Cornet, 1993: *Mytilus*), including the application of SCE, but without mitogens this approach is of limited value only. To date, the bulk of marine invertebrate cytogenetics has been conducted on early life-history stages which have intrinsically higher division rates and larger (i.e. less condensed) chromosomes than the cells of adult animals.

Micronucleus test

Developed originally for use on bone marrow and peripheral blood erythrocytes (Heddle, 1973; Schmid, 1977), the micronucleus (MN) test has become widely established in the field of mammalian genetic toxicology. In principle, MN is extremely simple to perform and depends upon fragments or entire chromosomes becoming detached from the mitotic spindle. These displaced fragments or chromosomes enter the cytoplasm where they assume the morphology of minute micronuclei (MNi) at the following interphase. In practice, distinguishing MNi from other cellular inclusions, e.g. viral plaques, can be difficult, but a more serious problem relates to the need to be able to distinguish cells which have undergone cell division from those which have not, without which the assay cannot strictly be used quantitatively. In mammalian *in vitro* studies, this problem of distinguishing divided from non-divided cells has been successfully overcome by the use of the cytokinesis-block technique (see Fenech & Morley, 1985), wherein cells in culture are treated with cytochalasin B, an inhibitor of actin polymerisation, which acts to inhibit the formation of the cell membrane between the two daughter nuclei at the end of telophase (Carter, 1967). Cells that have undergone nuclear division can be distinguished from those yet to divide by the presence of two daughter nuclei within the same bounding membrane. It is this subset of binucleate cells which is scored for MNi. Without the benefit of these refinements, the application of MN to aquatic invertebrate studies remains prone to error in estimating the frequency of damage, particularly when dealing with heterogeneous tissues with differing cell turnover rates. The use of centromeric and telomeric probes has aided the interpretation of the different types of chromosome damage leading to MN formation in mammalian studies (e.g. Russo et al., 1996).

The MN test has the advantage of being applicable to a wide range of different species (plants and animals) without any requirement for a detailed knowledge of the karyotype. The test has been applied to a number of marine invertebrate species, e.g. the larvae of echinoderms (Hose & Puffer, 1983) and blood cells of bivalve molluscs (Wrisberg et al., 1992; Burgeot & Galgani, 1995), although these studies reported finding high levels of inter-individual variation affecting this end-point. Burgeot & Galgani (1995) drew attention to the inability to distinguish between viral inclusion bodies and MN of animal origin. Using human lymphocytes, Fenech & Neville (1992) reported that MN was relatively insensitive to genotoxic agents which produce DNA adducts (i.e. monofunctional methylating agents) or base modifications (e.g. UV), rather than strand breakage or chromosome loss. Improvements in sensitivity (up to forty-fold) were obtained for agents which induce DNA base damage when excision-repair sites were converted to double strand breaks using cytosine arabinoside (an inhibitor of excision repair). However, without the benefit of *in vitro* methods, these refinements in the MN tech-

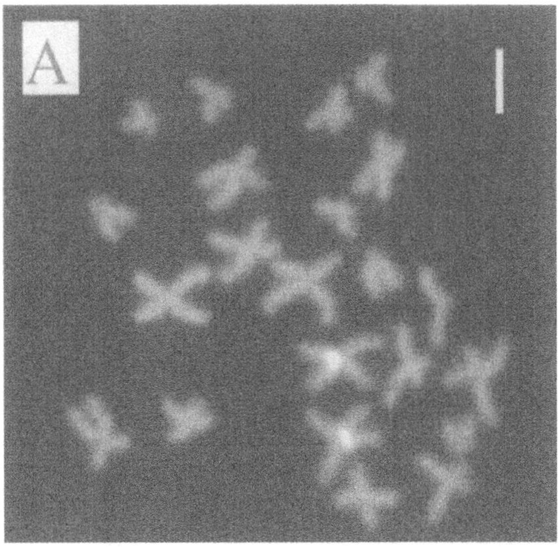

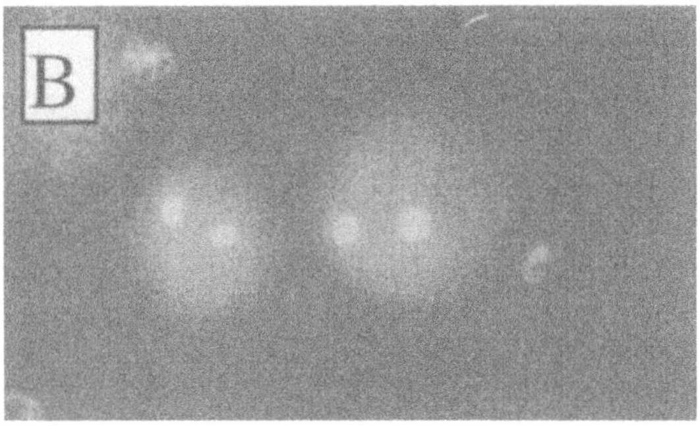

Figure 1. (A) Partial metaphase spread from *Pomatoceros lamarckii* showing the fluorescently-labelled rDNA probe on chromosome pair 2; (B) ribosomal DNA loci can also be seen in the interphase nuclei. Scale bar=5-μm.

nique remain theoretical, at least for the present time, in relation to marine genotoxicology.

Fluorescence in situ hybridisation (FISH)

This technique allows sites of specific nucleic acid sequences to be detected in metaphase spreads or interphase nuclei. Developed independently by Pardue & Gall (1969) and John et al. (1969), using radiolabelled probes, and modified more recently to allow the use of fluorescently-labelled probes (Bauman et al., 1980), the method facilitates the identification of specific chromosomes or chromosome arm regions in species, such as marine invertebrates, which have proved largely intractable to DNA banding. An example of this approach was the study carried out by Pascoe et al. (1995), who investigated the phenomenon of Robertsonian polymorphism in the common European dog-whelk *Nucella lapillus*, and revealed evidence of inversions/centric shifts otherwise undetectable using conventional giemsa staining. The implications for mutation assays such as MN are obvious. By introducing a FISH step, by using a centromeric probe, into the staining procedure, the specificity of the MN test could be dramatically increased. Since this technique can also be applied to interphase nuclei, it is possible to score large numbers of cells for aneuploidy relating to specific chromosomes, e.g. using rDNA probes (Figure 1).

Anaphase analysis

Anaphase analysis in its simplest form can be used to screen for lagging chromosomes and bridges. This technique has the advantage of displaying both the effects of aneugenicity (lagging chromosomes and multipolar anaphases) and clastogenicity (bridges and fragments). Anaphase analysis has been used over the past two decades in a number of marine invertebrate studies (e.g. Hose & Puffer, 1983; Hose et al., 1983; Anderson et al., 1994). Recently, we have applied this technique to the developing embryos of *Pomatoceros lamarckii* as an integrator of genotoxin exposure (Dixon et al., 1999) (described later).

Molecular approaches to the detection of DNA damage

The advent of molecular techniques over the past decade has had a profound effect on mutation testing, although, not surprisingly, most of the advances to date have been restricted to the field of mammalian genotoxicology. At last it has become possible to detect individual DNA base changes so that gene mutations (point mutations and frameshift mutations) can be measured directly. The development of automated DNA sequencers allows the rapid screening of samples for known and unknown mutations, but these methods have yet to make an impact in the marine field. DNA strand breaks represent a convenient quantifiable response to genotoxic insult. Strand breaks occur under normal conditions, but exposure to genotoxicants can increase the levels of strand breakage, either directly through chemical impact, or indirectly through the process of excision repair (Shugart & Theodorakis, 1994). There are a number of techniques that have been developed to investigate DNA strand breakage; these involve both electrophoretic and non-electrophoretic approaches.

The SCGE (comet) assay

First introduced by Ostling & Johanson (1984), the single cell gel electrophoresis (SCGE) or comet assay is a method for detecting DNA strand breaks in isolated cell nuclei. Originally applied to irradiated mammalian cells, the method has since undergone a series of technical modifications (Singh et al., 1988; Olive et al., 1990; Klaude et al., 1996), and is now used increasingly in chemical studies. In principle, the technique is relatively simple to perform: isolated cell nuclei previously exposed *in vitro* or *in vivo* to a suspected mutagen are suspended in a sandwich of agarose on a microscope slide and then lysed using strong detergent followed by treatment with a strong alkali in order to relax and unwind the DNA. The cell nuclei are then microelectrophoresed in a horizontal gel tank, where any broken negatively charged DNA strands migrate towards the anode in the electric field. Subsequent staining with a fluorescent DNA stain (e.g. ethidium bromide or propidium iodide) and visualisation under a fluorescent light source reveals any DNA single strand breakage as a comet tail protruding from the nucleus (hence the assay's name). Image analysis (e.g. Komet-Kinetic Imaging, Liverpool) allows semi-automated, rapid and objective determination of the levels of DNA damage. Other refinements of the method include the adjustment of the pH of the gel-tank buffer to visualise either predominantly single (alkaline buffer) or double (neutral buffer) strand breaks (Olive et al., 1991).

Some progress has recently been made in applying the comet assay to the cells of selected marine invertebrates. For example, Steinert (1996) used the technique to demonstrate DNA damage in the blood cells of *Mytilus edulis* and reported finding high levels of apoptosis affecting these cells which appeared to be a major component of the animal's response to chemical insult. Nacci et al. (1996) reported finding high levels of intra- and inter-individual variation affecting the comet response in *Crassostrea*. More recently, Wilson et al. (1998), while applying the comet assay to the gill cells of *Mytilus edulis*, showed that a reduction in the levels of inter-individual variation could be achieved by exposing the animals to vitamin E, an antioxidant, thus improving the sensitivity of the assay for laboratory screening purposes.

The comet assay offers considerable advantages over cytogenetic assays since it is rapid and inexpensive to perform, and there is no requirement for the detailed knowledge of the karyotype, thus enabling it to be easily transferred between species. The comet assay also has the added advantage of being applicable to non-dividing cells, an important consideration when working with poikilotherms where the natural rates of cell division are extremely low. An obvious requirement is that there should be a tissue that can easily be reduced to a single cell suspension without inflicting significant damage to the cell nuclei. For this reason, studies have tended to focus on blood cells and simple epithelial tissues.

Electrophoresis of extracted DNA

Agarose gel electrophoresis is a technique which has been used successfully for determining DNA damage in animal tissues (e.g. Drouin et al., 1996). In brief, the method involves incorporating cells into agarose plugs, followed by lysis and then subjecting the partially purified DNA to electrophoresis in a cooled tank (to prevent DNA damage through excessive heat). The gels are then run under alkaline or neutral conditions, which, after staining with ethidium bromide and UV fluorescence, allows the visualisation of single and double strand breaks respectively. The average break frequency is then calculated from the ethidium-bromide fluorescence intensity by comparison with a DNA weight standard.

To date the method has not been applied to a marine invertebrate but Theodorakis et al. (1994) carried out a successful study using agarose gel electrophoresis which enabled the detection of significant levels of DNA damage in fish blood cells. This method is currently being developed in our laboratory to allow the detection of DNA damage in marine invertebrates using a modification of the standard comet image analysis software.

Polymerase chain reaction (PCR)

The polymerase chain reaction was first used in the detection of DNA damage by Govan et al. (1990). The theory behind the assay is simple and is based on the concept that the efficiency of primer binding is affected by the amount of DNA template and the level of similarity between the PCR primers and the target sequence. Any change in these parameters may result in an alteration in the quantity or quality of the amplification products, thus giving a potential index of DNA damage. However, given the extreme sensitivity of PCR, and its susceptibility to a wide variety of non-genotoxic influences, this apparently simple molecular approach needs careful validation before any meaningful results can be obtained.

Atienzar et al. (1998) used a PCR approach to detect xenobiotic-induced DNA damage in the marine macroalga, *Palmaria palmata*. They used a modification of the technique known as arbitrarily primed polymerase chain reaction (AP-PCR) which relies on a comparison between amplification products from control and exposed DNA templates produced by random oligonucleotide primers (see Welsh et al., 1995). In their study significant problems were encountered, in particular, spurious amplification products were observed in the negative controls and there was a general lack of reproducibility of the amplification patterns. It was concluded that these were a result of aberrant primer-induced polymerisation which might be controlled by optimisation of the deoxynucleotide triphosphates (dNTPs) concentration. Clearly, further work is required before this approach can be considered suitable for routine analysis.

Other methods for measuring DNA strand breakage

Alkaline elution

The alkaline elution technique was first described by Kohn et al. (1976). The technique measures single strand breaks (SSBs) and no attempt is made to discriminate between DNA strand breaks and alkali labile sites. Cells are placed on a fine millipore filter, lysed, and then treated with high alkaline conditions to unwind their DNA. Alkaline washes are then carried out and the proportion of DNA retained by filter is then measured microfluorometrically. The proportion of DNA passing through the filter corresponds to the amount of DNA damage.

The alkaline elution technique has been applied in a number of mutation studies based on the Mediterranean mussel *Mytilus galloprovincialis* using haemolymph (Bihari et al., 1992; Vukmirovic et al., 1994) and gill cells (Bolognesi et al., 1992). These demonstrated the applicability of the technique for detecting DNA damage caused by *in vivo* exposure to different test compounds.

Alkaline unwinding assay

This assay was developed by Ahnstrom & Erixon (1973) and is similar to the method described above. It has been used successfully for the detection of chemically-induced DNA strand breaks. The technique is based on the time-dependent partial unwinding of DNA, which allows the determination of the double-stranded (dsDNA) to single-stranded DNA (ssDNA) ratio. In brief, genomic DNA is extracted from tissue and then allowed to partially denature under alkaline conditions. The degree of molecular unwinding is then estimated, using fluorometry, from the amount of fluorescence of either ethidium bromide or Hoechst 33258 stain. These DNA stains fluoresce with greater intensity in the presence of dsDNA compared to ssDNA. The rationale is that the lower the proportion of dsDNA to ssDNA, the greater the degree of DNA damage in the form of SSB.

The alkaline unwinding assay has proved to be a rapid and sensitive method of assessing DNA strand breakage in the cells of marine invertebrates, under both laboratory and field conditions. Nacci & Nelson (1992) used this assay successfully to detect DNA damage in gill tissues of marine bivalves, *Mytilus edulis* and *Crassostrea virginica*, transplanted into an urban estuary contaminated with PCBs, PAHs and heavy metals; their findings were supported by the results of *in vivo* and *in vitro* experiments using reference mutagens. Other studies using this method include those on *Mytilus galloprovincialis* (Accomando et al., 1991; Viarengo et al., 1991), and the starfish, *Asteria rubens* (Everaarts, 1995).

DNA adducts

A DNA adduct is formed when an exogenously-derived electrophilic molecule binds covalently to DNA. There are a number of immunological methods available for the detection of these so-called DNA adducts (e.g. the enzyme-linked immunosorbent assay; ELISA), although a major disadvantage of immunological methods is the time and expense involved in developing individual antibodies for each specific type of adduct. For this reason, the most widely used technique for detecting DNA adducts is ^{32}P-postlabelling, which has the advantage of being sensitive (one adduct in 10^9 to 10^{10} nucleotides) without requiring the characterisation of individual adducts before their measurement. This makes the method suitable for application under field conditions where complex mixtures are common. For ^{32}P-postlabelling (see Gupta & Randerath, 1988), DNA is first isolated from cells or tissues and then purified to remove RNA and protein, before being hydrolysed to the 3′-monophosphate form. Adducts are then enriched relative to normal nucleotides by enzyme action or chemical extraction. Finally, the adducts are radiolabelled using ^{32}P phosphate. The radiolabelled adducts are detected and quantified using a combination of thin layer chromatography (TLC), autoradiography and liquid scintillation.

The ^{32}P-postlabelling technique is quite laborious, and, because of the use of a high-energy radionuclide, there is a requirement for additional safety precautions in the laboratory. Nevertheless, its high sensitivity and potential value in biomarker studies has made ^{32}P-postlabelling a popular choice amongst workers interested in quantifying the effects of chemical exposures on exposed organisms, particularly under field conditions. The difficulty of distinguishing between natural and exposure-derived DNA adducts can, however, sometimes make it difficult to interpret results.

The DNA adduct technique was first validated in the context of environmental monitoring using fish species (e.g. Stein et al., 1990). In that study the levels of adducts in English sole from Puget Sound, Washington were compared with sediment levels of high molecular weight PAHs and bile levels of PAH metabolites. The level of adducts correlated well with the degree of toxicant exposure. There have been a number of investigations involving the use of ^{32}P-postlabelling for the detection of DNA adducts in marine invertebrates with particular emphasis on *Mytilus* species (e.g. Sole et al., 1996; Venier & Canova, 1996; Harvey et al., 1997). These studies demonstrated the sensitivity of the assay and also considered the possible biological implications. The relationship between environmental contamination, DNA adducts and the consequences in terms of organism health remains largely to be determined. As referred to above, there is some evidence to suggest a link between high levels of PAHs in the environment and the incidence of DNA adducts in the tissues of marine organisms; similarly there is circumstantial evidence linking high levels of PAHs and incidence of tumours in marine organisms (e.g. Gardner et al., 1992). However, a causal link between the presence of DNA adducts and the initiation of tumours in exposed individuals remains to be demonstrated.

A naturally polluted marine environment

Life has evolved to cope with a wide range of physical and chemical conditions. However, deep-sea hydrothermal vents are undoubtedly the most hostile and toxic environment on the face of the planet. Vent emissions contain high levels of toxic substances which have the potential to damage DNA. Apart from the extreme pressure (200 atmospheres) and temperature (400 °C), the highly acidic vent emissions (pH 2) contain a rich cocktail of heavy metals, arsenic, radon and other radionuclides, substances with well-documented mutagenic and carcinogenic properties. In order to exploit the rich supply of geothermal-derived energy at the vents, the specialised vent fauna has evolved ways to cope with the damaging effects of these harmful substances. Since these adaptations have undoubtedly evolved over millions of years of evolutionary time, they are of relevance to our understanding, and possible management, of contaminated sites elsewhere. In addition, these studies may lead to discoveries which are of considerable biotechnological importance.

Dixon et al. (1998b) have shown that due to the high temperature (20–30 °C) and rich food supply in the zone in which they live, vent species exhibit naturally high rates of cell division, thus making them suitable candidates for cytogenetic analysis. Given their extremely harsh environment, it is of great interest to see whether the specialised vent fauna is at risk from genetic disease and birth defects induced by environmental contaminants. Recent studies have revealed evidence of shell abnormalities in juvenile gastropods similar to those recorded in areas of low water quality elsewhere (Thorson, 1946; Dixon & Pollard, 1985).

Recent work using the comet assay (Dixon et al., unpublished data) has revealed surprisingly high levels of DNA damage in the cells of vent mussels (*Bathymodiolus*), which correlates with the emission rates of specific vent fields on the mid-Atlantic ridge. It would appear that, despite their long association with environmental toxicity (fossil evidence can be traced back to the Mesozoic), vent organisms are not fully resistant to the damaging effects of their environment. Older (larger) mussels exhibited higher levels of DNA damage compared to younger (smaller) animals, indicating that the latter are more efficient at repairing DNA lesions. The hydrothermal environment is characterised by extemely high productivity as a result of bacterial chemosynthesis linked with the vent chemistry (hydrogen sulphide and methane), and as a consequence vent organisms have growth rates among the fastest of any metazoan group. Rapid growth and early reproduction, previously thought of as adaptations to temporal instability at the vents, can equally well be seen as important adaptations to cope with the damaging effects of environmental exposure to toxic compounds through the restorative effects of DNA

Figure 2. Adult specimens of *P. lamarckii* releasing their gametes. Scale bar=5 mm.

repair, particularly that achieved through telomerase activity during embryogenesis.

Anaphase aberrations in embryos of the marine polychaete *Pomatoceros lamarckii*

In our laboratory, a new *in vivo* assay has been developed using the eggs, embryos and larvae of the common marine tubeworm, *Pomatoceros lamarckii* (Figure 2) (Dixon et al., 1999). This organism has advantages over other marine invertebrates, such as the widely used *Mytilus edulis* (e.g. Bayne, 1976), because of the extended period over which gametes are available, and its well-documented karyotype (Dixon et al., 1998a). *Pomatoceros* lacks the synchrony in gamete development which limits the temporal availability of eggs and sperm when using other marine invertebrate species. Given that the gametes are retained within the bodies of adult worms over many months (natural spawning takes place in SW England in July-August), this species offers the potential for examination of the effects of long-term exposures to low levels of mutagens both in the field and under laboratory conditions. In the marine environment, apart from pollution hotspots, contaminants usually occur at low concentration, at or below the chemical detection limit. Since little is known about the effects these may be having on exposed populations in terms of fitness, this aspect is currently under investigation.

Protocol for the Pomatoceros anaphase aberration test

- De-tube adult worms using stout forceps (chip away posterior end of tube and ease the worm out in the anterior direction using a blunt needle applied to the centre of the toughened operculum; for more details see Dixon et al., 1999). Place the worms individually in watch glasses containing clean filtered seawater (FSW) and leave them to release their gametes over 10–15 mins. Select only gametes that appear in good condition, i.e. regularly shaped oocytes and actively swimming spermatozoa.
- Expose oocytes (unfertilised eggs), at a concentration of approximately 1000 ml^{-1}, to the test compounds in 10-ml volumes of FSW at 15 °C for 3 h prior to fertilisation (for acute studies).

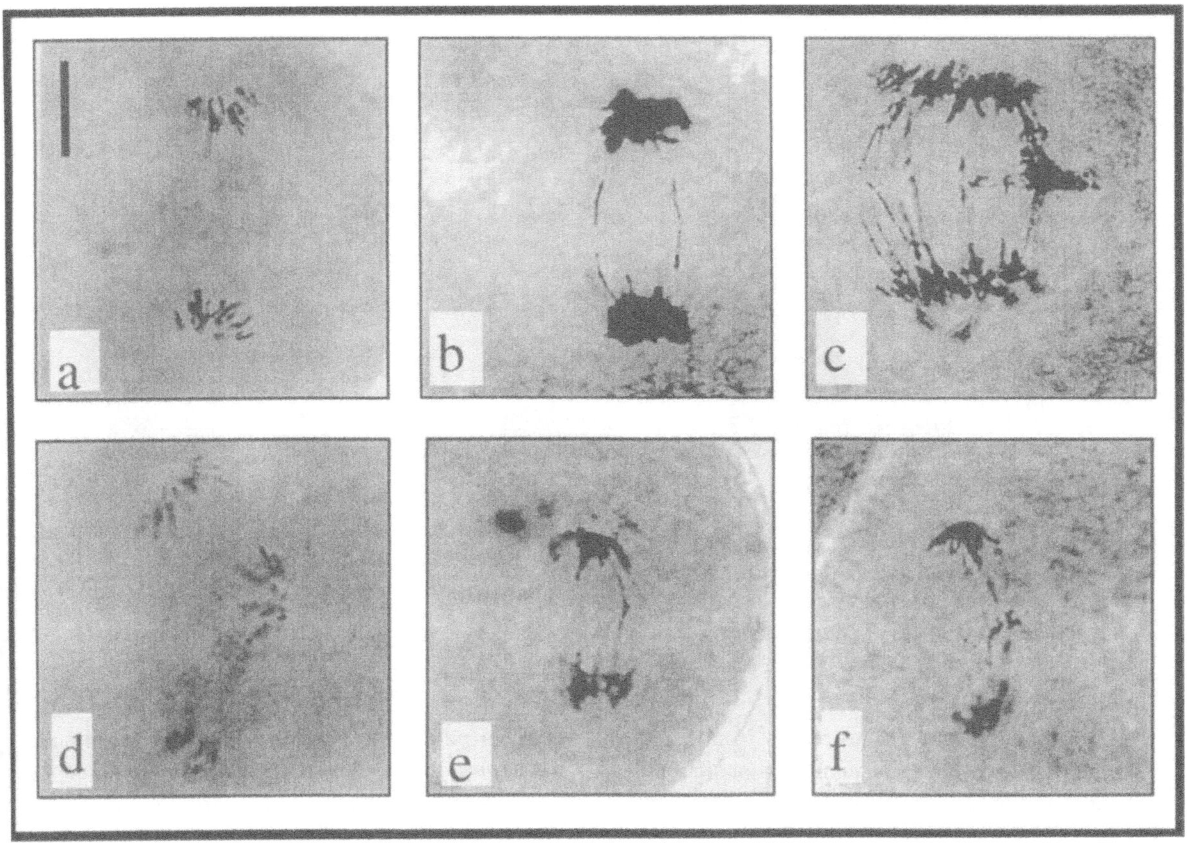

Figure 3. Typical anaphase aberrations: (a) normal anaphase; (b) lagging chromosomes; (c) lagging metaphase; (d) tripolar anaphase; (e) chromosome bridge; (f) chromosome fragments. Scale bar: 10 μm.

For chronic studies, expose the adults (still attached to their stones) in 2-litre volumes of FSW for 7, 14 and 21 days. If desired, retain samples of the animal tissues and gametes for chemical and histological analyses.

- Following exposure, fertilise the oocytes by adding fresh sperms at a concentration of approximately $10^6 ml^{-1}$ (as previously determined by haemocytometer counts). In the experiments, use only gametes which yield negative controls where the level of fertilisation reaches or exceeds 95%.
- Monitor embryo development at 20-minute intervals. When approximately 20% of the embryos reach the 8-cell stage fix a sample in cold Carnoy's fixative (3:1 methanol:glacial acetic acid), with 3 changes.
- Rinse the remaining embryos in FSW (using a 30-μm nitex mesh sieve) and allow to continue their development at 15 °C until the 48 h larval stage. Fix the larvae in 4% formalin in FSW and score for abnormal development (small, non-ciliated or larvae with undeveloped gut structure).
- Stain the Carnoy's fixed embryos for anaphase analysis with aceto orcein as follows: Pipette approximately 300 embryos onto a clean microscope slide. Add a few drops of aceto-orcein stain (Gurrs) and apply a large coverglass. Subject the slide to moderate heat (either by using a hotplate or by placing in contact with a light bulb) to intensify the staining reaction, blot to remove excess stain and seal the coverglass with clear nail varnish or glycerol gelatin.
- Score 2×100 embryos for the presence of aneugenic effects (i.e. lagging chromosomes/multipolar anaphases) and clastogenicity (chromosome bridges and fragments) (Figure 3). Score an additional 100 embryos for other cellular and chromosomal effects including asynchronous cell divisions and haplo/polyploidy.

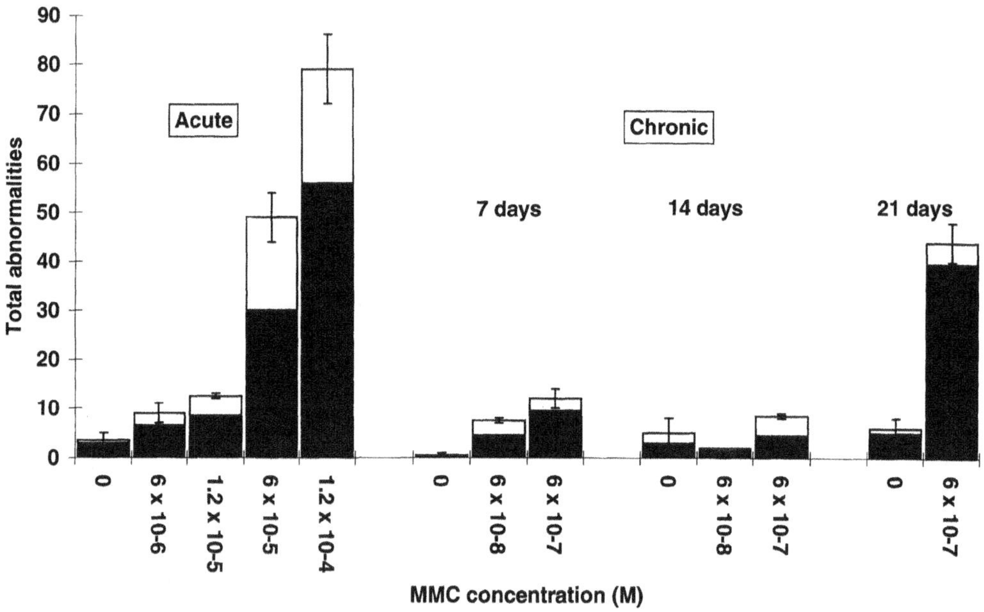

Figure 4. Summary of the results obtained using the *Pomatoceros* anaphase aberration test following both acute and chronic exposure to the reference mutagen mitomycin c (MMC). Black bars = aneugenic; white bars = clastogenic.

For example, a summary of results from chronic and acute exposure studies conducted in our laboratory (e.g. Dixon et al., 1999) is shown in Figure 4. Results from chronic exposures have shown that genotoxic agents, or their effects, will accumulate in the developing oocytes of *P. lamarckii* so that, after spawning and fertilisation, damage can be detected in the early cleavage (4–8 cell stage) embryos, at exposure concentrations below that which will induce a positive response under acute exposure conditions.

Future developments

Apart from improvements in monitoring and detection, one issue which will undoubtedly attract increasing attention in the future is the consequences of genotoxin exposure in marine species. The role played by genetic damage in human disease has long been appreciated. In the human context, effects on individuals (i.e. cancer and birth defects) are considered important. However, when dealing with animal populations, the focus switches from the individual to the population. An aspect that makes interpretation particularly difficult in the case of marine invertebrates is that there is a tendency to vastly over-produce (e.g. *Mytilus edulis* releases on average eight million eggs each time it spawns; Bayne, 1976). With such excessive over-production of gametes, it becomes extremely difficult to predict how an increase in the number of genetically defective offspring will affect population fitness (e.g. Dixon, 1982); a problem made even more difficult by recognition that only a small proportion of individuals in a mussel bed may actually contribute to the effective reproductive output (Li & Hedgecock, 1998).

Advances in molecular biology will revolutionise mutation testing, and will ultimately lead to a greater understanding of the effects of genetic damage on the individual and, ultimately, the population. With molecular techniques becoming ever more powerful, while simultaneously becoming simpler and cheaper, there will probably soon come a time when molecular approaches take over completely from cytological methods for routine analysis and testing. Advances in transgenics have already had a major impact on mammalian genetic toxicology where there has been a long-standing gap in strategies to evaluate mutagenic hazards *in vivo* (Boer et al., 1996). Using the Big Blue® system (Kohler et al., 1991) and MutaMouse™ (Gossen et al., 1989), where *E. coli lacI* and *lacZ*, respectively, act as reporter genes, transgenic rodents equipped with bacterial plasmids provide important, short term, *in vivo* assays for the measurement of species- and tissue-specific rates of mutation. These assays, due to a shortage of endogenous animal genes amenable to quantitative analysis of somatic mutation (Vijg & Douglas, 1996), depend upon the use of surrogate bacterial genes and detection via the mutation-induced expression of β-galactosidase, whose activity is used to detect bacterial clones containing mutated copies of the gene after plasmid rescue from the host genomic DNA. It will not be long before we see transgenic assays applied to non-mammalian subjects. Marine bivalves have already been identified as suitable subjects for transgenic manipulation in a mutation detection context (Boer et al., 1996), although public concern about the release of genetically-modified organisms (GMOs) into the environment may restrict their use to the laboratory for the foreseeable future.

Acknowledgements

Part of the work described (anaphase aberration assay) was carried out during a NERC ROPA project: GR3/RO22.

References

Accomando, R., A. Viarengo, R. Bordone, M. Taningher, L. Canesi & M. Orunesu, 1991. A rapid method for detecting DNA strand breaks in *Mytilus galloprovincialis* Lam. induced by genotoxic xenobiotic chemicals. Int. J. Biochem. 23: 227–229.

Ahnstrom, G. & K. Erixon, 1973. Radiation induced strand breakage in DNA from mammalian cells. Int. J. radiat. Biol. 23: 285–289.

Anderson, S. L., J. E. Hose & J. P. Knezovich, 1994. Genotoxic and developmental effects in sea urchins are sensitive indicators of effects of genotoxic chemicals. Envir. Toxicol. Chem. 13: 1033–1041.

Atienzar, F., P. Child, A. Evenden, A. N. Jha, D. Sawa, C. Walker & M. H. Depledge, 1998. Application of the arbitrarily primed polymerase chain reaction for the detection of DNA damage. Mar. envir. Res. 46: 331–335.

Bauman, J. G. J., J. Wiegant, P. Borst & P. Van Duijn, 1980. A new method for fluorescence microscopical localisation of specific DNA sequences by *in situ* hybridisation of fluorochrome-labeled RNA. Exp. Cell Res. 138: 485–490.

Bayne, B. L., 1976. Marine mussels: their ecology and physiology. Cambridge University Press, London.

Bihari, N., R. Batel & R. K. Zahn, 1992. Fractionation of DNA from marine invertebrate, (*Mytilus galoprovincialis*) hemolymph by alkaline elution. Comp. Biochem. Physiol. 102B: 419–424.

Boer, J. G., H. L. Erfle, D. Walsh, J. Holcroft & B. W. Glickman, 1996. The use of *lacI* transgenic mice in genetic toxicology. In Pfeifer, G. F. (ed.), Technologies for Detection of DNA Damage and Mutations. Plenum Press, New York: 411–429.

Bolognesi, C., M. Parrini, P. Roggieri, C. Ercolini & C. Pellegrino, 1992. Carcinogenic and mutagenic pollutants: impact on marine organisms. Map. Tech. Rep. Ser. 69: 113–121.

Burgeot, T., F. Galgani & E. His, 1995. The micronucleus assay in *Crassostrea gigas* for the detection of seawater genotoxicity. Mutat. Res. 342: 125–140.

Carter, S. B., 1967. Effects of cytochalasins on mammalian cells. Nature 213: 261–264.

Cornet, M., 1993. A short-term method for chromosome preparation from somatic tissues of adult mussel (*Mytilus edulis*). Experientia 49: 87–90.

Dixon, D. R., 1982. Aneuploidy in mussel embryos (*Mytilus edulis* L.) originating from a polluted dock. Mar. Biol. Letts. 3: 155–161.

Dixon, D. R., 1985. *Pomatoceros triqueter*: a test system for environmental mutagenesis. In Bayne, B. L. (ed.), The Effects of Stress and Pollution on Marine Animals. Praeger Publishers, New York: 205–214.

Dixon, D. R. & K. R. Clarke, 1982. Sister chromatid exchange – a sensitive method for detecting damage caused by exposure to environmental mutagens in the chromosomes of adult *Mytilus edulis*. Mar. Biol. Letts. 3: 163–172.

Dixon, D. R., L. R. J. Dixon, P. L. Pascoe & J. T. Wilson, 1998b. Chromosome numbers and genome size in deep-sea hydrothermal vent organisms. InterRidge News 7: 20–22.

Dixon, D. R. & N. Flavell, 1986. A comparative study of the chromosomes of *Mytilus edulis* and *Mytilus galloprovincialis*. J. mar. biol. Ass. U.K. 66: 219–228.

Dixon, D. R. & I. R. B. McFadzen, 1987. Heterochromatin in the interphase nuclei of the common mussel *Mytilus edulis* L. J. exp. mar. Biol. Ecol. 112: 1–9.

Dixon, D. R. & P. L. Pascoe, 1994. Mussel eggs as indicators of mutagen exposure in coastal and estuarine marine environments. In Sutcliffe, D. W. (ed.), Water Quality and Stress Indicators in Marine and Freshwater Systems: Linking Levels of Organisation. Freshwater Biological Association, Cumbria: 124–137.

Dixon, D. R., P. L. Pascoe & L. R. J. Dixon, 1998a. Karyotypic differences between two species of *Pomatoceros*, *P. triqueter* and *P. larmarckii* (Polychaeta: Serpulidae). J. mar. biol. Ass. U.K. 78: 1–14.

Dixon, D. R. & D. Pollard, 1985. Embryo abnormalities in the periwinkle, *Littorina saxatilis*, as indicators of stress in polluted marine environments. Mar. Poll. Bull. 16: 29–33.

Dixon, D. R. & H. Prosser, 1986. An investigation of the genotoxic effects of an organotin antifouling compound (bis(trybutyltin)oxide) on the chromosomes of the edible mussel, *Mytilus edulis*. Aquat. Toxicol. 8: 185–195.

Dixon, D. R., J. T. Wilson, P. L. Pascoe & J. M. Parry, 1999. Anaphase aberrations in the embryos of the marine tubeworm *Pomatoceros lamarckii* (Polychaeta: Serpulidae): a new *in vivo* test assay for detecting aneugens and clastogens in the marine environment. Mutagenesis 14: 375–383.

Drouin, R. et al., 1996. Cupric ion/ascorbate/hydrogen peroxide-induced DNA damage: DNA-bound copper ion primarily induces base modification. Free Radic. Biol. Med. 21: 261–273.

Everaarts, J. M., 1995. DNA integrity as a biomarker of marine pollution: strand breaks in seastar (*Asterias rubens*) and dab (*Limanda limanda*). Mar. Poll. Bull. 31: 431–438.

Fenech, M. & A. A. Morley, 1985. Cytokinesis-block micronucleus method in human lymphocytes: effect on aging and low-dose x-irradiation. Cytobios 43: 233–246.

Fenech, M. & S. Neville, 1992. Conversion of excision-repairable DNA lesions to micronuclei within one cell-cycle in human lymphocytes. Envir. Mol. Mutagen. 19: 27–36.

Gardner, G. R., R. J. Pruell & A. R. Malcolm, 1992. Chemical induction of tumors in oysters by a mixture of aromatic and chlorinated hydrocarbons, amines and metals. Mar. envir. Res. 34: 59–63.

Gossen, J. A., W. J. F. De Leeuw, C. H. T. Tan, E. C. Zwarhoff, F. Berends, P. H. M. Lohman, D. L. Knook & J. Vijg, 1989. Efficient rescue of integrated shuttle vectors from transgenic mice: A model for studying mutation *in vivo*. Proc. natnl. Acad. Sci. U.S.A. 86: 7971–7975.

Govan, H. L., Y. Vallesayoub & J. Braun, 1990. Fine mapping of DNA damage and repair in specific genomic segments. Nucl. Acids Res. 18: 3823–3830.

Gupta, R. C. & K. Randerath, 1988. Analysis of DNA adducts by ^{32}P-postlabelling and Thin Layer Chromatography. In Friedberg, E. & P. H. Hanawalt (eds), DNA Repair. Marcel Dekker, New York: 399–418.

Hagmar, L. et al., 1994. Cancer risk in humans predicted by increased levels of chromosomal aberrations in lymphocytes – nordic study group on the health risk of chromosome damage. Cancer Res. 54: 2919–2922.

Harrison, F. L. & I. M. Jones, 1982. An *in vivo* sister chromatid exchange assay in the larvae of the mussel *Mytilus edulis* – response to three mutagens. Mutat. Res. 105: 235–242.

Harvey, J. S., B. P. Lyons, M. Waldock & J. M. Parry, 1997. The application of the ^{32}P-postlabelling assay to aquatic biomonitoring. Mutat. Res. 378: 77–88.

Hassold, T. D., 1986. Chromosome abnormalities in human reproductive wastage. Trends Genet. 2: 105–110.

Heddle, J. A., 1973. A rapid *in vivo* test for chromosome damage. Mutat. Res. 18: 187–192.

Hook, E. B., 1983. Contribution of chromosome abnormalities to human morbidity and mortality and some comments upon surveillance of chromosome mutation rate. Mutat. Res. 114: 389–423.

Hose, J. E. & H. W. Puffer, 1983. Cytologic and cytogenetic anomalies induced in purple sea urchin embryos (*Strongylocentrotus purpuratus* S.) by parental exposure to benzo(a)pyrene. Mar. Biol. Letts. 4: 87–95.

Hose, J. E., H. W. Puffer, P. S. Oshida & S. M. Bay, 1983. Developmental and cytogenetic abnormalities induced in the purple sea urchin by environmental levels of benzo(a)pyrene. Arch. envir. Contam. Toxicol. 12: 319–325.

Jha, A. N., T. H. Hutchinson, J. M. Mackay, B. M. Elliott & D. R. Dixon, 1996. Development of an *in vivo* genotoxicity assay using the marine worm *Platynereis dumerilii* (Polychaeta: Neridae). Mutat. Res. 359: 141–150.

John, H., M. Birnstiel & K. Jones, 1969. RNA-DNA hybrids at the cytological level. Nature 223: 582–587.

Jones, I. M. & F. L. Harrison, 1987. Variability in the frequency of sister-chromatid exchange in larvae of *Mytilus edulis*: implications for field monitoring. J. exp. mar. Biol. Ecol. 113: 283–288.

Klaude, M., S. Eriksson, J. Nygren & G. Ahnstrom, 1996. The comet assay: mechanisms and technical considerations. Mutat. Res. 363: 89–96.

Kohler, S. W., G. S. Provost, A. Fieck, P. L. Kretz, W. O. Bullock, D. L. Putman, J. A. Sorge & J. M. Short, 1991. Analysis of spontaneous and induced mutations in transgenic mice using a lambda ZAP/*lacI* shuttle vector. Envir. Mol. Mutagen. 18: 316–321.

Kohn, K. W., C. K. Erickson, A. G. Ewig & L. A. Zwelling, 1976. Fractionation of DNA from mammalian cells by alkaline elution. Biochemistry 15: 4629–4637.

Korenberg, J. R. & E. F. Freelander, 1974. Giemsa technique for the detection of sister chromatid exchanges. Chromosoma 48: 355.

Li, G. & D. Hedgecock, 1998. Genetic heterogeneity, detected by PCR-SSCP, among samples of larval pacific oysters (*Crassostrea gigas*) supports the hypothesis of large variance in reproductive success. Can. J. Fish. aquat. Sci. 55: 1025–1033.

Levan, G. & F. Mitelman, 1977. Chromosomes and the aetiology of cancer. In de la Chapelle, A. & M. Sorsa (eds), Chromosomes Today. Elsevier, Amsterdam: 363–371.

Martinez-Lage, A., A. Gonzalez-Tizon & J. Mendez, 1994. Characteristics of different chromatin types in *Mytilus galloprovincialis* L. after C-banding, fluorochrome and restriction-endonuclease treatments. Heredity 72: 242–249.

Mix, M. C., 1986. Cancerous diseases in aquatic animals and their association with environmental pollutants: a critical literature review. Mar. envir. Res. 20: 1–141.

Moore, M. N., D. M. Lowe, D. R. Livingstone & D. R. Dixon, 1986. Molecular and cellular indices of pollutant effects and their use in environmental impact assessment. Wat. Sci. Tech. 18: 223–232.

Nacci, D. E., S. Cayula & E. Jackim, 1996. Detection of DNA damage in individual cells from marine organisms using the single cell gel assay. Aquat. Toxicol. 35: 197–210.

Nacci, D. E. & S. Nelson, 1992. Application of the alkaline unwinding assay to detect DNA strand breaks in marine bivalves. Mar. envir. Res. 33: 38–100.

Olive, P. L. & J. P. Banath, 1993. Induction and rejoining of radiation-induced DNA single-strand breaks: 'tail moment' as a function of position in the cell cycle. Mutat. Res. 294: 275–283.

Olive, P. L., J. P. Banath & R. E. Durand, 1990. Heterogeneity in radiation-induced DNA damage and repair in tumour and normal cells measured using the 'comet' assay. Radiation Res. 122: 86–94.

Olive, P. L., D. Wlodek & J. P. Banath, 1991. DNA double-strand breaks measured in individual cells subjected to gel electrophoresis. Cancer Res. 51: 4671–4676.

Ostling, O. & K. J. Johanson, 1984. Microelectrophoretic study of radiation-induced DNA damage in individual mammalian cells. Biochem. Res. Communicat. 123: 291–298.

Pardue, M. L. & J. G. Gall, 1969. Molecular hybridisation of radioactive DNA to the DNA of cytological preparations. Proc. natnl. Acad. Sci. U.S.A. 64: 600–604.

Pascoe, P. L., S. J. Patton, R. Critcher & D. R. Dixon, 1995. Robertsonian polymorphism in the marine gastropod *Nucella lapillus*: advances in karyology using rDNA loci and NORs. Chromosoma 104: 455–460.

Perry, P. & S. Wolff, 1974. New giemsa method for the differential staining of sister chromatids. Nature 251: 156.

Pesch, G. G., C. E. Pesch & A. R. Malcolm, 1981a. *Neanthes arenaceodentata*, a cytogenetic model for marine genetic toxicology. Envir. Mutagen. 3: 386–387.

Pesch, G. G., C. E. Pesch & A. R. Malcolm, 1981b. *Neanthes arenaceodentata*, a cytogenetic model for marine genetic toxicology. Aquat. Toxicol. 1: 301–311.

Russo, A., G. Priante & A. M. Tommasi, 1996. PRINS localization of centromeres and telomeres in micronuclei indicates that in mouse splenocytes chromatid non-disjunction is a major mechanism of aneuploidy. Mutat. Res. 372: 173–180.

Savage, J. R. K., 1993. Update on target theory as applied to chromosome aberrations. Envir. Mol. Mutagen. 22: 198–207.

Schmid, W., 1977. The micronucleus test. In Kilbey, B. J., M. Legator, W. Nichols & C. Ramel (eds), Handbook of Mutagenicity Test Procedures. Elsevier, Amsterdam: 235–242.

Shugart, L. R. & C. W. Theodakaris, 1994. Environmental genotoxicity: probing the underlying mechanisms. Envir. Health Perspect. 102: 13–17.

Singh, N. P., M. T. McCoy, R. R. Tice & E. L. Schneider, 1988. A simple technique for quantitation of low levels of DNA damage in individual cells. Exp. Cell Res. 175: 184–191.

Sole, M., C. Porte, X. Biosca, C. L. Mitchelmore, J. K. Chipman, D. R. Livingstone & J. Albaiges, 1996. Effects of the 'Aegean Sea' oil spill on biotransformation enzymes, oxidative stress and DNA-Adducts in digestive gland of the mussel (*Mytilus edulis* L). Comp. Biochem. Physiol. 113B: 257–265.

Stein, J. E., W. L. Reichert, M. Nishimoto & U. Varanasi, 1990. Overview of studies on liver carcinogenesis in English Sole from Puget Sound – evidence for a xenobiotic chemical etiology. Sci. total Envir. 94: 51–69.

Steinert, S. A., 1996. Contribution of apoptosis to observed DNA damage in mussel cells. Mar. envir. Res. 42: 253–259.

Taylor, J. H., P. S. Woods & W. L. Hughes, 1957. The organisation and duplication of chromosomes as revealed by autoradiographic studies using tritium-labelled thymidine. Proc. natnl. Acad Sci. U.S.A. 43: 122–128.

Theodorakis, C. W., S. J. Dsurney & L. R. Shugart, 1994. Detection of genotoxic insult as DNA strand breaks in fish blood cells by agarose-gel electrophoresis. Envir. Toxicol. Chem. 13: 1023–1031.

Thorson, G., 1946. Reproduction and larval development of Danish marine bottom invertebrates with special reference to the planktonic larvae in the sound (Oresund). Medd. Komm. Haverundersog (ser. Plankton) 4: 170–173.

Tsutsui, T., H. Maizumi, J. A. McLachlan & J. C. Barrett, 1983. Aneuploidy induction and cell transformation by diethylstilbestrol. Cancer Res. 43: 3814–3821.

Venier, P. & S. Canova, 1996. Formation of DNA adducts in the gill tissue of *Mytilus galloprovincialis* treated with benzoapyrene. Aquat. Toxicol. 34: 119–133.

Viarengo, A., L. Canesi, M. Pertica & D. R. Livingstone, 1991. Seasonal variations in the antioxidant defence systems and lipid peroxidation of the digestive gland of mussels. Comp. Biochem. Physiol. 100B: 187–190.

Vijg, J. & G. R. Douglas, 1996. Bacteriophage lambda and plasmid *lacZ* transgenic mice for studying mutations *in vivo*. In Pfeifer, G. F. (ed.), Technologies for Detection of DNA Damage and Mutations. Plenum Press, New York: 391–410.

Vukmirovik, M., N. Bihari, R. K. Zahn, W. E. G. Muller & R. Batel, 1994. DNA damage in marine mussel *Mytilus galloprovincialis* as a biomarker of environmental contamination. Mar. Ecol. Prog. Ser. 109: 165–171.

Welsh, J., D. Ralph & M. McClelland, 1995. In Innis, M. A., D. H. Gelfand & J. J. Sninsky (eds), PCR Strategies. Academic Press, London: 249–276.

Wilson, J. T., P. L. Pascoe, J. M. Parry & D. R. Dixon, 1998. Evaluation of the comet assay as a method for the detection of DNA damage in the cells of a marine invertebrate, *Mytilus edulis* L (Mollusca: Pelecypoda). Mutat. Res. 399: 87–95.

Wrisberg, M. N., C. M. Bilbo & H. Spliid, 1992. Induction of micronuclei in hemocytes of *Mytilus edulis* and statistical analysis. Ecotoxicol. envir. Safety 23: 191–205.

Zakour, H. R., M. L. Landolt & R. M. Kocan, 1984. Sister chromatid exchange analysis in cultured peripheral blood leukocytes of the coldwater marine fish. Mar. envir. Res. 14: 499–500.

Hydrobiologia **420:** 45–47, 2000.
A. M. Solé-Cava, C. A. M. Russo & J. P. Thorpe (eds), Marine Genetics.

Biotechnology and fish culture

F. Foresti
Departamento de Morfologia, Instituto de Biociências, Universidade Estadual Paulista, 18618-000 Botucatu, São Paulo, Brazil

Key words: aquaculture, chromosome manipulation, ploidy

Abstract

Biotechnology can currently be considered of importance in aquaculture. The increase in the production of aquatic organisms over the last two decades through the use of biotechnology indicates that in a few generations biotechnology may overtake conventional techniques, at least for the commercially more valuable species. In the last few years, genetics has contributed greatly to fish culture through the application of the more recent techniques developed in biotechnology and in genetic engineering. At present, the most commonly used methods in fish biotechnology are chromosome manipulation and hormonal treatments, which can be used to produce triploid, tetraploid, haploid, gynogenetic and androgenetic fish. These result in the production of individuals and lineages of sterile, monosex or highly endogamic fish. The use of such strategies in fish culture has as a practical objective the control of precocious sexual maturation in certain species; other uses are the production of larger specimens by control of the reproductive process and the attainment of monosex lines containing only those individuals of greater commercial value. The use of new technologies, such as those involved in gene transfer in many species, can result in modified individuals of great interest to aquaculturists and play important roles in specific programmes of fish production in the near future.

Introduction

The role of aquaculture in increasing fish production is well recognised today (Donaldson, 1997). Although the basis of fish genetic improvement is roughly similar to that employed for animals, and appreciable progress has been made in recent years in the fields of selection, inbreeding, hybridisation and sex control, the application of such technologies to aquaculture has been limited. One important difference between fish and terrestrial animals for cultivation and genetic improvement is that, usually, fish have higher levels of genetic variation, and hence more scope for selection, than most mammals or birds.

Chromosome manipulation

Fish species are generally very tolerant of artificial manipulation of their chromosomes during early development and this property has been exploited for the production of inbred lines, monosex populations, and the control of ploidy. A variety of different techniques has been applied to obtain polyploids, gynogenetics and androgenetics, sex-reversed individuals and transgenics (Purdom, 1993; Toledo-Filho et al., 1996). Methods used include interspecific hybridisation or the control of sex by the administration of sex steroids to larvae or juveniles and even surgical or autoimmune castration.

The artificial modification of the chromosome set of an organism permits the production of monosex and sterile individuals, while gynogenesis and androgenesis provide methods for the rapid production of inbred populations which can be used in crossbreeding programs.

Manipulation of chromosome number to give polyploid individuals is common in aquaculture. The retention of the second polar body during the second meiotic division in the oocyte results in triploid individuals, which have two chromosome sets from the mother and one from the father. Unlike mammals, where chromosomal rearrangements of this magnitude are usually fatal, many fish with three sets of chromosomes survive quite readily. In addition to triploids, individuals with four sets of chromosomes (tetraploids) or those with both of their chromosome sets derived from a single parent can be produced.

Triploid fish are of interest because their sterility is useful in aquaculture and fisheries management (Lutz, 1998a). Sterility caused by genetic and physiological factors leads to different characteristics in male and female sexual development. A number of studies have shown that triploid rainbow trout are sterile, whether spontaneously occurring in a population or artificially induced through heat or pressure shocks.

Male triploid rainbow trout, however, typically undergo considerable changes in secondary sex characteristics and gonad development at the time of maturation, affecting the carcass appearance and reducing meat quality. Female triploids, on the other hand, normally have minimal gonad development and maintain carcass quality throughout the period of maturation of their diploid counterparts. The performance of triploid fish in production situations has been found to be comparable to that of diploid individuals, and the use of triploid females should be considered when they are to be grown past the time of normal sexual maturation.

Gynogenesis is the production of offspring with genes from the mother only. In practice, in fish and some other organisms, it is possible to produce offspring from a mature female with no paternal genetic contribution (Purdom, 1993). The technique for producing gynogenetic individuals requires the inactivation of the male genome and the diploidisation of the female genetic material in the zygote, in a process induced by physical or chemical agents. Gynogenesis can be achieved easily in rainbow trout and other species of fish by fertilising eggs with irradiated sperm and inducing polar body retention through the same type of treatments (pressure or temperature shocks) used to produce triploids (Parsons, 1998). In the same way, if the eggs are irradiated and then fertilised by normal spermatozoa, androgenetic individuals can be produced under certain conditions, with only a set of parental chromosomes. These haploids are then changed to diploids by heat shock treatment before the first cell division.

The technique for inducing gynogenesis has been usefully applied to different species of fish (Lutz, 1997a,b; 1998b). Gynogenetic individuals are used to produce all-female fish populations. The reasons for inducing gynogenesis range from the production of monosex populations to the development of partially or completely inbred organisms. Partially inbred offspring (from meiotic gynogenesis) may be useful in genome studies for examining the map position of different loci in relation to the centromeres of their chromosomes. If completely inbred animals (from mitotic gynogenesis) survive to maturity, their eggs can be subjected to a second round of gynogenesis, producing true clones in large numbers.

In a successful long-term program based on genetic selection and hybridisation for the development of inbred pure lines of carp in Hungary (Bakos, 1979; Bakos & Gorda, 1995), the use of such a technique resulted in very productive hybrids (Nagy et al., 1979). Although the first gynogenetic individuals produced in a line typically perform rather poorly due to inbreeding effects, in trout they can be hormonally sex-reversed to functional males that produce only X-bearing sperm. This sperm is then used to fertilize eggs from normal females, producing all-female populations (Donaldson, 1997). All-female rainbow trout and salmon have shown a variety of advantages in aquaculture, including faster and more uniform growth and better carcass yield, particularly in stocks with high rates of male precocity; they have found favour with aquaculturists throughout the world.

Concluding remarks and notes on emerging technologies

As seen in the above discussions, genetic technology is rapidly being applied in aquaculture. Hormonal treatments that regulate the action of genes and modify the sex of offspring are now widely used in fish culture programs. Chromosome manipulation, resulting in sterile polyploid individuals or monosex inbred lines, constitutes an important tool. DNA markers are being used to study stock identification and population differences, in gene mapping studies, and potentially as aids to selective breeding programs. Gene transfer work in many species of fish, stimulated by the possibility of producing rapidly-growing individuals through the introduction of foreign growth hormone genes, has produced modified fish in numerous laboratories around the world.

Evidence has been documented of the expression of foreign genes introduced into embryonic fish, as well as transmission of the introduced genes by transgenic individuals to their progeny (Iyengar et al., 1996; Dunham, 1999). Once this technique has proved to be effective, and difficulties overcome in the optimal genome incorporation and expression of genetic characteristics, a variety of genes with the potential for growth promotion or disease resistance will be of great interest to aquaculturists. To date, a large number of species of fish have been genetically modified using different DNA constructs, reflecting the wide application of this emerging technology in fish culture. However, as might be expected, considerable concern has been expressed from philosophical and ethical points of view. Also, the potential negative impact that transgenic fish could have should be considered (Kapuscinski & Hallerman, 1991); this includes negative market impacts (Berkowitz & Kryspin-Sorensen, 1994) and possible interaction of transgenic fish with wild populations. Regulation has been proposed in many countries attempting to control the production and use of transgenic animals (Hallerman & Kapuscinski, 1995). These concerns will cause significant delay in the direct adoption of gene transfer technologies in fish culture. In the meantime, other less controversial biotechnological methodologies could play a role.

While these biotechnologies are being developed and adopted by aquaculturists, new developments in molecular genetics point to the evolution of stem cell tissue cultures and even the growing of totipotent cells, such as egg blastomeres in culture, as future possibilities. Thus, the production of cells containing new gene arrangements, some of them introduced by chromosome manipulation or by gene transfer, could be undertaken in tissue culture to give new combinations of genetic material. Furthermore, the organisation of gene banks based on collections of frozen sperm and cells or on purified DNA molecules, obtained from different fish species, should be considered and could play an important role in the future.

References

Bakos, J., 1979. Crossbreeding Hungarian races of common carp to develop more productive hybrids. In Pillay, T. V. R. & W. Dill (eds), Advances in Aquaculture. Fishing News Books, Farnham, Surrey, UK: 633–635.

Bakos, J. & S. Gorda, 1995. Genetic improvement of common carp strains using intraspecific hybridization. Aquaculture 129: 183–186.

Berkowitz, D. B. & I. Kryspin-Sorensen, 1994. Transgenic fish: safe to eat? A look at the safety considerations regarding food transgenics. Biotechnology 12: 247–252.

Donaldson, E. M., 1997. The role of biotechnology in sustainable aquaculture. In Bardach, J. E. (ed.), Sustainable Aquaculture. John Wiley & Sons, New York.

Dunhan, R. A., 1999. Utilization of transgenic fish in developing countries: potential benefits and risks. J. Word aquacult. Soc. 30: 1–11.

Hallerman, E. M. & A. R. Kapuscinki, 1995. Incorporating risk assessment and risk management into public policies on genetically modified finfish and shellfish. Aquaculture 137: 9–17.

Iyengar, A., F. Muller & N. Maclean, 1996. Regulation and expression of trangenes in fish – a review. Transgen. Res. 5: 147–166.

Kapuscinski, A. R. & E. M. Hallerman, 1991. Implications of introduction of transgenic fish into natural ecosystems. Can. J. Fish. aquat. Sci. 48: 99–107.

Lutz, C. G., 1997a. Tinkering with chromosomal complements: gynogenesis – an introduction. Aquacult. Magaz. 23: 67–71.

Lutz, C. G., 1997b. Gynogenesis part II: examining recent work. Aquacult. Magaz. 23: 73–78.

Lutz, C. G., 1998a. Triploidy: potential and problems. Aquacult. Magaz. 24: 92–95.

Lutz, C. G., 1998b. Gynogenesis part III. Aquacult. Magaz. 24: 75–78.

Nagy, A., K. Rajki, J. Bakos & V. Csányi, 1979. Genetic analysis in carp (*Cyprinus carpio*) using gynogesesis. Heredity 43: 35–40.

Parsons, J., 1998. Status of genetic improvement in commercially reared stocks of rainbow trout. World Aquacult. 29: 44–47.

Purdom, C. E., 1993. Genetics in fish breeding. Chapman & Hall, London, 277 pp.

Toledo-Filho, S. A., F. Foresti & L. F. Almeida-Toledo, 1996. Biotecnologia genética aplicada à piscicultura. Cadernos de Genética 3, CCS/USP, São Paulo, Brazil.

Hydrobiologia **420:** 49–54, 2000.
A. M. Solé-Cava, C. A. M. Russo & J. P. Thorpe (eds), Marine Genetics.

Genetic differentiation in Chilean hake *Merluccius gayi gayi* (Pisces: Merlucciidae)

R. Galleguillos[1], L. Troncoso[1], C. Oyarzún[1], M. Astorga[2] & M. Peñaloza[2]
[1]*Dept. of Oceanography, Faculty of Natural Sciences and Oceanography, University of Concepcion, P.O. Box 160-C, Concepcion, Chile*
E-mail: rgalleg@udec.cl, ltronco@udec.cl, coyarzun@udec.cl
[2]*Graduate Program in Zoology, Dept. of Zoology, Faculty of Natural Sciences and Oceanography, University of Concepcion, P.O. Box 160-C, Concepcion, Chile*

Key words: allozymes, stock, Pacific Ocean, Chile, electrophoresis

Abstract

The genetic structure of Chilean hake *Merluccius gayi gayi*, was analyzed using starch gel protein electrophoresis. Samples were collected from four localities along the coast off Chile. A total of 1500 specimens sampled from Coquimbo, San Antonio, Talcahuano and Puerto Montt were used in the study. Genetic information was obtained for six allozyme loci *Pgi-1, Pgi-2, Pgm, Idh-1, Idh-2* and *Aat*. The results of the analysis showed no significant allelic differences among samples from various localities, with an average F_{ST} value of 0.007 for all loci and samples. A heterogeneity test for all loci and specimens from the four localities showed only two significant values. Thus, Chilean hake populations along the coast of Chile appear to be genetically homogeneous.

Introduction

Most fish species do not live in continuous, homogeneous populations, but as groups of populations with various characteristics. If gene flow is minimal or almost non-existing between populations, they can exhibit significant levels of differentiation, which may be maintained through time under certain conditions (Spanakis et al., 1989). Research on genetic variability in species with wide geographic distributions has produced useful information for estimating levels of genetic differentiation between allopatric species. The use of a number of molecular techniques, such as allozyme polymorphism, has produced genetic markers which have been used to investigate migration patterns, reproductive isolation and stock composition in fish (Carvalho & Hauser, 1995).

From the point of view of fishery management, the most useful definition of a stock of marine animals should have solid genetic basis. Lack of knowledge of the number of populations and level of interbreeding in a species subject to exploitation, may lead to management policies and to conservation objectives that are not properly fulfilled (Ovenden, 1990).

One of the most important fishery resources in Chile is the hake *Merluccius gayi gayi*, landings of which reached 90 000 tons in 1997. It is a benthic-demersal fish that lives over the continental shelf from Antofagasta (23° 38′ S) to Chiloé Island (43° 00′ S). According to Balbontín & Fisher (1981) it spawns the year round, with peaks in August and November and makes large migrational movements. In mark and recapture experiments carried out by Villegas & Saetersdal (1968) fish marked in the San Antonio area were found more than 500 km away in landings at Talcahuano.

The first study using protein genetic markers in Chilean hake was carried out by Sepúlveda & Rojas (1973), who established phenotypic patterns and intraspecific variability by analyzing muscle and blood proteins. Stepien & Rosenblatt (1996) used isozymes to genetically characterize hake and estimated the degree of heterozygosity and genetic polymorphism.

Among studies on hake populations, George-Nascimento (1996), analyzed hake parasite community assemblages and, based on differences of four parasite species among four length classes of fish, suggested the existence of at least two ecological stocks of hake along the Chilean coast. On the other hand, Oyarzún (1997) analyzed the control region of mitochondrial DNA between positions L-19 and 12 Sar-H and found no significant differences between hake samples from the latitudinal extremes of its distribution.

In other *Merluccius* species Grant et al. (1988), using allozymes, showed weak population differentiation between two *M. capensis* populations, while no geographic differences were detected for *M. paradoxus*. A different situation has been indicated by Roldan et al. (1998) who found evidence of separation between samples of *Merluccius merluccius* from the North Atlantic and the Mediterranean Sea.

This work attempts to describe allozyme variability for Chilean hake (*Merluccius gayi gayi*), using samples collected in several localities off the Chilean coast.

Materials and methods

Hake samples were obtained (July 1995, August 1996, December 1996) from the commercial fishery that operates in Chile in the areas of Coquimbo (29° 50′ S; 71° 15′ W), San Antonio (33° 34′ S; 71° 36′ W), Talcahuano (36° 41 ′ S; 73° 06′ W) and Puerto Montt (41° 30′ S; 72° 50′ W) (see Figure 1). A total of about 1500 individuals from the four areas was analyzed using horizontal starch gel protein electrophoresis. Portions of muscle and liver tissue were dissected from fresh fish and stored at −20 °C. Each tissue was homogenized in distilled water and centrifuged at 3000 r.p.m. for 5 min at 5 °C. The supernatant was used as the protein source. The enzyme systems analyzed were Isocitrate dehydrogenase, *Idh* (E.C. 1.1.1.42), Aspartate amino transferase, *Aat* (E.C. 2.6.1.1), Phosphoglucomutase, *Pgm* (E.C. 2.7.5.1) and Phosphoglucose isomerase *Pgi*, (E.C. 5.3.1.9). The buffers used were Tris citric pH 6.9 and Tris discontinuous pH 5.1–6.0, the enzymes were specifically stained following Harris & Hopkinson (1976). Allele frequencies and values of deviation from Hardy–Weinberg (D), Nei's identity and genetic distance (1978) and Wright's F_{ST} were estimated with the Program BYOSIS 1.7 (Swofford & Selander, 1989).

The *G*-test (Sokal & Rohlf, 1995) was used to test allelic frequency homogeneity among samples and standardized deviation values from the frequency of the most common allele were estimated for each locus in the twelve samples analyzed. The basic procedure for the latter considers calculating the mean of the allele frequencies for the most common allele and standardizing the frequencies of the other alleles of the same locus with regard to an average for all samples and then, analyzing whether the standardized values are within the limit of ±2 standard deviations from the mean (Christiansen et al., 1976).

Results

Allele frequencies and sample sizes for each polymorphic locus and for each sample obtained from Coquimbo, San Antonio, Talcahuano and Puerto Montt are shown in Table 1.

The mean observed heterozygosity per sample was estimated from direct counts and ranged between 0.21 and 0.16 (Table 1). No significant differences were found between localities. These estimates indicate a high level of intra- and inter-locality homogeneity. D values, which describe heterozygote deficiency, were low indicating that most samples did not differ from Hardy-Weinberg equilibrium (Table 1). Two samples (Puerto Montt-2 and San Antonio-2) were the only exceptions, with significant values for the *Idh-1* (18.49, $P < 0.01$) and *Aat-1* (7.11, $P < 0.05$) loci. Since genetic identity and distance show values close to 1 and 0 respectively, the groups could not be differentiated by cluster analysis.

For all loci and samples analyzed Wright's F_{ST} index, which estimates genetic differentiation between subpopulations (Hartl, 1980), showed a maximum of 0.015 at the *Aat-1* locus and a minimum of 0.004 at the *Pgi-2* locus with an average of 0.007. For each locus considered these results show the lack of significant genetic differentiation among hake from the different areas in the study.

A heterogeneity test for each locus and locality produced significant results for *Idh-2* in the 3 samples from Coquimbo (G=15.44, $P < 0.01$) and for *Pgm* at San Antonio (G=19.30, $P < 0.01$). The analysis of the standard deviation of the frequency of the most common allele, for all loci, shows that almost all groups are within the range of ±2 standard deviations. The only two exceptions found were *Pgi-1* in samples from Talcahuano (Figure 2) and *Idh-1* from San Antonio samples (Figure 3).

Discussion

Protein electrophoresis techniques have been extensively used in population genetic studies. They have

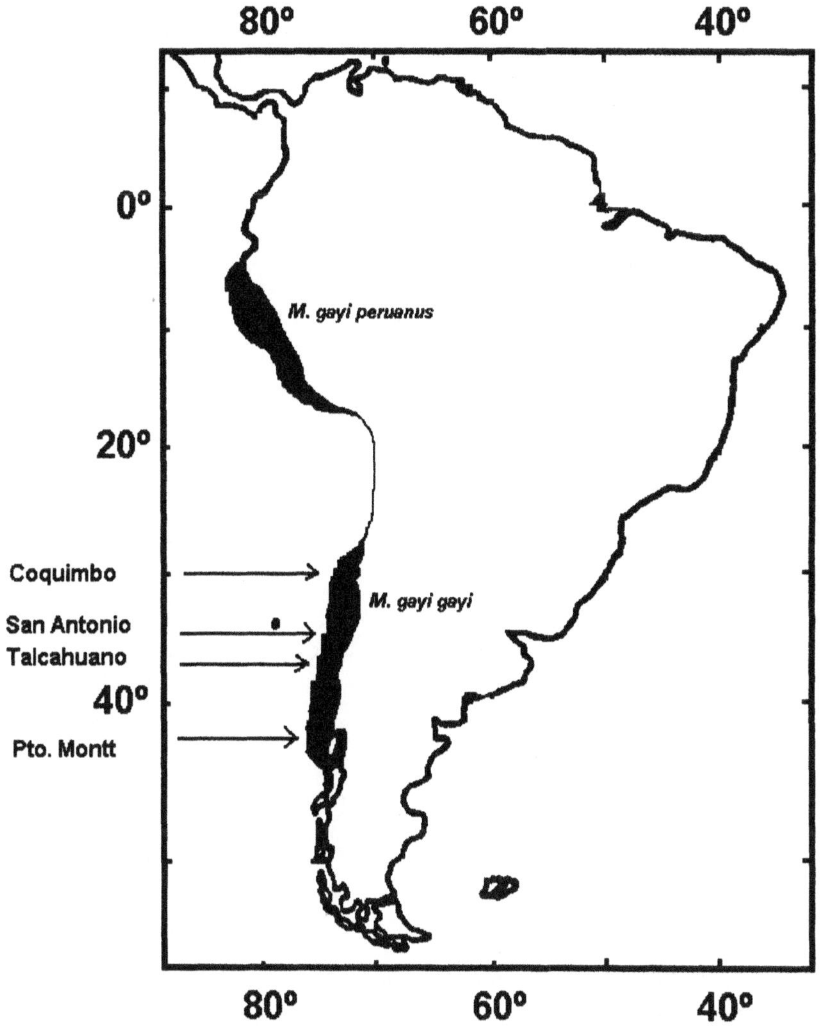

Figure 1. Geographical distribution and sampling sites of *Merluccius gayi* over the southeastern Pacific continental shelf.

been mostly applied to establish taxonomic status, to estimate levels of genetic differentiation among subpopulations and for stock identification of commercially important resources (Ward et al., 1994).

The results of this research show a high degree of genetic homogeneity for *M. g. gayi* in its distribution area along the Chilean coast: Wright's F_{ST} values were very small (between 0.004 and 0.015). Identity values were all very close to one and the tests applied to allele frequencies showed almost no significant differences between samples from the different localities. Thus, allozyme analysis indicates that this species has a single population or stock within the area in study and represents an intraspecific group with random mating and continued integrity in space and time.

According to Sinclair & Iles (1989) *Merluccius* species could be weakly structured at the population level, because of the effect the current gyres and larval retention areas might have on population structure. However, in Chilean hake the lack of strong geographic barriers to migration and extensive larval dispersion appear to produce high levels of genetic homogeneity over its wide geographic ranges.

Mean heterozygosity per locus for *M. gayi* (0.051 ± 0.015) is similar to the levels of genetic variability found for *M. productus*, *M. hubbsi* and *M. capensis* (Stepien & Rosenblat, 1996). Grant et al. (1988), used allozymes to estimate the population

Table 1. Allele frequencies for six loci and twelve samples of Chilean hake (*Merluccius gayi gayi*). Each allele is named in lowercase and samples are in uppercase. (A=Coquimbo 1; B=Coquimbo 2; C=Coquimbo 3; D=San Antonio 1; E=San Antonio 2; F=San Antonio 3; G=Talcahuano 1; H=Talcahuano 2; I=Talcahuano 3; J=Puerto Montt 1; K=Puerto Montt 2; L=Puerto Montt 3). H_e=Mean Hardy–Weinberg (HW) expected heterozygosity, D=Deviation from HW

Locus	Sample											
	A	B	C	D	E	F	G	H	I	J	K	L
Pgi-1												
(N)	109	134	115	130	120	118	127	116	122	129	123	94
a	0.624	0.687	0.709	0.654	0.642	0.644	0.681	0.685	0.799	0.663	0.715	0.665
b	0.358	0.299	0.287	0.342	0.338	0.352	0.303	0.289	0.201	0.326	0.272	0.314
c	0.018	0.015	0.004	0.004	0.021	0.004	0.016	0.026	0.000	0.012	0.012	0.021
Pgi-2												
(N)	109	134	115	132	123	127	131	124	122	131	121	92
a	0.991	0.989	0.996	0.992	1.000	1.000	1.000	1.000	0.992	0.996	1.000	0.995
b	0.005	0.011	0.004	0.004	0.000	0.008	0.000	0.000	0.008	0.004	0.000	0.005
c	0.005	0.000	0.000	0.004	0.000	0.000	0.000	0.000	0.000	0.000	0.000	0.000
Pgm-1												
(N)	109	142	116	132	56	124	131	114	122	149	80	94
a	0.954	0.944	0.957	0.928	0.911	0.911	0.927	0.934	0.922	0.940	0.931	0.952
b	0.018	0.039	0.034	0.057	0.080	0.065	0.046	0.048	0.066	0.047	0.038	0.043
c	0.023	0.014	0.009	0.011	0.000	0.000	0.027	0.013	0.012	0.013	0.000	0.005
d	0.005	0.004	0.000	0.004	0.009	0.024	0.000	0.004	0.000	0.000	0.031	0.000
Idh-1												
(N)	109	115	116	92	40	66	131	104	122	111	59	74
a	0.968	0.957	0.987	0.967	0.913	0.985	0.977	0.976	0.975	0.986	0.966	0.993
b	0.032	0.043	0.013	0.033	0.087	0.015	0.023	0.024	0.025	0.014	0.034	0.007
Idh-2												
(N)	104	120	116	135	99	128	105	123	121	110	117	94
a	0.798	0.712	0.728	0.759	0.727	0.758	0.790	0.752	0.707	0.800	0.761	0.702
b	0.202	0.287	0.272	0.241	0.273	0.242	0.210	0.248	0.293	0.200	0.239	0.298
Aat-1												
(N)	79	40	41	71	90	62	131	40	59	131	65	94
a	0.981	1.000	1.000	0.986	0.961	1.000	1.000	0.988	0.992	0.992	1.000	1.000
b	0.019	0.000	0.000	0.014	0.039	0.000	0.000	0.013	0.008	0.008	0.000	0.000
H_e	0.169	0.178	0.155	0.178	0.213	0.171	0.160	0.170	0.161	0.157	0.163	0.166
D	0.012	0.028	0.052	0.056	0.019	0.029	0.006	0.053	0.037	0.051	0.169	0.030

structure of *M. capensis* and *M. paradoxus*. These researchers found weak differentiation between two groups of populations of the former species and no significant differences between populations within its geographic distribution for the latter. For *M. merluccius* along the west European continental shelf the analysis of genotype frequency distribution for three loci showed a high degree of genetic similarity (Mangaly & Jamieson, 1978).

Genetic heterogeneity has been reported for some *Merluccius* species; Utter & Hodgins (1971) found differences in allelic frequencies for *M. productus* populations that appear to be isolated within Puget Sound. Roldán (1991) also found evidence of genetic

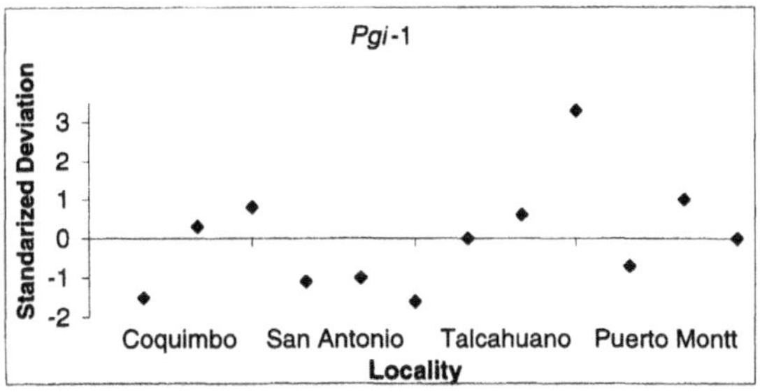

Figure 2. Standardized deviation from the most common allele for the *Pgi-1* locus in samples of hake, *Merluccius gayi gayi*.

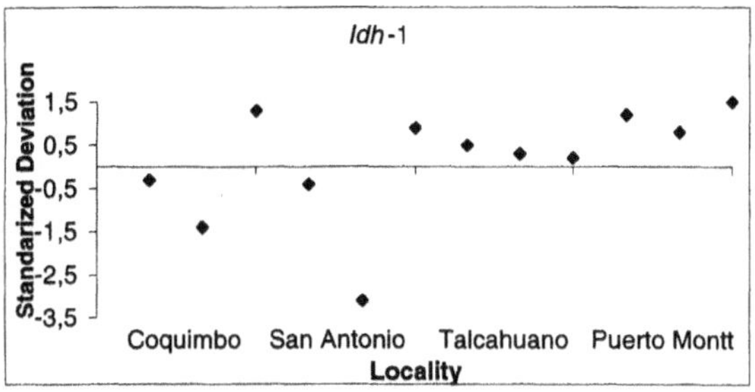

Figure 3. Standardized deviation from the most common allele for the *Idh-1* locus in samples of hake, *Merluccius gayi gayi*.

heterogeneity at the *Est-1* locus, among ten samples of *M. hubbsi* from the continental shelf of Argentina. It is not clear, however, which mechanisms have produced geographic heterogeneity at this locus. More recently, considering a bigger area Roldán et al. (1998) found evidence of subdivision between North Atlantic and Mediterranean populations of *M. merluccius* and concluded that the populations structure within the Atlantic is more complex than the discrete northern and southern stocks proposed by ICES.

Molecular techniques, based on the study of nuclear and mitochondrial DNA, have also been used to analyse population and stock structure. The analysis of mtDNA may be more powerful than enzyme analysis because mtDNA has an effective population size that is only one quarter of that for nuclear DNA thus increasing the effects of genetic drift (Ward & Grewe, 1994).

Oyarzún (1997) carried out the first study of nucleotide sequences in the D-loop of mtDNA in Chilean hake (*M. gayi gayi*) samples from the South eastern Pacific. Results after the application of restriction enzymes showed that the most common haplotypes had similar frequencies at the two extremes of the distribution of the species: Coquimbo (29° 50′ S; 71° 15′ W) and Valdivia (39° 48′ S; 73° 14′ W).

The information from allozyme markers and mitochondrial DNA disagrees with the results of George-Nascimento (1996), who postulated the existence of two ecological stocks of hake off Chile based on parasite analysis. Even though he recognized that differences were evident in small and medium length classes only, he argued that this difference reflected isolation in different nursery areas. It is not known whether the convergence in parasite composition in larger size classes implies a mixture of older age groups.

In many instances low genetic differentiation among marine fish populations is the result of high levels of gene flow between geographic areas, because of planktonic larval drift and active migration of adults (Shaklee, 1984). In addition, the usually large population sizes of marine fish reduce the importance of genetic drift in producing genetic differences between populations. Migration of *M. g. gayi* over great latitudinal distances on the Chilean continental shelf was shown by the re-capture of tagged individuals over 500 km away from the area of their release in catch-recapture experiments (Villegas & Saetersdal, 1968). In terms of management of Chilean hake as a fishery resource, our results suggest that *M. gayi gayi* should be treated as a single population.

Acknowledgments

This work was financed by FONDECYT Grant 19050057 and FIP 94/18.

References

Balbontín, F. & W. Fischer, 1981. Ciclo sexual y fecundidad de la merluza, *Merluccius gayi gayi*, en la costa de Chile. Rev. Biol. mar. 28: 111–132.

Carvalho, G. R. & L. Hauser, 1995. Molecular genetics and the stock concept in fisheries. In Carvalho, G. R. & T. J. Pitcher (eds), Molecular Genetics and Fisheries. Chapman & Hall, London: 54–79.

Christiansen, F. B., O. Frydenberg, J. P. Hjorth & V. Simonsen, 1976. Genetics of *Zoarces* populations. IX. Geographic variation at the three phosphoglucomutase loci. Hereditas 83: 245–256.

George-Nascimento, M., 1996. Populations and assemblages of parasites in hake, *Merluccius gayi*, from the southeastern pacific ocean: stock implications. J. Fish. Biol. 48: 557–568.

Grant, W. S., R. Leslie & I. Becker, 1988. Genetic stock structure of the southern African hakes: *Merluccius capensis* and *Merluccius paradoxus*. Heredity 61: 13–20.

Harris, H. & D. A. Hopkinson, 1976. Handbook of Enzyme Electrophoresis in Human Genetics. North Holland, Amsterdam.

Hartl, D., 1980. Principles of population genetics. Sinauer Associates, Sunderland, Massachusetts.

Mangaly, G. & A. Jamieson, 1978. Genetic tags applied to the European hake, *Merluccius merluccius* (L.). Anim. Blood Grps. biochem. Genet. 9: 39–48.

Nei, M., 1978. Genetic distance between populations. Am. Nat. 106: 283–292.

Ovenden, J. R., 1990. Mitochondrial DNA and Marine Stock Assessment: A Review. Aust. J. mar. Freshwat. Res. 41: 835–853.

Oyarzún, C., 1997. Estudios microevolutivos en *Merluccius gayi gayi*. Unpublished Ph.D thesis. University of Concepción, Chile.

Roldán, M. I., 1991. Enzymatic polymorphism in the Argentinean Hake, *Merluccius hubbsi* Marini, of the Argentinean continental shelf. J. Fish. Biol. 39: 53–59.

Roldán, M. I., J. L. García-Marín, F. M. Utter & C. Pla, 1998. Population genetic structure of European hake, *Merluccius merluccius*. Heredity 81: 327–334.

Sepúlveda, J. I. & P. Rojas, 1973. Comportamiento electroforético de algunas proteínas tisulares de la merluza (*Merluccius gayi*, Guichenot). Inv. mar. 4: 113–136.

Shaklee, J. B., 1984. Genetic variation and population structure in the damselfish, *Stegastes fasciolatus*, throughout the Hawaiian Archipelago. Copeia 3: 629–640.

Sinclair, M. & T. D. Iles, 1989. Population regulation and speciation in the oceans. J. Cons. int. Explor. Mer. 45: 165–175.

Sokal, R. R. & F. J. Rohlf, 1995. Biometry: the Principles and Practice of Statistics in Biological Research. W.H. Freeman, San Francisco.

Spanakis, E., N. Tsimenides & E. Zouros, 1989. Genetic differences between populations of sardine, *Sardina pilchardus* and anchovy, *Engraulis encrassicolus*, in the Aegean and Ionian seas. J. Fish Biol. 35: 417–437.

Stepien, C. A. & R. H. Rosenblatt, 1996. Genetic divergence in antitropical pelagic marine fishes (*Trachurus, Merluccius* and *Scomber*) between north and south America. Copeia 3: 586–598.

Swofford, D. L. & R. B. Selander, 1989. BIOSYS 1.7. A Computer program for the analysis of allelic variation in population genetics and biochemical systematics. Smithsonian Illinois Natural History Survey.

Utter, F. M. & H. O. Hodgins, 1971. Biochemical polymorphism in the Pacific hake (*Merluccius productus*). J. Conseil. 161: 87–89.

Villegas, L. & G. Saetersdal, 1968. Informe preliminar sobre experimentos de marcación de merluza (*Merluccius gayi*) en Chile. Publ. Inst. Fomento Pesquero 37: 1–7.

Ward, R. D. & P. M. Grewe, 1994. Appraisal of molecular genetic techniques in fisheries. Rev. Fish. Biol. Fish. 4: 300–325.

Ward, R. D., M. Woodwark & D. O. F. Skibinski, 1994. A comparison of genetic diversity levels in marine, freshwater, and anadromous fishes. J. Fish. Biol. 44: 213–232.

Hydrobiologia **420:** 55–62, 2000.
A. M. Solé-Cava, C. A. M. Russo & J. P. Thorpe (eds), Marine Genetics.

An overview of marine fish cytogenetics

P. M. Galetti Jr.[1,*], C. T. Aguilar[2] & W. F. Molina[3]
[1]*Departamento de Genética e Evolução, Universidade Federal de São Carlos, 13565-905 São Carlos, SP, Brazil*
E-mail: galettip@power.ufscar.br
[2]*Departamento de Genética – Instituto de Biologia, Universidade Federal do Rio de Janeiro, CCS – Bloco A, Caixa Postal 68011, Cidade Universitária, 21944-970 Rio de Janeiro, RJ, Brazil*
E-mail: aguilar@acd.ufrj.br
[3]*Departamento de Biologia Celular e Genética, Universidade Federal do Rio Grande do Norte, 59078-970 Natal, RN, Brazil*
E-mail: molinawf@lineu.cb.ufrn.br

Key words: chromosomes, evolution, nucleolar organizer region

Abstract

The number of cytogenetic studies of marine fish has increased in recent years. Fish groups, such as Perciformes, which comprises many of the extant marine teleosts of economic importance, show little divergence in chromosome number and most species display a diploid number of 48 acrocentric chromosomes. In the Serranidae, Sparidae, Sciaenidae (Perciformes) and Mugilidae (Mugiliformes) small chromosome variations are restricted to subtle heterochromatin or nucleolar organizer region (NOR) modifications. There appears to exist a strict relationship between both absence of geographic barriers throughout the marine environment and high mobility of these animals (eggs, larvae, or adults), with a rarity of chromosome rearrangement at the macrostructural level. Moreover, a cellular homeostasis might also be important to karyotype maintenance among these fishes, limiting changes in the chromosome complement to cryptic chromosome rearrangements. Other groups, such as Blenniidae, Gobiidae and Scorpaenidae, for instance, show more extensive chromosome diversity, which is probably related to limited mobility. Numerical and structural chromosome polymorphisms and several sexual chromosome systems are recurrent among these fishes. A wide karyotypic diversification also characterizes the Tetraodontiformes, an interesting fish group with peculiar morphological, physiological and ecological characteristics.

Introduction

About 13 000 marine fish species exist (Nelson, 1994) and so far fewer than 2% of these have been studied cytogenetically (Brum, 1996). Although most of these fishes share a common environment lacking apparent physical barriers, they do not show a general chromosome evolutionary trend. The diploid chromosome number varies from $2n$=22–26, in some species of Nototheniidae (Ozouf-Costaz et al., 1997), an Antarctic fish group, to $2n$=240–260, in some anadromous Acipenseridae, which show several microchromosomes (Fontana et al., 1997). Most marine fishes studied have a diploid complement of 48 acrocentric chromosomes (Sola et al., 1981; Klinkhardt et al., 1995; Brum, 1996). However, in some taxa, close species have been reported showing changes in chromosome numbers and formula (Cataudella et al., 1973; Sola et al., 1978; Thode et al., 1983; Amores et al., 1990; Yokoyama et al., 1992; Caputo et al., 1996a, 1997; Corrêa & Galetti, 1997) and different sex chromosome systems (Nogusa, 1960; Murofushi et al., 1980; Ojima & Ueda, 1982; Murofushi & Yosida, 1984; Carbone et al., 1987; Brum et al., 1992; Morescalchi et al., 1992; Gomes et al., 1994; Cano et al., 1996). Moreover, some species show supernumerary chromosomes (Pauls et al., 1996). Cryptic chromosome rearrangements, particularly those involving the heterochromatin segments and/or the nucleolar organizer sites, have often been described in groups that

* Author for correspondence

show an apparently conservative chromosome macrostructure among species (Cano et al., 1981; Martinez et al., 1989; Morescalchi et al., 1992; Ozouf-Costaz et al., 1996; Aguilar & Galetti, 1997).

Karyotype diversification

Despite the variation in chromosome number observed among marine fishes, some groups such as Perciformes, with almost 7000 marine species and comprising many living marine teleosts of economic importance, show little chromosome divergence. Approximately 60% of the perciform species thus far studied show a karyotype characterized by 48 uniarmed (acrocentric) chromosomes including, for example, some Pomacentridae (Alvarez et al., 1980), Sparidae (Cano et al., 1981), Sciaenidae (LeGrande & Fitzsimons, 1988), Blenniidae (Brum et al., 1995a), Channichthyidae, Nototheniidae (Ozouf-Costaz et al., 1996, 1997), and Serranidae (Aguilar & Galetti, 1997). It is believed that this chromosome complement is a plesiomorphic character (Ozouf-Costaz et al., 1997; Brum & Galetti, 1997). In the Serranidae, for instance, 18 species have been chromosomally characterized and all show a diploid number of 48 and a majority of uniarmed chromosomes (Aguilar & Galetti, 1997). However, striking differences involving the heterochromatin and nucleolar organizer regions (NORs) have been reported among these fishes (Takai & Ojima, 1986; Martinez et al., 1989; Rodriguez-Daga et al., 1993; Vitturi et al., 1993; Aguilar & Galetti, 1997) and may have played an important role in their karyotypic evolution.

In most serranids, the NOR sites are interstitially located in the long arm of a chromosome pair, which may vary between species (e.g. pair 1 in *Paracentropistis hepatus*; pair 2 in *Serranus cabrilla* and *Cromileptis altivelis*; pair 16 in *Serranus scriba*; and pair 24 in *Epinephelus alexandrinus*, *E. caninus* and *E. guaza*; Aguilar & Galetti, 1997). The variation in location of NOR sites may constitute a strong cytotaxonomic character between *Diplectrum radiale* and *D. formosum*. In the former, NOR sites were reported to be intercalated, relatively centromere-proximal on the third acrocentric chromosome pair, while in the latter these sites were observed on the telomere of the long arm in the largest acrocentric chromosome pair (Aguilar & Galetti, 1997). A similar trend is shown by other perciform groups. The cytogenetic pattern is also uniform in the Sparidae, Sciaenidae and Carangidae, which are characterized by the same karyotypic pattern (Cano et al., 1981; LeGrande & Fitzsimons, 1988; Caputo et al., 1996b; Sola et al., 1997). In the Blenniidae ($2n$=48, a higher variability in the number of chromosome arms (FN=48–72) is observed, especially among the species of Blenniini, in which local populations may show diverse FN values, such as in *Blennius pavo*, FN=48, 52 and 64, in *B. sanguinolentus*, FN=48 and 68, in *B. gattorugine*, FN=52 and 54, and in *Scartella cristata*, with FN=48 and 50 (e.g. Cataudella & Civitelli, 1975; Cano et al., 1982; Vitturi et al., 1986; Garcia et al., 1987, Arefjev, 1991; Brum et al., 1995a). The distribution of heterochromatin also shows differentiation among the blenniids. Two C-banding patterns have been described. While *Blennius gattorugine*, *B. ponticus incognitus*, *B. pavo* and *B. sanguinolentus* show a homogeneous distribution of the heterochromatin in paracentromeric and telomeric regions, others such as *B. galerita*, *B. pholis* and *B. trigloids* have most of the heterochromatin concentrated on the entire arm of two or three chromosome pairs. Also the NOR bearing chromosomes differ between karyotypes, revealing extensive intra-karyotypic changes involving heterochromatin and NOR sites among this fish group (Garcia et al., 1987). In the Labridae, although most species have been reported with $2n$=48, the number of chromosome arms (FN) is often higher (FN=48–90), indicating a predominant occurrence of pericentric inversions within this group (Alvarez et al., 1986).

The family Channichthyidae (suborder Notothenioi dei) includes a large proportion of endemic Antarctic fish species. These have received much attention because of their morphological and physiological adaptations, such as their antifreeze glycopeptides and the hemoglobinless blood, although their evolution and phylogeny are still largely unknown. Their karyotypes are highly uniform ($2n$=48, FN=48) and exhibit little interspecific macrochromosomal variation, making it difficult to identify possible chromosome rearrangements involved in the evolution of this group. NOR sites are restricted to only one chromosome pair in all species, except for *Champsocephalus hematus* (Ozouf-Costaz et al., 1991, 1996) and *Pagetopsis macropterus* (Morescalchi et al., 1992), in which an additional site was detected on a third chromosome.

Similarly, the Mugilidae (Mugiliformes) seem to have retained the same basic karyotypic structure shared by all Perciformes, reinforcing their strict phylogenetic relationship. Different species of *Lisa*, *Chelon* and *Mugil* show $2n$=48 and FN=48, although a few biarmed chromosomes have also been described in *Lisa ramada*, *L. aurata*, *L. saliens* and *Chelon labrosus* (Cataudella & Capanna, 1973; Rossi et al., 1996, 1997). Even molecular cytogenetic data showed no chromosome differences between several distinct populations of *Mugil cephalus* all over the world (Rossi et al., 1996), while mtDNA and allozyme analyses revealed an appreciable degree of genetic divergence between these populations (Crosetti et al., 1994; Rossi et al., 1998). Recently, a significant reduction in the number of chromosomes was described in *Mugil incilis* and *M. curema*, 2*n*=28 (Pauls et al., 1998), suggesting a rare occurrence of extensive chromosomal rearrangements among the mugilids.

Other marine fishes show more striking chromosome diversification. The Gadiformes show high karyotypic variability, ranging from 2*n*=26 in the genus *Eleginus* to 2*n*=48 in *Raniceps* and *Trisopterus*. It has been suggested that pericentric inversions and Robertsonian fusions are the main mechanisms of karyotypic evolution in this group (Klinkhardt, 1994). In Anguilliformes, the diploid number varies from 2*n*=26 in *Synaphobranchus kaupi*, to 2*n*=50 in *Echelus uropterus*, while *E. myrus* shows 2*n*=38 (Salvadori et al., 1994; Amores et al., 1995). Among the Scorpaeniformes, even though different genera have been reported with a karyotypic pattern similar to the putative ancestral (2*n*=48, most chromosomes uniarmed), a conspicuous reduction of chromosome number appears to occur in some genera (e.g. Nogusa, 1960; Cataudella et al., 1973; Sola et al., 1978; Yokoyama et al., 1992; Corrêa & Galetti, 1997). In *Scorpaena* (Scorpaenidae), *S. notata* shows 2*n*=34 uniarmed chromosomes (FN=34), *S. isthimensis* has 2*n*=40 and FN=54, *S. porcus* has 2*n*=42 and FN=48, and *S. brasiliensis* shows 2*n*=46 and FN=60 (Sola et al., 1978; Corrêa & Galetti, 1997). *Prionotus punctatus* (Triglidae) shows the largest chromosome number among Scorpaeniformes, with 2*n*=100–102, all acrocentrics (Corrêa, 1995). NOR analyses revealed the presence of two nucleolar sites in the short arms of distinct chromosome pairs in the different species (Thode et al., 1985a; Yokoyama et al., 1992; Corrêa 1995). NOR size heteromorphisms were detected in *Scorpaena porcus*, *S. notata* and *S. brasiliensis* and may be related to regional chromosome duplications or deletions (Thode et al., 1985a; Corrêa & Galetti, 1997). Major chromosome changes (inversions, tandem fusions and Robertsonian mechanisms) and minor NOR and heterochromatin rearrangements are involved in the chromosome diversification of these fishes. Recently, fluorescence *in situ* hybridization (FISH) revealed ectopic telomeric sequences in some chromosomes of *S. notata*, suggesting that tandem fusions played a substantial role in the reduction of both chromosome number and FN values observed in this species (Caputo et al., 1998).

Wide karyotype diversity has also been reported in Tetraodontiformes, a highly specialized teleost group, which shows the smallest amount of DNA per cell among fishes (Hinegardner & Rosen, 1972). While species of Triacanthidae present 2*n*=48 and FN=48 (Rishi, 1973; Choudhury et al., 1982), other families (Balistidae, Ostraciidae, Tetraodontidae, Diodontidae) are represented by derived karyotypes with diploid numbers ranging from 28 to 50 and FN values from 33 to 78 (e.g. Natarajan & Subrahmanyam, 1974; Arai & Nagaiwa, 1976; Murofushi et al., 1980; Choudhury et al., 1982; Arai, 1983; Takai & Ojima, 1987; Brum et al., 1995b). Among the Siluriformes, only the family Ariidae is marine, although a few others are estuarine. Representatives of this family show a karyotypic pattern of 2*n*=54–58 and high chromosome arm numbers, FN>80 (Rishi & Singh, 1983; Gomes et al., 1994), which is very similar to the most probable ancient karyotype of the Siluriformes (LeGrande, 1981).

Among non-teleostean fish, the Chondrostei, which include the paddlefishes (Polyodontidae) and the sturgeons (Acipenseridae), show the highest diploid chromosome numbers among fishes. Polyodontids with 2*n*=112–120 (Dingerkus & Howell, 1976) and acipenserids showing 2*n*=240–260 (Fontana et al., 1997) both present a large number of microchromosomes. These karyotypic patterns might have arisen by tetraploidy (Ohno et al., 1969). Microchromosomes were also described in several sharks (*Etmopterus spinax*, *Cephaloscyllium umbratile*, *C. ventriosum*, *Scyliorhinus torazame*, *S. canicula* and *S. stellaris*, for instance) and rays (*Raja clavata*), although other species of these groups only show macrochromosomes in their karyotypes (Nygren & Jahnke, 1972; Asahida et al., 1987; Asahida & Ida, 1989). In fact, the diploid number in Carcharhiniformes (Elasmobranchii), which comprises about 55% of all shark species, ranges from 2*n*=64 to 86, showing a large number of meta- and submeta-centric chromosomes (FN=90–124), denoting extensive karyotypic evolution (Asahida & Ida, 1989; Asahida et al., 1995).

As expected in a classical population genetic model (e.g. Lande, 1979) there appears to be a strong relationship between both absence of geographic barriers in the marine environment and high mobility of marine fish groups (eggs, larvae, or adults) with a rarity of chromosome rearrangement, particularly at the macrostructural level. Moreover, karyotypic macrostructure might also be constrained by cellular homeostasis and only cryptic chromosomal rearrangements can occur (Venere & Galetti, 1989). In contrast, other groups with more limited distributions and exposure to greater environmental heterogeneity, may show more extensive chromosome diversity including interpopulational differentiation.

Chromosome polymorphisms

Chromosome polymorphisms appear to be related to the formation in some fish groups of isolated populations and genetic differentiation. The Gobiidae is the most speciose family of marine fishes comprising about 212 genera and roughly 1875 species, occurring mainly in tropical and subtropical areas. Gobies are often the most abundant freshwater fish on oceanic islands. This family contains the world's smallest fishes and the world's smallest vertebrates (Nelson, 1994).

About one hundred gobiid species have been analyzed cytogenetically and show great karyotypic diversity. Diploid number varies from 30 to 62 (probably due to Robertsonian chromosome rearrangements), while variation in chromosome morphology (FN=40–98) might additionally be related to non-Robertsonian mechanisms (e.g. Cataudella et al., 1973; Thode et al., 1983; Amores et al., 1990; Klinkhardt, 1992; Caputo et al., 1996a, 1997). High intraspecific chromosomal variability has been detected in some species. Chromosomal polymorphisms have been extensively examined in populations of some *Gobius* species. For example *G. paganellus* presents several chromosome rearrangements, including tandem and multiple centric fusions, pericentric inversions, and deletions (Giles et al., 1985; Thode et al., 1985b; Amores et al., 1990). In *G. fallax* (Thode et al., 1988), *G. bucchichii* (Thode et al., 1983) and *G. niger* (Vitturi & Catalano, 1989) multiple Robertsonian fusions have also been described. In addition, heterochromatic heterogeneity has been revealed in *Gobius paganellus*, *G. niger*, *G. cobitis* and *Zosterisessor ophiocephalus*, suggesting that the chromosomal variability observed in gobiids may be also related to the structural and compositional characteristics of the centromeric and telomeric heterochromatin (Caputo et al., 1997).

Inter and intrapopulation chromosome polymorphisms have also been described among different salmonid species, including some anadromous species, for instance, in *Oncorhynchus gorbuscha*, *O. keta*, *O. mykiss*, *O. nerka*, *Salmo aguabonita*, *S. salar* and *S. trutta* (Roberts, 1968; Muramoto et al., 1974; Thorgaard, 1976, 1978; Hartley & Horne, 1982), probably as a result of extensive Robertsonian translocations. Changes in the heterochromatin segments and NOR sites might also be important sources of chromosome variation among these fishes (Sánchez et al., 1990; Pendás et al., 1993; Castro et al., 1994). In addition, polymorphism within individuals has been observed in some of these species, reflecting the possibility of a higher degree of cytogenetic flexibility in these fishes (Hartley & Horne, 1982).

Sex chromosomes

Chromosome analyses in phylogenetically distant fish groups suggest that the presence of functional, but morphologically cryptic, sex chromosomes is a general characteristic of teleost karyotypes (Wiberg, 1983; Ewulonu et al., 1985). In fact, there are many different evolutionary strategies concerning sexuality in fishes, represented by either synchronous or sequential hermaphroditism and gonocoristic species. In gonocoristic species cytologically undistinguishable sex chromosomes are frequent, with their genetic differences limited to the occurrence of a few sex determination loci (Ohno, 1974).

There are many sex chromosome system among fishes. Heterologous sex chromosomes may be related to males in some fish groups and to females in other ones. In addition, different system types such as XX/XY, ZZ/ZW, $X_1X_1X_2X_2/X_1X_2Y$, XX/XY_1Y_2 and ZZ/ZW_1W_2 have been reported (see Moreira-Filho et al., 1993 for review). In a few cases, within a single taxon (a family, for instance), sex chromosomes may appear associated with males in some species or females in others (e.g. Andreata et al., 1992, 1993). It has been suggested that, in contrast to higher vertebrates (birds and mammals), heterogamety, *per se*, is not a fixed condition in lower vertebrates such as lizards (Olmo et al., 1987) and possibly fishes.

In marine fishes, despite of the large number of living species, the occurrence of cytologically differentiated sex chromosomes appears to be rare. Few fish species have been described as having sex chromosomes and only three types of systems, XX/XY, ZZ/ZW and $X_1X_1X_2X_2/X_1X_2Y$, have been identified to date (e.g. Nogusa, 1960; Murofushi et al., 1980; Ojima & Ueda, 1982; Murofushi & Yosida, 1984; Carbone et al., 1987; Brum et al., 1992; Morescalchi et al., 1992; Gomes et al., 1994; Cano et al., 1996). In some families, such as Serranidae (Perciformes) and Mugilidae (Mugiliformes), which have been studied extensively, no sex chromosomes have been reported (e.g. Aguilar & Galetti, 1997; Rossi et al., 1997). The absence of sex chromosomes in hermaphroditic groups such as the serranids, is expected and may be explained by their peculiar sexual differentiation. In the protogynous hermaphrodite *Anthias squamipinnis*, two sex-specific probes (BKm and ZFY) failed to detect any sex chromosome differentiation at the molecular level (Wachtel et al., 1991).

For fishes, two alternative mechanisms of sex chromosome differentiation have been proposed:

1. initial heterochromatinization; and
2. the occurrence of a structural chromosome rearrangement (e.g., Haaf & Schmid, 1984; Moreira-Filho et al., 1993).

Initial heterochromatinization appears to be the first step of sex chromosome differentiation in some freshwater fishes, as *Poecilia sphenops* (Haaf & Schmid, 1984), *Eigenmannia* (Almeida-Toledo et al., 1988), *Leporinus* (Galetti & Foresti, 1986). This mechanism has also been postulated for a few marine fishes, such as in *Conger myriaster* (Congridae, Anguilliformes), $2n$=38 (FN=56) with a ZZ/ZW system (Ojima & Ueda, 1982), in *Netuma barba* (Ariidae, Siluriformes), $2n$=56 (FN=92) and XX/XY (Gomes et al., 1994) and in *Dicentrarchus labrax* (Moronidae, Perciformes), with $2n$=48 (FN=48) and an early differentiated XX/XY system (Cano et al., 1996). Less well defined sex chromosomes have been described in *Mogrunda obscura* (Gobiidae), $2n$=62, in which an atypical bivalent in male meiosis suggests a XX/XY system (Nogusa, 1960), but confirmation is still needed.

Multiple sex chromosomes have been reported in several marine fish groups, such as in *Stephanolepis cirrhifer*, Balistidae (Tetraodontiformes) (Murofushi et al., 1980), *Muraenichthys gymnotus*, Ophichthidae (Anguilliformes) (Murofushi & Yosida, 1984), *Blennius tentacularis*, Blenniidae (Perciformes) (Carbone et al., 1987) and in *Brevoortia aurea*, Clupeidae (Clupeiformes) (Brum et al., 1992). An $X_1X_1X_2X_2/X_1X_2Y$ sex determination system is shared by all of these fishes, although independent structural chromosome rearrangements may have occurred as the first step of differentiation of each species. Similarly, among 13 endemic Antarctic Channichthyidae species thus far analyzed, five (*Chaenodraco wilsoni*, *Chionobathyscus dewitti*, *Chionodraco hamatus*, *C. myersi* and *Pagetopsis macropterus*) have shown an $X_1X_1X_2X_2/X_1X_2Y$ system, which arose by diverse translocation events involving autosomes and a morphologically cryptic Y chromosome (Morescalchi et al., 1992).

It appears quite probable that when a new sex chromosome arises it can contribute to the establishment of genetically isolated populations. It is possible that some cases of sex chromosome differentiation may have played a decisive role in speciation of fishes in the marine environment, where geographic isolation is less common than in freshwater systems.

Acknowledgements

The authors are grateful to Conselho de Desenvolvimento Científico e Tecnológico (CNPq) and Fundação Coordenação de Aperfeiçoamento de Pessoal de Nível Superior (CAPES) for financial support.

References

Aguilar, C. T. & P. M. Galetti Jr., 1997. Chromosomal studies in South Atlantic serranids (Pisces, Perciformes). Cytobios 89: 105–114.

Almeida-Toledo, L. F., F. Foresti & S. A. Toledo-Filho, 1988. An early stage of sex chromosome differentiation in the fish *Eigenmannia virescens* (Sternopygidae). Genome 30: 258.

Alvarez, M. C., J. Cano & G. Thode, 1980. DNA content and chromosome complement of *Chromis chromis* (Pomacentridae, Perciformes). Caryologia 33: 267–274.

Alvarez, M. C., E. Garcia & G. Thode, 1986. Contribution to the karyoevolutive study of the Labridae (Perciformes). The karyotypes of *Ctenolabrus rupestris* and *Symphodus ocellatus*. Caryologia 39: 353–357.

Amores, A., J. Bejar & M. C. Alvarez, 1995. Replication, C and Ag-NOR chromosome-banding in two anguilliform fish species. Mar. Biol. 123: 845–849.

Amores, A., V. Giles, G. Thode & M. C. Alvarez, 1990. Adaptive character of a Robertsonian fusion in chromosomes of the fish *Gobius paganellus* (Pisces, Perciformes). Heredity 65: 151–155.

Andreata, A. A., L. F. Almeida-Toledo, C. Oliveira & S. A. Toledo-Filho, 1992. Chromosome studies in Hypoptopomatinae (Pisces, Siluriformes, Loricariidae). I. XX/XY sex chromosome heteromorphism in *Pseudotocinclus tietensis*. Cytologia 57: 369–372.

Andreata, A. A., L. F. Almeida-Toledo, C. Oliveira & S. A. Toledo-Filho, 1993. Chromosome studies in Hypoptopomatinae (Pisces, Siluriformes, Loricariidae). II. ZZ/ZW sex-chromosome system, B chromosomes and constitutive heterochromatin differentiation in *Microlepidogaster leucofrenatus*. Cytogenet. Cell Genet. 63: 215–220.

Arai, R., 1983. Karyological and osteological approach to phylogenetic systematics of tetraodontiform fishes. Bull. natnl. Sci. Mus., Tokyo 9A: 175–210.

Arai, R. & K. Nagaiwa, 1976. Chromosomes of tetraodontiform fishes from Japan. Bull. natnl. Sci. Mus., Tokyo 2A: 59–72.

Arefjev, V. A., 1991. Different rate in intrapopulation chromosome polymorphism in two species of Black Sea *Blennius* (Blenniidae, Pisces). Genetica 83: 181–187.

Asahida, T. & H. Ida, 1989. Karyological notes on four sharks in the order Carcharhiniformes. Jap. J. Ichthyol. 36: 275–280.

Asahida, T., H. Ida & K. Hayashizaki, 1995. Karyotypes and cellular DNA contents of some sharks in the Order Carcharhiniformes. Jap. J. Ichthyol. 42: 21–26.

Asahida, T., H. Ida & S. Inoue, 1987. Karyotypes of three rays in the Order Myliobatiformes. Jap. J. Ichthyol. 33: 426–430.

Brum, M. J. I., 1996. Cytogenetic studies of Brazilian marine fish. Braz. J. Genet. 19: 421–427.

Brum, M. J. I., M. M. O. Corrêa, C. C. Oliveira & P. M. Galetti Jr., 1995a. Cytogenetic studies on the Perciformes *Orthopristis ruber* (Haemulidae) and *Scartella cristata* (Blenniidae). Caryologia 48: 309–318.

Brum, M. J. I. & P. M. Galetti Jr., 1997. Teleostei ground plan karyotype. J. Comp. Biol. 2: 91–102.

Brum, M. J. I., P. M. Galetti Jr., M. M. O. Corrêa & C. T. Aguilar, 1992. Multiple sex chromosomes in South Atlantic fish, *Brevoortia aurea*, Clupeidae. Braz. J. Genet. 15: 547–553.

Brum, M. J. I., C. C. Oliveira & P. M. Galetti Jr., 1995b. Cytogenetic studies of two puffer species (*Sphoeroides*, Tetraodontidae) from Rio de Janeiro coast, Brazil. Cytologia 60: 369–374.

Cano, J., M. C. Alvarez, G. Thode & E. Muñoz, 1982. Phylogenetic interpretation of chromosomal and nuclear-DNA-content data in the genus *Blennius* (Blenniidae: Perciformes). Genetica 58: 11–16.

Cano, J., A. Pretel, S. Melendez, F. Garcia, V. Caputo, A. S. Fenocchio & L. A. C. Bertollo, 1996. Determination of early stages of sex chromosome differentiation in the sea bass *Dicentrarchus labrax* L. (Pisces: Perciformes). Cytobios 87: 45–59.

Cano, J., G. Thode & M. C. Alvarez, 1981. Analisis cariologico de seis especies de esparidos del Mediterraneo. Genét. Ibérica 33: 181–188.

Caputo, V., F. Marchegiani & E. Olmo, 1996b. Karyotype differentiation between two species of carangid fishes, genus *Trachurus* (Perciformes: Carangidae). Mar. Biol. 127: 193–199.

Caputo, V., F. Marchegiani, M. Sorice & E. Olmo, 1997. Heterochromatin heterogeneity and chromosome variability in four species of gobiid fishes (Perciformes: Gobiidae). Cytogenet. Cell Genet. 79: 266–271.

Caputo, V., M. Sorice, R. Vitturi, H. Magistrelli & E. Olmo, 1998. Cytogenetic studies in some species of Scorpaeniformes (Teleostei: Percomorpha). Chrom. Res. 6: 255–262.

Carbone, P., R. Vitturi, E. Catalano & M. Macaluso, 1987. Chromosome sex determination and Y-autosome fusion in *Blennius tentacularis* Brunnich, 1765 (Pisces, Blenniidae). J. Fish Biol. 31: 597–602.

Caputo, V., R. Vitturi, G. Odierna, J. Cano, E. Olmo & M. S. Colomba, 1996a. Characterization of mitotic chromosomes in the gobiid fish *Zosterisessor ophiocephalus* (Pallas, 1811) (Perciformes, Gobiidae). Biol. Zent. Bl. 115: 328–336.

Castro, J., S. Rodríguez., J Arias & L. Sánchez, 1994. A population analysis of Robertsonian and Ag-NOR polymorphisms in brown trout (*Salmo trutta*). Theor. appl. Genet. 89: 105–111.

Cataudella, S. & E. Capanna, 1973. Chromosome complements of three species of Mugilidae (Pisces, Perciformes). Experientia 29: 489–492.

Cataudella, S. & M. V. Civitelli, 1975. Cytotaxonomical consideration on the genus *Blennius* (Pisces, Perciformes). Experientia 31: 167–169.

Cataudella, S., M. V. Civitelli & E. Capanna, 1973. The chromosomes of some Mediterranean teleosts: Scorpaenidae, Serranidae, Labridae, Blenniidae, Gobiidae (Pisces – Scorpaeniformes, Perciformes). Boll. Zool. 40: 385–389.

Choudhury, R. C., R. Prasad & C. C. Das, 1982. Karyological studies in five tetraodontiform fishes from the Indian Ocean. Copeia 1982: 728–732.

Corrêa, M. M. O., 1995. Contribuição à citotaxonomia dos Scorpaeniformes (Osteichthyes – Teleostei): Estudos Citognéticos em espécies do litoral do Rio de Janeiro, Brasil. Unpublished M.Sc. thesis, Federal University of Rio de Janeiro, Brazil.

Corrêa, M. M. O. & P. M. Galetti Jr., 1997. Chromosomal diversity in Scorpaenidae (Teleostei, Scorpaeniformes): Cytogenetic studies in *Scorpaena brasiliensis* and *Scorpaena isthmensis* from the coast of Rio de Janeiro, Brazil. Cytologia 62: 397–404.

Crosetti, D., W. S. Nelson & J. C. Avise, 1994. Pronounced genetic structure of mitochondrial DNA among populations of the circumglobally distributed grey mullet (*Mugil cephalus*). J. Fish. Biol. 44: 47–58.

Dingerkus, G. & W. M. Howell, 1976. Karyotypic analysis and evidence of tetraploidy in the North American paddlefish, *Polyodon spathula*. Science 194: 842–844.

Ewulonu, U.K., R. Haas & B. J. Turner, 1985. A multiple sex chromosome system in the annual killifish, *Nothobranchius ghentheri*. Copeia 1985: 503–508.

Fontana, F., R. Rossi, M. Lanfredi, G. Arlati & P. Bronzi, 1997. Cytogenetic characterization of cell lines from three sturgeon species. Caryologia 50: 91–95.

Galetti Jr., P.M. & F. Foresti, 1986. Evolution of the ZZ/ZW system in *Leporinus* (Pisces, Anostomidae). Role of constitutive heterochromatin. Cytogenet. Cell Genet. 43: 43–46.

Garcia, E., M. C. Alvarez & G. Thode, 1987. Chromosome relationships in the genus *Blennius* (Blenniidae, Perciformes). C-banding patterns suggest two karyoevolution pathways. Genetica 72: 27–36.

Giles, V., G. Thode & M. C. Alvarez, 1985. A new Robertsonian fusion in the multiple chromosome polymorphism of a Mediterranean population of *Gobius paganellus* (Gobiidae, Perciformes). Heredity 55: 255–260.

Gomes, V., V. N. Phan, & M. J. A. C. Passo, 1994. Karyotypes of three species of marine catfishes from Brazil. Bolm. Inst. oceanogr. USP 42: 55–61.

Haaf, T. & M. Schmid, 1984. An early stage of ZW/ZZ sex chromosome differentiation in *Poecilia sphenops* var. *melanistica* (Poeciliidae, Cyprinodontiformes). Chromosoma 89: 37–41.

Hartley, S. E. & M. T. Horne, 1982. Chromosome polymorphism in the rainbow trout (*Salmo gairdneri* Richardson). Chromosoma 87: 461–468.

Hinegardner, R. & D. E. Rosen, 1972. Cellular DNA content and the evolution of teleostean fishes. Am. Nat. 106: 621–644.

Klinkhardt, M. B., 1992. Chromosome structures of four Norwegian gobies (Gobiidae, Teleostei) and a hypothetical model of their karyo-evolution. Chromatin 1: 169–183.

Klinkhardt, M. B., 1994. Karyotypic divergence between species of Gadidae (Pisces, Gadiformes). Cytobios 77: 207–214.

Klinkhardt, M. B., M. Tesche & H. Greven, 1995. Database of fish chromosomes. Westarp Wissenschaften, Magdeburg.

Lande, R., 1979. Effective deme sizes during long-term evolution estimated from rates of chromosomal rearrangement. Evolution 33: 234–251.

LeGrande, W. H., 1981. Chromosomal evolution in North American catfishes (Siluriformes: Ictaluridae) with particular emphasis on the madtoms, *Noturus*. Copeia 1981: 33–52.

LeGrande, W. H. & J. M. Fitzsimons, 1988. Chromosome numbers of some gulf coast sciaenid fishes. Copeia 1988: 491–493.

Martinez, G., G. Thode, M. C. Alvarez & J. R. López, 1989. C-banding and Ag-NOR reveal heterogeneity among karyotypes of serranids (Perciformes). Cytobios 58: 53–60.

Moreira-Filho, O., L. A. C Bertollo & P. M. Galetti Jr., 1993. Distribution of sex chromosome mechanisms in neotropical fish and description of a ZZ/ZW system in *Parodon hilarii* (Parodontidae). Caryologia 46: 115–125.

Morescalchi, A., J. C. Hureau, E. Olmo, C. Ozouf-Costaz, E. Pisano & R. Stanyon, 1992. A multiple sex-chromosome system in antarctic ice-fishes. Polar Biol. 11: 655–661.

Muramoto, J., J. Azumi & H. Fukuoka, 1974. Karyotypes of nine species of salmonidae. Chrom. Inf. Serv. 17: 20–23.

Murofushi, M., S. Oikawa, S. Nishikawa & T. H. Yosida, 1980. Cytogenetical studies on fishes. III. Multiple sex chromosome mechanism in the filefish *Stephanolepis cirrhifer*. Jap. J. Genet. 55: 127–132.

Murofushi, M. & T. H. Yosida, 1984. Cytogenetical studies on fishes. VIII. XX-Y sex chromosome mechanism newly found in the snake eel, *Muraenichthys gymnotus* (Anguilliformes, Pisces). Proc. jap. Acad. 60B: 21–23.

Natarajan, R. & K. Subrahmanyam, 1974. A karyotype study of some teleosts from Portonovo waters. Proc. ind. Acad. Sci. 79: 173–196.

Nelson, J. S., 1994. Fishes of the world. 3rd edition. Wiley, New York.

Nogusa, S., 1960. A comparative study of the chromosomes in fishes with particular considerations on taxonomy and evolution. Mem. Hyogo Univ. Agric. (Biol. Ser.) 3: 1–62.

Nygren, A. & M. Jahnke, 1972. Microchromosomes in primitive fishes. Swed. J. agric. Res. 2: 229–238.

Ohno, S. 1974. Sex chromosomes and sex determining mechanisms. In John, B. (ed.), Animal Cytogenetics 4. Gebrüder Borntraeger, Berlin: 46–63.

Ohno, S., J. Muramoto, C. Stenius, L. Christian, W. A. Kittrell & N. B. Atkin, 1969. Microchromosomes in holocephalian, chondrostean and holostean fishes. Chromosoma 26: 35–40.

Ojima, Y. & H. Ueda, 1982. A karyotypical study of the conger eel (*Conger myriaster*) in *in vitro* cells, with special regard to the identification of the sex chromosome. Proc. jap. Acad. 58B: 56–59.

Olmo, E., G. Odierna & T. Capriglione, 1987. Evolution of sex-chromosomes in lacertid lizards. Chromosoma 96: 33–38.

Ozouf-Costaz, C., J-C. Hureau & M. Beaunier, 1991. Chromosome studies on fish of the Suborder Notothenioidei collected in the Weddell Sea during EPOS 3 cruise. Cybium 15: 271–289.

Ozouf-Costaz, C., E. Pisano, C. Bonillo & R. Williams, 1996. Ribosomal RNA location in the Antarctic fish *Champsocephalus gunnari* (Notothenioidei, Channichthyidae) using banding and fluorescence *in situ* hybridization. Chrom. Res. 4: 557–561.

Ozouf-Costaz, C., E. Pisano, C. Thaeron & J-C. Hureau, 1997. Antarctic fish chromosome banding: significance for evolutionary studies. Cybium 21: 399–409.

Pauls, E., P. R. A. M. Affonso, M. R. C. B. Netto & M. L. Pacheco, 1996. Supernumerary chromosomes on marine fish *Upeneus parvus* (Poy 1853, Mullidae) from Atlantic Ocean. Arch. Zootec. 45: 295–299.

Pauls, E., I. A. Coutinho, P. R. A. M. Affonso, M. R. C. B. Netto, A. S. S. Oliveira, W. Guedes & A. Afonso, 1998. Characterization of fish diversity along the coast of the state of Rio de Janeiro: a cytogenetic approach. First International Workshop on Marine Genetics – Rio 98, Rio de Janeiro, Brazil, book of abstracts: 58.

Pendás, A. M., P. Morán & E. García-Vázquez, 1993. Multi-chromosomal location of ribosomal RNA genes and heterochromatin association in brown trout. Chrom. Res. 1: 63–67.

Rishi, K. K., 1973. A preliminary report on the karyotypes of eighteen marine fishes. Res. Bull (N.S.) Panjab Univ. 24: 161–162.

Rishi, K. K. & J. Singh, 1983. Karyological studies on two Indian estuarine catfishes, *Plotosus canius* Ham. and *Pseudeutropius atherinoides* (Bloch). Caryologia 36: 139–144.

Roberts, F. L., 1968. Chromosomal polymorphism in North American landlocked *Salmo salar*. Can. J. Gen. Cytol. 10: 865–875.

Rodriguez-Daga, R., A. Amores & G. Thode, 1993. Karyotype and nucleolus organizer regions in *Epinephelus caninus* (Pisces, Serranidae). Caryologia 46: 71–76.

Rossi, A. R., M. Capula & D. E. Campton, 1998. Allozyme variation in global populations of striped mullet, *Mugil cephalus* (Pisces: Mugilidae). Mar. Biol. 131: 203–212.

Rossi, A. R., D. Crosetti, E. Gornung & L. Sola, 1996. Cytogenetic analysis of global populations of *Mugil cephalus* (striped mullet) by different staining techniques and fluorescent in situ hybridization. Heredity 76: 77–82.

Rossi, A. R., E. Gornung & D. Crosetti, 1997. Cytogenetic analysis of *Liza ramada* (Pisces, Perciformes) by different staining techniques and fluorescent *in situ* hybridization. Heredity 79: 83–87.

Salvadori, S., A. Cau, E. Coluccia, A. Milia & A. M. Deiana, 1994. Karyotype, C- and G-banding and nucleolar organizer regions of *Conger con*ger (Osteichthyes, Anguilliformes). Boll. Zool. 61: 59–63.

Sánchez, L., P. Martínez, A. Viñas & C. Bouza, 1990. Analysis of the structure and variability of nucleolar organizer regions of *Salmo trutta* by C-, Ag-, and restriction endonuclease banding. Cytogenet. Cell Genet. 54: 6–9.

Sola, L., S. Cataudella & E. Capanna, 1981. New developments in vertebrate cytotaxonomy III. Karyology of bony fishes: a review. Genetica 54: 285–328.

Sola, L., S. Cataudella & A. Stefanelli, 1978. I cromosomi di quattro specie di Scorpaenidae Mediterranei (Pisces, Scorpaeniformes). Acc. Lincei-Rend. Sc. fis. mat. nat. 64: 393–397.

Sola, L., O. Cipelli, E. Gornung, A. R. Rossi, F. Andaloro & D. Crosetti, 1997. Cytogenetic characterization of the greater amberjack, *Seriola dumerili* (Pisces, Carangidae), by different staining techniques and fluorescence *in situ* hybridization. Mar. Biol. 128: 573–577.

Takai, A. & Y. Ojima, 1986. Some features on nucleolus organizer regions in fish chromosomes. In Uyeno, T. R., R. Arai & K. Matsura (eds), Proceedings of the Second International Conference on Indo-Pacific Fishes. Icthyological Society of Japan, Tokyo: 899–909.

Takai, A. & Y. Ojima, 1987. Comparative chromosomal studies in three balistid fishes. Kromosomo (Tokyo) 47-48: 1545–1550.

Thode, G., M. C. Alvarez, E. Garcia & V. Giles, 1985a. Variation in C-banding patterns and DNA values in two scorpion-fishes (*Scorpaena porcus* and *S. notata*, Teleostei). Genetica 68: 69–74.

Thode, G., J. Cano & M. C. Alvarez, 1983. A karyological study on four species of Mediterranean gobiid fishes. Cytologia 48: 131–138.

Thode, G., V. Giles & M. C. Alvarez, 1985b. Multiple chromosome polymorphism in *Gobius paganellus* (Teleostei, Perciformes). Heredity 54: 3–7.

Thode, G., G. Martinez, J. L. Ruiz & J. R. Lopez, 1988. A complex chromosomal polymorphism in *Gobius fallax* (Gobiidae, Perciformes). Genetica 76: 65–71.

Thorgaard, G. H., 1976. Robertsonian polymorphism and constitutive heterochromatin distribution in chromosomes of the rainbow trout (*Salmo gairdneri*). Cytogenet. Cell Genet. 17: 174–184.

Thorgaard, G. H., 1978. Sex chromosomes in the sockeye salmon: a Y-autosome fusion. Can. J. Genet. Cytol. 20: 349–354.

Venere, P. C. & P. M. Galetti Jr., 1989. Chromosome evolution and phylogenetic relationships of some Neotropical Characiformes of the family Curimatidae. Braz. J. Genet. 12: 17–25.

Vitturi, R., P. Carbone, E. Catalano & M. Macaluso, 1986. Karyotypes of five species of Blennioidea (Pisces, Perciformes). Caryologia 39: 273–279.

Vitturi, R. & E. Catalano, 1989. Multiple chromosome polymorphism in the gobiid fish *Gobius niger jozo* L. 1758 (Pisces, Gobiidae). Cytologia 54: 231–235.

Vitturi, R., D. Colombera & E. Catalano, 1993. Intra-populational chromosome polymorphisms in four teleost species. Cytobios 75: 171–182.

Wachtel, S., S. Demas, T. Tiersch, P. Pechan & D. Shapiro, 1991. BKm satellite DNA and ZFY in the coral reef fish *Anthias squamipinnis*. Genome 34: 612–617.

Wiberg, U. H., 1983. Sex determination in the European eel (*Anguilla anguilla*, L.). A hypothesis based on cytogenetic results, correlated with the findings of skewed sex ratios in eel culture ponds. Cytogenet. Cell Genet. 36: 589–598.

Yokoyama, T., N. Ebitani & T. Kubo, 1992. Karyotypes and banding patterns in eight species of the scorpionfish (Scorpaenidae). Zool. Sci. 9: 1210.

Hydrobiologia **420:** 63–71, 2000.
A. M. Solé-Cava, C. A. M. Russo & J. P. Thorpe (eds), Marine Genetics.

Genetics of marine bioinvasions

B. S. Holland
Kewalo Marine Laboratory, Pacific Biomedical Research Center, University of Hawaii, 41 Ahui Street, Honolulu, HI 96813, U.S.A.
E-mail: bholland@hawaii.edu

Key words: ballast water, conservation, invasive species, population structure

Abstract

Anthropogenic biological introductions have captured the attention of marine scientists and resource managers in recent years. Human-mediated marine bioinvasions are presently acknowledged as often ecologically and financially devastating events. Despite recent increases in scientific interest and financial resources devoted to nonindigenous nuisance species globally, fundamental questions pertaining to taxonomic identity, geographic source, introduction vector and invasive population dynamics frequently remain unanswered. Ecological surveys based on photometric and observational methods are unable to identify cryptogenic taxa, source populations, multiple introductions, or genetic diversity. The twofold goal of this paper is to discuss the application of molecular genetic techniques to fundamental ecological questions pertaining to bioinvasions and to demonstrate the utility of DNA technology in providing data useful in the development of predictive models for marine bioinvasion science.

Introduction

A global perspective

One of the most pervasive, irreversible and devastating impacts of human activity on natural ecosystems is the introduction of nonindigenous species (Elton, 1958). Introductions of nonindigenous species have occurred for centuries on a global scale, often resulting in deleterious ecological, economic and social consequences (Elton, 1958; Mooney & Drake, 1989). The natural distributions of more than 1000 marine taxa have been directly altered by human activity (Carlton, 1989). Biological introductions constitute an extreme form of range expansion, attributable to human agency rather than to natural dispersal (Vermeij, 1992). Ballast water discharge and permanent biological introductions that periodically result have been identified as one of the five principal threats to global marine biodiversity (Norse, 1993) and are cited as the primary causative mechanism in the break-down of natural 'faunal realms' (Elton, 1958). The inadvertent spread of exotic marine taxa may also occur via hull fouling.

Ballast water is carried by large commercial ocean-going vessels that frequently require substantial amounts of weight in addition to cargo in order to impart trim, or stability, for open ocean voyages. The amount of ballast water carried by a single large cargo ship can exceed 150 000 tons (Schormann et al., 1990). Ballast water transport refers to the unintended transfer of living organisms contained in seawater carried in specialized ballast tanks by commercial vessels. Ballast water introductions occur when aquatic organisms, typically planktonic flora and fauna, are pumped into ballast tanks at one location, usually within their natural range, transported alive within the ship and released at a location outside of the natural range. If the new habitat has favorable ecological conditions, the exotic organism becomes a permanent resident and the introduction event becomes an anthropogenic range expansion.

Following aquatic invasions of Australia by toxic Japanese dinoflagellates and of Canada by the zebra mussel in the late 1980s (Table 1), both countries initiated extensive ballast water studies and began exploring regulatory measures. By 1989, Canada had instituted voluntary measures requesting that international vessels originating at foreign ports exchange their ballast water in the open ocean prior to entering the St. Lawrence Seaway and the Great Lakes. One

Table 1. Global examples of recent ballast water introductions. This table is not intended as a comprehensive list. In the 15-year period between 1979 and 1993, it is likely that hundreds of ballast water introductions occurred globally (from a Ballast Water Work Group Report, Chesapeake Bay Commission and National Academy of Science, 1995)

Year of Discovery	Species and source region	Introduced to:
1979–1983	American razor clam – *Ensis directus*	Germany/Denmark
	Asian copepod – *Oithona davisae*	San Francisco Bay
	Chinese copepod – *Limnoithona sinensis*	San Francisco Bay
	Japanese mussel – *Musculista senhousia*	New Zealand
	American comb jelly – *Mnemiopsis leidyi*	Black and Azov Seas
	Indo-Pacific goby – *Butis koilomatodon*	Nigeria/Cameroon
	American polychaete – *Marenzelleria viridis*	Germany
	European seaslug – *Tritonia plebeia*	Massachusetts
1984–1988	European water flea – *Bythotrephes cederstroemi*	Great Lakes
	European seasquirt – *Ascidiella aspersa*	New England
	Asian clam – *Potamocorbula amurensis*	San Francisco Bay
	Japanese seastar – *Asterias amurensis*	Australia
	Asian copepod – *Pseudodiaptomus marinus*	Southern California
	Japanese red alga – *Antithamnion nipponicum*	Long Island Sound
	Japanese dinoflagellate – *Alexandrium catenella*	Australia
	Japanese dinoflagellate – *Gymnodinium catenatum*	Australia
	Indo-Pacific crab – *Charybdis helleri*	Colombia
	Chinese copepod – *Pseudodiaptomus forbesi*	California
	European bryozoan – *Membranipora membranacea*	New England
	Japanese goby – *Rhingobius brunneus*	Arabian Gulf
	Philippine goby – *Mugilogobius parvus*	Hawaii
	Japanese crab – *Hemigrapsus sanguineus*	U.S. Atlantic coast
	Eurasian zebra mussel – *Dreissena polymorpha*	Great Lakes
1989–1993	Eurasian zebra mussel – *Dreissena bugensis*	Great Lakes
	Brown Mussel – *Perna perna*	Texas/Mexico
	Indo-Pacific green mussel – *Perna viridis*	Southern Caribbean
	Asian copepod – *Pseudodiaptomus inopinus*	Columbia River
	Black Sea goby – *Neogobious melanostomus*	Great Lakes
	South American Cholera – *Vibrio cholerae*	Alabama
	New Zealand seaslug – *Philine auriformis*	San Francisco Bay

year later, in 1990, Australia issued similar voluntary guidelines and Canada passed the Nonindigenous Aquatic Nuisance Species Prevention and Control Act. In 1991, the United Nations' International Maritime Organization (IMO) issued a set of voluntary international guidelines for ballast water exchange. All United Nations member countries were requested to follow the guidelines, which included running ballast pumps to effect a complete exchange of ballast water in advance of docking or entering a foreign port. An Australian review in 1993, however, showed that very few countries had implemented the procedures and few vessels were found to be in compliance. This prompted a more flexible 1994 version of the original set of guidelines, in the hope that it would increase compliance. Currently a formal annex to the International Convention on Marine Pollution (MARPOL 73/78) is being considered, which would require that all 157 member states of the IMO follow ballast control guidelines, including open ocean exchange of ballast water.

A North American perspective

Over 4500 species of established non native organisms have been identified in North America (Mills et al., 1994) and the frequency of inadvertent human-mediated marine introductions is apparently increasing (Hamer, 1991). Two contributing factors in the increasing incidence of ballast water mediated introductions are the presence of larger, faster and more numerous commercial vessels than ever before in history and in the eutrophication of coastal waters due to increasing environmental pressure by human populations in coastal cities. Nutrient rich ballast water provides nourishment for plankton in transit and may enhance life expectancy of exotic plankton species during trans-oceanic voyages. It is also likely that the increase in the number of recorded biological introductions is due in part to increased scientific attention to the subject in the past 20 yrs.

Historically, biological invasions in North America have often proved both economically and ecologically disastrous (Hamer 1991; Carlton & Geller, 1993; Hedgpeth, 1993; Locke et al., 1993; Mills et al., 1994; Quammen, 1996). Approximately 15% of the recorded nonindigenous species in North America have caused severe negative impacts to agriculture, industry, human health and the natural environment (OTA, 1993).

Studies involving biological sampling of planktonic organisms from ballast tanks of commercial vessels entering U.S. waters have documented the transport of large numbers of living exotic species (Carlton 1989; Carlton et al., 1990). Ballast water surveys aboard cargo vessels from international ports have revealed hundreds of identifiable exotic marine taxa, primarily larvae of marine invertebrates. For example, in one such study Carlton & Geller (1993) identified 367 species of living organisms in ballast water of cargo vessels originating in Asia and arriving in Oregon. At least 32 species of established exotic marine and estuarine organisms have been identified in the South Slough National Estuarine Research Reserve in Coos Bay, Oregon. These exotic taxa derive from natural populations from the North Atlantic Ocean and from the Northwestern Pacific Ocean and the presence of all 32 taxa is attributed to ballast water discharge. On the northern California coast, the ecology of the San Francisco Bay has been irreversibly altered by introductions and subsequent establishment of at least 255 exotic marine invertebrates (Hedgpeth, 1993), including the Chinese clam *Potamocorbula amurensis*, currently the most abundant infaunal bivalve in the Bay (Duda, 1994).

Ballast water surveys conducted on the east coast of North America have revealed a similar situation in Atlantic harbors and estuaries, where numerous foreign invertebrates have become established, including the European tunicate (*Ascilla aspersa*), European bryozoan (*Membranipora membranacea*) and the European seaslug (*Tritonia plebia*) and additional exotic marine taxa are discovered each year (G. Ruiz, pers. comm.). In the Midwest, the cumulative effects of at least 139 established nonindigenous aquatic taxa have severely impacted the natural structure of the Great Lakes ecosystem (Mills et al., 1994). Individual invasive species such as the zebra mussel (*Dreissena polymorpha*), the spiny water flea (*Bythotrephes cederstroemi*) and the sea lamprey (*Petromyzon marinus*), are examples of taxa that have caused substantial economic hardships and ecological instability. Here again, ballast water transport is cited as the primary vector of introduction. The zebra mussel, along with more than 100 other exotic aquatic taxa, was introduced to the Great Lakes via ballast water discharge following the opening of the St. Lawrence Seaway to international commercial vessel traffic. The economic burden due to these organisms is over 100 million dollars per yr. Local fisheries are threatened with collapse, water quality throughout much of the watershed, by far the largest in North America, has been altered and municipal water and power generating facilities have been periodically shut down due to fouling of intake pipes. Numerous native species are considered severely threatened with extinction since the arrival and establishment of these invasive organisms.

Theoretical invasion genetics

Theoretical models of genetic organization and population structure following a founding event predict two very different scenarios, one model predicting subpopulations to show strong genetic structuring and clinal genetic variation, while others suggest that extinction and recolonization enhance gene flow and reduce interpopulation differentiation (Alvarez-Buylla & Garay, 1994). In the context of ballast water mediated marine introductions the genetic character of the founding population may be important to its ability to adapt to its new environment. The genetic structure of an invasive population depends on several factors, including the effective population size of the introduction event and the genetic diversity of the source population.

A founder event which results in a reduction in population size may exert a severe impact on natural genetic variation (e.g. Hartl & Clark, 1997). Chance plays an important role in determining which genes are present among the founders and this has a profound effect on the gene pool of subsequent generations. Genetic drift is the most powerful force acting on small founding populations and can result in drastic reductions in population level genetic variation.

Another potential factor leading to drastic decreases in genetic diversity is inbreeding. In instances where a bottleneck event results in a founding population composed of related individuals, such that the probability of reproductive pairing of relatives is higher than in the large, panmictic source population, inbreeding may be unavoidable. In outcrossing species inbreeding is generally detrimental to viability and fecundity. Inbreeding depression, or the often severe decrease in fitness associated with matings between related individuals, is due to the increase in homozygosity of rare recessive deleterious alleles (e.g. Hartl & Clark, 1997).

If an introduction occurs as a single event involving a limited number of founders, population genetic theory predicts that alleles will be fixed and lost at an accelerated rate relative to the source population (Mayr, 1963; Hartl & Clark, 1997). The gene pool of the introduced population is expected to be limited as a result of the stochastic process of the introduction mechanism. Consequently, in relatively few generations large amounts of genetic diversity can be lost via fixation and loss of founder alleles, resulting in a significantly different allelic composition from that of the gene pool of the source population. If, on the other hand, the introduction involves a large (>1000), genetically diverse assortment of individuals, we might expect to see little or no reduction in heterozygosity and allelic diversity relative to the gene pool of the source population.

The size of a population plays an important role in evolutionary theory. The effective size (N_e) of a population may differ significantly from the actual size and is usually much less (Wright, 1931). Several measures of effective population size (N_e) can be useful in predicting the magnitude of the random forces acting on the gene pool of invasive species following a fluctuation in population level (Hartl & Clark, 1997).

If the ship happens to take on organisms from more than one population, i.e. from several geographic source populations, founding organisms will derive from various sources and isolate breaking may result causing a Wahlund effect. The Wahlund effect is characterized by an excess of heterozygous genotypes in offspring resulting from isolate breaking. A founding population which derives from numerous previously isolated populations has the potential to produce a genetically highly diverse assortment of offspring.

F-statistics can be used to describe how genetic variation is partitioned among and within populations (Wright, 1965; Ayala, 1982) and when calculated using microsatellite data, give the highest possible resolution in elucidating population structure (Slatkin, 1995; Hartl & Clark, 1997). *F*-statistics are useful statistical tools in the estimation of departures from expected panmictic proportions, as due to inbreeding or genetic drift.

Data concerning molecular genetic structure of nonindigenous marine populations during initial phases of invasion are largely unavailable. Genetic surveys estimating levels of population variation and structure within natural distributions of marine invertebrates are available, but the majority that have been published to date are based on protein electrophoresis rather than molecular techniques (Bayne, 1976; Nevo et al., 1984; McLeod, 1986; Meehan et al., 1989; Garton & Haag, 1991; Koehn, 1991; Grant et al., 1992; Duda, 1994). Widely varying results have been obtained by comparing molecular and electrophoretic population data (Hare & Avise, 1996), and markers such as RAPDs, RFLPs and microsatellites have revealed high levels of genetic diversity in cases where mtDNA and allozyme variability is extremely low (Davies et al., 1999).

Due to commercial value, availability of large sample sizes and wide geographic distributions, *Mytilus* spp. and *Crassostrea virginica* represent two of the most intensively investigated marine taxa and are certainly the most intensively studied bivalve taxa. Over the past two decades, numerous studies have been published with the principal aim of investigating the relationship between genotype and metabolic energy demand and productivity in these economically important bivalves, with particular emphasis on growth rate (e.g. Koehn et al., 1976; Levinton & Koehn, 1976; Koehn, 1991). In the late 1970s, several researchers studying *Crassostrea virginica*, began to examine multilocus genotypes in bivalves. Singh & Zouros (1978) examined the effects of individual heterozygosity (the proportion of loci of an individual that are heterozygous) and metabolism and demonstrated a positive correlation between heterozygosity reported as multiple locus heterozygosity or MLH and growth rate. This observation was repeated and confirmed for both *Mytilus* and*C. virginica* (Zouros &

Foltz, 1984; Zouros, 1987). These data appear to demonstrate balancing selection, showing a correlation between heterozygosity and fitness enhancement in bivalve molluscs. It would follow that for an invading population, an adaptive advantage might be realized by a population having an allelically diverse gene pool. Indeed, Ehrlich (1986) has suggested that high levels of genetic diversity should be possessed by successful invading populations.

In natural populations of marine molluscs numerous genetic studies based on allozymes have reported that genotypic distributions at several loci deviate from Hardy-Weinberg expectations due to heterozygote deficiencies (Koehn, 1991). Lower than expected genetic variability in introduced populations has been reported for a number of mollusc species, including *Achatina fulica* (Selander & Ochman, 1983), *Cepea nemoralis* (Selander & Ochman, 1983), *Corbicula fluminea* (Smith et al., 1979; McLeod, 1986), *Littorina saxatalis* (Knight et al., 1987) and *Theba pisana* (Johnson, 1988). The origin of this pattern remains unclear, despite the presence of at least seven proposed hypotheses, including

1. undetected self-fertilization;
2. molecular imprinting;
3. null alleles;
4. mis-scoring of isozyme gels;
5. population mixing (the Wahlund effect);
6. aneuploidy; and
7. selection (Zouros & Foltz, 1984; Zouros, 1987; Gaffney et al., 1990; Koehn, 1991).

On the other hand, a number of studies of invasive mollusc populations have shown higher than expected levels of heterozygosity, including *Biomphalaria straminea* (Woodruff et al., 1985), *Crepidula onyx* (Woodruff et al., 1986), *Dreissena polymorpha* (Hebert et al., 1989), *Macoma balthica* (Meehan et al., 1989), *Mytilus galloprovincialis* (Grant & Cherry, 1985), *Potamocorbula amurensis* (Duda, 1994) and *Perna perna* (Holland, 1997) (Table 2).

Thus, no clear model has emerged to encompass the range of different genetic characteristics exhibited by invasive populations. Predictions of genetic diversity and population structure of nonindigenous taxa are, therefore, impossible in the absence of genetic data regarding the level of polymorphism and population structure of source populations and effective population size of the introduction.

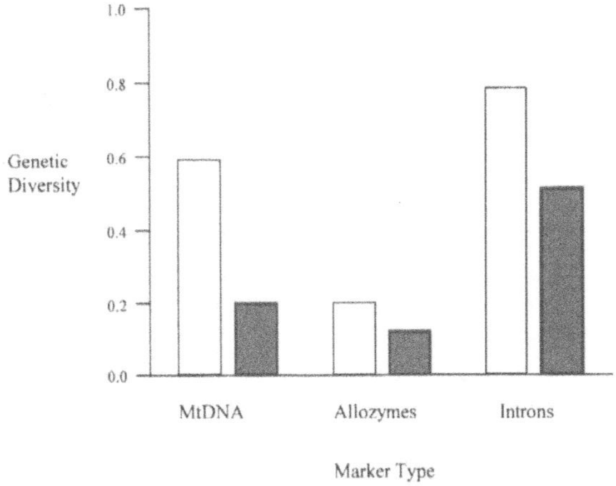

Figure 1. Genetic variation in ancestral African (open bars) and invading non-African (filled bars) populations of medflies from mitochondrial DNA (mtDNA), allozymes and introns (reproduced with permission, from Davies et al., 1999).

Methods for investigating genetic structure of invasive populations

Invasive populations present unique challenges to population geneticists. Recently founded populations have usually undergone a population bottleneck associated with the colonization event, resulting in low genetic diversity. Therefore selection of genetic markers that assay nuclear variation directly may be the most appropriate alternative. For example although mitochondrial DNA (mtDNA) loci are commonly used for population genetic studies, they may not be appropriate in invasive populations. Reductions in genetic variation following bottleneck events are amplified for mitochondrial loci due to the reduced effective population size relative to nuclear genes (Hartl & Clark, 1997). Although allozymes tend to retain more genetic diversity through population bottlenecks than mtDNA, they tend to underestimate genetic variation, since only a fraction of the actual genetic diversity at any given locus can be detected using protein markers. Therefore neutral nuclear markers such as microsatellites, introns, randomly amplified polymorphic DNAs (RAPDs) and restriction fragment length polymorphisms (RFLPs) may be more informative than either mtDNA or allozyme based assays (Figure 1).

According to Davies et al. (1999), traditional population-level approaches such as use of F_{ST} are inappropriate when the assumptions of equilibrium are violated, such as following a recent biological invasion. The authors go on to address the issue of nonequilibrium population genetics using multilocus

Table 2. Comparison of studies of genetic variation in bivalves. H_{dc}: mean direct-count heterozygosity; *P*: proportion of polymorphic loci; A/L number of alleles per locus

Organism	H_{dc}	*P*	A/L	Author
46 mollusc taxa	0.148	0.468	Nv	Nevo et al. (1984)
Dreissena polymorpha				
Lake St. Clair	0.316	0.739	3.1	Hebert et al. (1989)
Lake Erie	0.307	–	3.43	Garton & Haag (1991)
The Great Lakes	0.347	–	2.58	May & Marsden (1992)
The Great Lakes	0.464	–	2.75	Boileau & Hebert (1993)
Macoma balthica				
San Francisco Bay	0.395	0.818	4.4	Meehan et al. (1989)
Corbicula fluminea				
San Francisco Bay	0.000	0.143	1.14	McLeod (1986)
Perna perna				
Venezuela	0.351	0.125	1.16	Pompa et al. (1990)
Perna perna				
Western Gulf of Mexico	0.85	1.00	21	Holland (1997)
Perna viridis				
Trinidad	Het. Def.	–	–	Agard et al. (1992)
Potamocorbula amurensis				
San Francisco Bay	0.295	0.750	3.80	Duda (1994)

genotyping and discuss useful statistical approaches to multilocus genotyping in the context of recent nonequilibrium situations. One such statistical approach that has been used in determining the origin of an individual with multiple genetically similar candidate sources is mixed stock analysis (MSA). MSA has been demonstrated to be useful in assigning ocean-caught chinook salmon (*Oncorhynchus tshawytscha*) to different temporally isolated spawning groups (or runs) (Waser & Strobeck, 1998). MSA uses a maximum likelihood approach to identify most likely source populations by estimating the fraction of each source most likely to yield the genotype frequencies observed in the catch (for a review see Utter & Ryman, 1993). MSA methods also show potential for application to questions involving invasive taxa and may prove useful in identifying source populations for marine invaders.

Another promising method is the application of assignment tests to invasion genetic studies. Employed at the population level assignment tests can be used to assign individual genotypes to source populations. They provide a method for identifying source populations in the absence assumptions of equilibrium. For example, assignment test have been used successfully in human forensics to assign particular genotypes to particular ethnic groups. When multiple markers are used this test has been demonstrated to provide a means of reliably assigning individuals to their most likely sources and estimating dispersal and gene flow (Davies et al., 1999). Issues concerning the appropriate number of loci needed as well as limitations in terms of assessing statistical significance have been addressed recently and the results take into account most of the sources of error which weakened early versions of the test (Rannala & Mountain, 1997; Marshall et al., 1998; Davies et al., 1999).

Similarities between population level assignment tests discussed in this paper and likelihood-based paternity tests are many. A computer program designed to assess the significance of paternity tests based on multilocus genetic data is presented by Marshall et al. (1998) and is available at http://helios.bto.ed.ac.uk/evolgen/. Marshall and collaborators consider two sources of error, experimental error associated with scoring individual genotypes (thought to be around 1% for allozyme markers) and the number of potential sources in relation to the number that have been sampled. Davies et al. (1999) add that by taking into account these two sources of bias together with the effects of genetic drift, assignment tests for assessing likelihood of source populations can be greatly strengthened.

Conclusions

One of the principal goals of the emerging field of marine invasion science is to develop predictive models for marine invasion risk assessment. In order to accomplish this goal, researchers require fundamental information regarding the taxonomic status, source population (s) and population structure of nonindigenous organisms. According to Kuris (1999) an integrated pest management (IPM) approach to the mitigation of marine invasions includes the following principles:

1. Early detection of the invasive species;
2. Rapid evaluation of the risk of establishment and spread;
3. Evaluation of potential impact (s);
4. Prompt detection of natural predators and other limiting factors from within the native range.

I would argue that through the application of genetic techniques, many or all of these issues can be addressed.

Despite the promise and potential power which DNA technology can bring to the field of marine introduction biology, currently very little emphasis is being given to such techniques within the field. For example, of the more than 80 oral presentations given at a major recent conference on the subject (Marine Bioinvasions, 1999), only three papers mention the word 'genetics' in the title, and only one of the more than 30 poster presentations. Recurrent themes at this recent conference included risk assessment, predicted impacts, determination of introduction mechanism and species identification, all of which can be addressed using genetic techniques.

Informative genetic markers are currently available for the investigation of recent range expansions by marine taxa. Recently developed statistical approaches, such as MSA and assignment tests are proving to be informative and appropriate for nonequilibrium population genetics and hold enormous potential for application to anthropogenic marine introductions. Genetic data may provide the best means for identifying systematically troublesome species. For example, in cases where the karyotype of a species is a diagnostic or species specific feature, determination of the diploid number can be a quick and inexpensive means of taxonomic verification (Holland et al., 1999), and may also be a valuable tool in differentiating between cryptogenic taxa. In addition to taxonomic information, genetic markers can help to identify most likely source populations, which can be crucial to the process of determining the most likely introduction vector. Assignment tests provide a method for the identification of recent gene flow in the absence of assumptions of equilibrium, as in recently founded invasive populations. Assignment of invasive species to their most likely geographic source may be important in predicting the impacts and extent of invasions as well as whether future invasions can be expected. Such data could prove vital to resource managers and conservation biologists faced with managing exotic nuisance species. In the event that more stringent international legislation is eventually adopted, compliance could be monitored via genetic surveys of taxa from incoming ballast water. Techniques such as MSA and assignment tests may play a key role in enforcement efforts should such legislation require scientifically and statistically robust evidence regarding source and identity of nonindigenous organisms sampled from ballast tanks of international cargo vessels. Molecular genetic techniques may provide a powerful tool in the current and upcoming efforts of those of us concerned with marine conservation and the integrity of natural marine communities.

References

Alvarez-Buylla, E. R. & A. A. Garay, 1994. Population genetic structure of *Cecropia obtusifolia*, a tropical pioneer tree species. Evolution 48: 437–453.

Ayala, F. J., 1982. Population and Evolutionary Genetics: a Primer. Benjamin/Cummings Publishing, Menlo Park, California.

Bayne, B. L., 1976. Marine Mussels: their Ecology and Physiology. Cambridge University Press, Cambridge, Massachusetts.

Boileau, M. G. & P. D. N. Hebert, 1993. Genetics of the Zebra mussel (*Dreissena polymorpha*) in populations from the Great Lakes Region and Europe. In Nalepa, T. F. & D. W. Schloesser (eds), Zebra Mussels; Biology, Impacts and Control. Lewis Publishers, Boca Raton, Florida: 227–238.

Carlton, J. T., 1989. Man's role in changing the face of the ocean: biological invasions and implications for conservation of near-shore environments. Con. Biol. 3: 265–273.

Carlton, J. T. & J. B. Geller, 1993. Ecological roulette: The global transport of nonindigenous marine organisms. Science 261: 78–82.

Carlton, J. T., J. K. Thompson, L. E. Schemel & F. H. Nichols, 1990. Remarkable invasion of the San Francisco Bay by *Potamocorbula amurensis*. I Introduction and dispersal. Mar. Ecol. Prog. Ser. 66: 81–94.

Duda, T. F., 1994. Genetic population structure of the recently introduced Asian clam, *Potamocorbula amurensis*, in San Francisco Bay. Mar. Biol. 119: 235–241.

Davies, N., F. X. Villablanca & G. K. Roderick, 1999. Determining the source of individuals: multilocus genotyping in nonequilibrium population genetics. Trends Ecol. Evol. 14: 17–21.

Ehrlich, P. R., 1986. Which animals will invade? In Drake, J. A. & H. A. Mooney (eds), Ecology of Invasions of North America and Hawaii. Springer-Verlag, New York: 79–95.
Elton, C. S., 1958. The Ecology of Invasion by Plants and Animals. Methuen and Co. Ltd., London.
Garton, D. W. & W. R. Haag, 1991. Heterozygosity, shell length and metabolism in the European mussel, *Dreissena polymorpha*, from a recently established population in Lake Erie. Comp. Biochem. Physiol. 99A: 45–48.
Gaffney, P. M., T. M. Scott, R. K. Koehn & W. J. Diehl, 1990. Interrelationships of heterozygosity, growth rate and heterozygote deficiencies in the coot clam *Mulinia lateralis*. Genetics 124: 687–699.
Grant, W. S. & M. I. Cherry, 1985. *Mytilus galloprovincialis* (Lmk.) in southern Africa. J. exp. mar. Biol. Ecol. 90: 179–191.
Grant, W. S., A. C. Schneider, R. W. Leslie & M. I. Cherry, 1992. Population genetics of the brown mussel *Perna perna* in southern Africa. J. exp. mar. Biol. Ecol. 165: 45–58.
Hamer, M., 1991. Toxic hitchhikers conquer the world's oceans. New Scientist, March 23: 23 pp.
Hare, M. P. & J. C. Avise, 1996. Molecular genetic analysis of a stepped multilocus cline in the American oyster (*Crassostrea virginica*). Evolution 50: 2305–2315.
Hartl, D. L. & A. G. Clark, 1997. Principles of population genetics. Sinauer Associates, Sunderland, Massachusetts.
Hebert, P. D. N., B. W. Muncaster & G. L. Mackie, 1989. Ecological and genetic studies on *Dreissena polymorpha* (Pallas), a new mollusc in the Great Lakes. Can. J. Fish. aquat. Sci. 46: 1587–1591.
Hedgpeth, J. W., 1993. Foreign invaders. Science 261: 34–35.
Holland, B. S., 1997. Genetic architecture of a biological invasion: Geographic pattern of genetic variation in the brown mussel, *Perna perna*, from natural and nonindigenous populations. Unpublished Ph.D. thesis. Texas A&M University, College Station, Texas.
Holland, B. S., D. W. Hicks, D. Gallagher & S. K. Davis, 1999. Cytotaxonomic verification of a nonindigenous marine mussel in the Gulf of Mexico. Veliger 42: 281–283.
Johnson, M. S., 1988. Founder effects and geographic variation in the land snail *Theba pisana*. Heredity 61: 133–142.
Knight, A. J., R. N. Hughes & R. D. Ward, 1987. A striking example of the founder effect in the mollusc *Litorina saxatalis*. Biol. J. linn. Soc. 32: 417–426.
Koehn, R. K., 1991. The genetics and taxonomy of species in the genus *Mytilus*. Aquaculture 94: 125–145.
Koehn, R. K., R. Milkman & J. B. Mitton, 1976. Population genetics of marine pelecypods. IV. Selection, migration and genetic differentiation in the blue mussel *Mytilus edulis*. Evolution 30: 2–32.
Kuris, A., 1999. Panel discussion: comment on the premise. In Proceedings of the First National Conference on Marine Bioinvasions. Massachusetts Institute of Technology, Cambridge, Massachusetts.
Levinton, J. S. & R. K. Koehn, 1976. Population genetics of mussels. In Bayne, B. L. (ed.), Marine Mussels, their Ecology and Physiology. Cambridge University Press, Cambridge, Massachusetts: 357–384.
Locke, A., J. T. Carlton, D. M. Reid, W. G. Sprules & H. C. Van Leeuwen, 1993. Ballast water exchange as a means of controlling dispersal of freshwater organisms by ships. Can. J. Fish. aquat. Sci. 50: 2089–2093.
Marshall, T. C., J. Slate, L. E. B. Kruuk & J. M. Pemberton, 1998. Statistical confidence for likelihood-based paternity inference in natural populations. Mol. Ecol. 7: 639–655.
May, B. & J. E. Marsden, 1992. Genetic identification and implications of another invasive species of dreissenid mussel in the Great Lakes. Can. J. Fish. aquat. Sci. 49: 1501–1506.
Mayr, E., 1963. Animal Species and Evolution. Harvard University Press. Cambridge, Massachusetts.
McLeod, M. J., 1986. Electrophoretic variation in North American *Corbicula*. Am. malac. Bull. 2: 125–132.
Meehan, B. W., J. T. Carlton & R. Wenne, 1989. Genetic affinities of the bivalve *Macoma balthica* from the Pacific coast of North America: Evidence for recent introduction and historical distribution. Mar. Biol. 102: 235–241.
Mills, E. L., J. H. Leach, J. T. Carlton & C. L. Secor, 1994. Exotic species and the integrity of the Great Lakes. BioScience 44: 666–676.
Mooney, H. A. & J. A. Drake, 1989. Biological Invasions: A SCOPE Program Overview. Pages. In Drake, J. A. & H. A. Mooney (eds), Biological Invasions: A Global Perspective. Wiley, New York: 491–508.
Nevo, E., A. Beiles & R. Ben-Schlomo, 1984. The evolutionary significance of genetic diversity: ecological, demographic and life history correlates. In Mani, G. S. (ed.), Evolutionary Dynamics of Genetic Diversity. Springer, Heidelberg: 13–212.
Norse, E., 1993. Global marine biological diversity: a strategy for building conservation into decision making. Island Press, Washington, D.C.
Office of Technology Assessment (OTA) (U.S. Congress), 1993. Harmful nonindigenous species in the United States. U.S. Government Printing Office, D.C.
Quammen, D., 1996. The Song of the Dodo: Island Biogeography in an Age of Extinctions. Scribner, New York.
Rannala, B. & J. L. Mountain, 1997. Detecting immigration using multilocus genotypes. Proc. natnl. Acad. Sci. U.S.A. 94: 9197–9201.
Schormann, J., J. T. Carlton & M. R. Dochoda, 1990. The ship as a vector in biotic invasions. Mar. Eng. Digest. Oct 1990: 18–22.
Selander, R. K. & H. Ochman, 1983. The genetic structure of populations as illustrated by molluscs. Isozymes 10: 93–123.
Selander, R. K., M. H. Smith, S. Y. Yang, W. E. Johnson & J. R. Gentry, 1971. Biochemical polymorphism and systematics in the genus *Peromyscus*. Univ. Texas Publ. 7103: 49–90.
Singh, S. M. & E. Zouros, 1978. Genetic variation associated with growth rate in the American oyster (*Crassostrea virginica*). Evolution 32: 342–353.
Slatkin, M., 1985. Gene flow in natural populations. Ann. Rev. Ecol. Syst. 16: 393–430.
Smith, M. H., J. Britton, P. Burke, R. K. Chesser, M. W. Smith & J. Hagen, 1979. Genetic variability in *Corbicula*, an invading species. In Britton, J. C. (ed.), Proceedings, First International Corbicula Symposium. Texas Christian University Research Foundation, Fort Worth, Texas: 244–248.
Utter, F. & N. Ryman, 1993. Genetic markers and mixed stock fisheries. Fisheries 18: 11–21.
Vermeij, G. J., 1992. Trans-equatorial connections between biotas in the temperate eastern Atlantic. Mar. Biol. 112: 343–348.
Waser, P. M. & C. Strobeck, 1998. Genetic signatures of interpopulation dispersal. Trends Ecol. Evol. 13: 43–44.
Woodruff, D. S., M. Mulvey & M. W. Yipp, 1985. Population genetics of *Biomphalaria straminea* in Hong Kong. Heredity 76 355–360.
Woodruff, D. S., L. L. McMeekin, M. Mulvey & M. P. Carpenter, 1986. Population genetics of *Crepidula onyx*: variation in a Cali-

fornia slipper snail recently established in China. Veliger 29: 53–63.

Wright, S., 1931. Evolution in Mendelian populations. Genetics 16: 97–159.

Wright, S., 1965. The interpretation of population structure by F-statistics with special regard to systems of mating. Evolution 19: 395–420.

Zouros, E. & D. W. Foltz, 1984. Possible explanations of heterozygosity deficiency in bivalve molluscs. Malacologia 25: 583–591.

Zouros, E., 1987. On the relation between heterozygosity and heterosis: An evaluation of the evidence from marine mollusks. Isozymes: Curr. Top. biol. med. Res. 15: 255–277.

Hydrobiologia **420:** 73–90, 2000.
A. M. Solé-Cava, C. A. M. Russo & J. P. Thorpe (eds), Marine Genetics.

Molecular genetic analyses of species boundaries in the sea

N. Knowlton
Marine Biology Research Division – 0202; Scripps Institution of Oceanography; University of California San Diego; La Jolla, CA 92093-0202, U.S.A. and Smithsonian Tropical Research Institute; Apartado 2072; Balboa, Republic of Panama. E-mail: nknowlton@ucsd.edu

Key words: sibling species, taxonomy, DNA, electrophoresis, allozymes

Abstract

The tools of molecular genetics have enormous potential for clarifying the nature and age of species boundaries in marine organisms. Below I summarize the genetic implications of various species concepts, and review the results of recent molecular genetic analyses of species boundaries in marine microbes, plants, invertebrates and vertebrates. Excessive lumping, rather than excessive splitting, characterizes the current systematic situation in many groups. Morphologically similar species are often quite distinct genetically, suggesting that conservative systematic traditions or morphological stasis may be involved. Some reproductively isolated taxa exhibit only small levels of genetic differentiation, however. In these cases, large population sizes, slow rates of molecular evolution, and relatively recent origins may contribute to the difficulty in finding fixed genetic markers associated with barriers to gene exchange. The extent to which hybridization blurs species boundaries of marine organisms remains a subject of real disagreement in some groups (e.g. corals). The ages of recently diverged species are largely unknown; many appear to be older than 3 million years, but snails and fishes provide several examples of more recent divergences. Increasingly sophisticated genetic analyses make it easier to distinguish allopatric taxa, but criteria for recognition at the species level are highly inconsistent across studies. Future molecular genetic analyses should help to resolve many of these issues, particularly if coupled with other biological and paleontological approaches.

Introduction

Taxonomists have described species for centuries without the aid of protein electrophoresis, RAPDs, RFLPs, AFLPs or DNA sequencing. Indeed, given the scarcity of taxonomists relative to the number of undescribed species, most species descriptions still do not involve molecular genetic analyses, and will not for the foreseeable future. Such analyses are expensive, and thus their application is most justified when other types of data yield equivocal results. Nevertheless, genetics and systematics are increasingly intertwined at all levels of the Linnean hierarchy. In this paper I will explore what molecular genetics can tell us about diversity at the species level in the sea. In particular, I will focus on the role of molecular genetics in recognizing cryptic or sibling species (i.e. species that are difficult to distinguish using more traditional techniques). Before doing so, however, a few words are necessary concerning what is meant by a species.

A brief taxonomy of species concepts and how they apply to marine organisms

Genetics can contribute to our understanding of species as we define them, but does not itself provide the definition. Mayr (1963) developed a compelling argument for using reproductive criteria to define species, but the rise of cladistics has challenged the supremacy of his view. Many other species concepts have also been proposed, 22 according to a recent survey (Mayden, 1997). These different species concepts have distinct implications for the use and interpretation of genetic data, although some of the differences are more apparent than real (Avise & Wollenberg, 1997; De Queiroz, 1998).

Mayr's biological species concept defines species as groups of actually or potentially interbreeding individuals, with boundaries between species defined by intrinsic barriers to gene flow that have a genetic basis. These barriers may be expressed through their effects on: (1) key features of the mating system (e.g. by influencing mate choice, gamete attraction or fertilization, or the timing of reproduction); (2) other ecological characteristics, such as habitat preference, which influence the probability of mating; or (3) developmental pathways, so that inviability or sterility interfere with gene exchange in subsequent generations. Although the existence of hybridization is often viewed as a reason for discarding the biological species concept, this is not consistent with Mayr's recommendations (e.g. Mayr & Ashlock, 1991).

For the vast majority of marine organisms we do not know the genetic basis of reproductive barriers (Palumbi, 1994). The best studied examples concern the proteins involved in egg-sperm interactions in sea urchins (Metz & Palumbi, 1996; Metz et al., 1998a) and gastropods (Metz et al., 1998b; Swanson & Vacquier, 1998; Hellberg & Vacquier, 1999). Gastropods represent a particularly elegant model system because both the sperm protein and the egg receptor have been identified. However, this level of understanding is not a viable short-term goal for most marine organisms. Fortunately for systematists, reproductively isolated taxa will, with time, accumulate other fixed genetic differences that can be revealed using more standard and accessible genetic techniques. The concordance of several independent genetic characters provides particularly strong evidence for the existence of reproductive barriers in sympatry because other explanations for their concordance are generally less plausible (Avise & Ball, 1990).

Despite the intuitive appeal of the biological species concept, the phylogenetic species concept has become increasingly popular, in part associated with the success of cladistic methods within systematics as a whole. It has been presented under a variety of guises (De Queiroz, 1998) but, in its simplest and most extreme form, species are defined as minimum diagnosable units (Cracraft, 1989). Any type of diagnostic genetic difference can thus in principle be used to define a species, with the proviso that entities known to interbreed (e.g. males and females within a single population) do not qualify for separate species status. In sympatry, the biological and phylogenetic species concepts are equivalent, because fixed genetic differences imply barriers to gene exchange, and barriers to gene exchange imply some sort of genetic difference (Knowlton & Weigt, 1997).

In allopatry, however, biological and phylogenetic species can be potentially quite different (Knowlton & Weigt, 1997). This is because the biological species concept requires that genetic differences be such that isolated populations are not potentially interbreeding, while phylogenetic species can be defined by any fixed difference, regardless of its triviality. Avise and Wollenberg (1997) argue that any approach based on one or a small number of diagnostic genetic characters is nonsensical because of the enormous resolving power of genetic techniques and the diversity of possible genealogical pathways for individual loci. They suggest, however, that a phylogenetic species concept based on multiple loci is compatible with the biological species concept, because reproductive barriers will emerge during the lengthy period of geographic isolation that is required for many loci to acquire fixed (diagnostic) differences.

Of the remaining species concepts, Templeton's (1989) cohesion species concept most explicitly addresses the limitations of the biological species concept, which he summarizes as being associated with either too much sex (e.g. extensive hybridization) or too little sex (e.g. asexual taxa or species with highly fragmented populations). In both cases, he recommends use of the concept of 'demographic exchangeability', which in essence involves defining species based on the sharing of comparable ecological niches. For example, when discrete forms exist despite extensive hybridization, he suggests recognizing those units that can be defined by shared ecological constraints. Similarly, for taxa with many isolated populations due to limited dispersal or clonal propagation, ecological distinctiveness provides the criterion for recognition at the species level. Although he does not explicitly address problems of the phylogenetic species concept, trivial genetic differences with no obvious ecological significance would presumably go unrecognized using the cohesion species concept.

Templeton's demographic exchangeability is not an explicitly genetic concept, although genetics may be invaluable for identifying situations where it is particularly appropriate (e.g. extensive hybridization, highly subdivided population structure, or clonal reproduction). However, in some groups, the level of genetic differentiation may itself serve as a proxy for estimating likely ecological divergence. For example, the sexual habits of many microbes do not lend themselves to species definitions based on reproducve barriers and, for a number of groups, our only ıformation on these organisms are DNA sequences ecause they are unculturable at present. Thus Emb-y and Stackebrandt (1997) discuss a rule of thumb /hereby prokaryote taxa be recognized at the spe-ies level when their 16S rDNA sequences are less ıan 98% similar. Implicit in this approach is the idea ıat substantial genetic divergence implies ecological ifferentiation of the type specified by the cohesion pecies concept, but the converse (small genetic differ-ntiation) does not necessarily imply little ecological ivergence (see below).

Among marine metazoans, strictly clonal repro-luction is rare (Hughes, 1989), although the clam *asaea* provides a clear case of some niche diversi-ication among ancient asexual lineages (Ó Foighil & mith, 1996) for which the cohesion concept may be vell suited. A much more widespread situation favor-ng application of the cohesion species concept is that of extremely limited dispersal (Knowlton & Jackson, 993). When highly philopatric species live in discrete ıabitats separated by uninhabitable areas, large num-ers of ecologically similar but genetically distinctive opulations can result, some of which also exhibit re-productive incompatibilities. The copepod *Tigriopus californicus* (Burton et al., 1999) and the isopod *Excir-olana braziliensis* (Lessios & Weinberg, 1994; Lessios et al., 1994) provide compelling examples of the po-tential utility of the cohesion species concept for such situations, and there are undoubtably many others.

The extent of hybridization in the sea remains a subject of controversy (Gardner, 1997), so that the ex-tent to which it impinges on species concepts remains uncertain. Molecular genetic techniques are ideally suited for its detection (see Rieseberg & Linder, 1999 for a critical evaluation), and a number of marine examples are available (e.g. Liu et al., 1991; Bert & Arnold, 1995; Brown, 1995; Bert et al., 1996; Gardner, 1996; Lessios & Pearse, 1996; Foltz, 1997; Röhner et al., 1997; Quesada et al., 1998; Väinölä & Vainio, 1998; Comesaña et al., 1999). Although only 3 of the 108 cases summarized by Gardner (1997) employed genetic methods, he concludes that (1) rates of hybridization in the sea and on land are roughly comparable, and that (2) in the sea, sporadic hybridization is more common than either well-defined hybrid zones between distinct, allopatric taxa or hybrid swarms. The latter provide the biggest challenge for species concepts, a problem that has been explicitly discussed for corals (see below).

Examples of recent genetic analyses

Several recent reviews have listed cases where genetic analyses have played a critical role in the recognition of species boundaries (Knowlton, 1993; Thorpe & Solé-Cava, 1994). Rather than dwell on previously-covered examples, I will highlight others not covered in these reviews and, for a number of groups, attempt to summarize what genetic techniques have taught us to date. This summary is not intended to be comprehensive, but rather to point out interesting developments and give a feeling for the kinds of results and problems that have been encountered. I will focus primarily on marine invertebrates, with a few examples being drawn from other marine groups.

Bacteria and single-celled eukaryotes

Our understanding of the world of marine microbes has been transformed by the tools of molecular genetics. Major surprises include the discovery of entirely new major groups, never before cultured, based on rDNA sequences obtained from environmental samples (Embley & Stackebrandt, 1997; Potter et al., 1997; Rappé et al., 1998; Hinrichs et al., 1999). These findings are more akin to the discovery of new higher taxa than to recognition of diversity at the species level.

Nevertheless, in some cases, taxa formerly considered to be a single species have been shown to harbor enormous and unsuspected genetic diversity. One striking example concerns dinoflagellates associated with marine invertebrates, conventionally referred to as *Symbiodinium microadriaticum*. Only in the last decade has it been discovered that they exhibit genetic differences based on small subunit rDNA sequences comparable to those observed between different families or orders of free-living dinoflagellates. These genetically defined groups have characteristic differences in depth distribution (and other ecological attributes) as well (Rowan, 1998). Recently, taxa of cyanobacteria have also been shown to exhibit depth zonation (Moore et al., 1998; Ferris & Palenik, 1998), but in this case the genetic differences are so small that they might not normally be attributed to species level differentiation (*sensu* Embley & Stackebrandt, 1997). Cryptic species of planktonic foraminifera that differ in a variety of biological features have also been discovered through the use of genetic techniques (Huber et al., 1997). The cohesion species concept is in principle highly appropriate in many of these cases. However, the relationship between genetic and ecological divergence at various levels of genetic differentiation is largely unknown for most microbial groups.

Multicellular plants

Marine angiosperms have been examined genetically, but recent studies have focused primarily on population genetics and clonal structure rather than alpha taxonomy (e.g. Kirsten et al., 1998). They will not be discussed further here. Marine algae, in contrast, provide a number of examples where genetic analyses have played a major role in clarifying species boundaries. Genetic techniques are particularly useful because recognizing species of marine algae is often complicated by their morphological simplicity (Van Oppen et al., 1996), morphological plasticity (e.g. Bakker et al., 1995a; Benzie et al., 1997; Blomster et al., 1998), and heteromorphic life histories.

Genera recently examined using genetic techniques include *Porphyra* (Lindstrom & Cole, 1992; Stiller & Waaland, 1993; Brodie et al., 1996), *Phycodris* (Van Oppen et al., 1995a,b), *Gracilariopsis* (Goff et al., 1994), *Gracilaria* (Goff et al., 1994), *Pikea* (Maggs & Ward, 1996), *Cladophora* (Bakker et al., 1995a,b), *Caulerpa* (Benzie et al., 1997; Pillmann et al., 1997; Olsen et al., 1998) and *Enteromorpha* (Blomster et al., 1998). These taxa include economically important species (*Porphyra, Gracilaria, Gracilariopsis*), as well as bioindicators of pollution (*Enteromorpha*) and pest species (*Caulerpa*). Genetic techniques have been invaluable in evaluating the possibility of species introductions (Maggs et al., 1992; Van Oppen et al., 1995b; Maggs & Ward, 1996; Pillmann et al., 1997; Olsen et al., 1998), which are common among algae.

Although allozymes continue to be used in alpha systematics (e.g. Lindstrom & Cole, 1992; Van Oppen et al., 1995a; Benzie et al., 1997), DNA based techniques (especially sequencing of nuclear rRNA and chloroplast Rubisco genes) are becoming increasingly prominent. The ability to characterize dried herbarium material and compare it to other collections of uncertain status makes DNA-based techniques particularly attractive (e.g. Goff et al., 1994). On the other hand, in at least one case allozymes provided greater resolution than RAPDs (Van Oppen et al., 1995a). Diagnostic allozyme loci are often monomorphic within species, yielding low heterozygosities (e.g. Lindstrom & Cole, 1992; Van Oppen et al., 1995a; Benzie et al., 1997).

Many pairs of cryptic taxa are very distinct genetically (e.g. Olsen et al., 1987), and are not even necessarily sister species (Maggs et al., 1992; Goff et al., 1994; Bakker et al., 1995a; Blomster et al., 1998). In some of these cases, other kinds of differences had been noted prior to genetic studies (e.g. Brodie et al., 1996; Blomster et al., 1998), while, in other cases the genetic identification of cryptic species came as a complete surprise (Stiller & Waaland, 1993; Benzie et al. 1997; Pillman et al., 1997). Not all cryptic species are highly divergent genetically, however (Brodie et al., 1996), and genetic data have also been used to synonymize morphotaxa (Lindstrom & Cole, 1992; Blomster et al., 1998).

Porifera

Sponges are another group where morphological simplicity (Klautau et al., 1999) and morphological plasticity have led to uncertainties in systematics at the species level. Recent genetic studies include investigations of cryptic species in *Petrosia* (Bavestrello & Sarà, 1992), *Tethya* (Sarà et al., 1993), *Clathrina* (Klautau et al., 1994), *Cliona* (Barbieri et al., 1995), *Plakina* (Muricy et al., 1996), *Spirastrella* (Boury-Esnault et al., 1999), *Cinachyrella* (Lazoski et al., 1999), and *Chondrilla* (Klautau et al., 1999). Solé-Cava and Boury Esnault (1999) review the results of these and earlier studies (63 in total) reporting levels of interspecific genetic divergence based on allozymes. As they note, almost all of these studies focused on taxonomically problematic groups where species boundaries were deemed unclear.

Most studies involved no more than 12 scorable loci, and substantial differences (usually one or more diagnostic loci) were generally detected. In some cases these differences were larger than those often reported between different genera of sponges. Genetic differences were typically associated with morphological differences that were also diagnostic, although relatively subtle and/or generally not recognized as 'significant' (e.g. color) in traditional approaches to sponge taxonomy. One sympatric pair (two *Suberites* species that differ in color) were, however, exceptionally closely related (Nei's I of 0.98).

The overwhelming message from these studies is that sponge biodiversity at the species level has been seriously underestimated. So-called cosmopolitan taxa probably consist of complexes of regionally more restricted forms (e.g. Klautau et al., 1999), and sympatric morphotypes defined by colony morphology, color, habitat, or subtle skeletal differences are probably non-interbreeding. DNA analyses have not been applied to the problem of cryptic species in sponges, and some PCR-based techniques that are routinely used in other groups (e.g. RAPDs) are more difficult to use in sponges because of the symbionts embedded in the tissues. As technical difficulties are resolved, however, the extent of diversity may increase further, because the sometimes very large genetic differences observed with protein electrophoresis suggest that current approaches are touching only the tip of the iceberg.

Cnidaria

Several important anthozoan genera have been the subject of recent genetic studies of species boundaries: the anemones *Epiactis* (Edmands, 1996), *Anthopleura* (McFadden et al., 1997), and *Actinia* (Monteiro et al., 1997); the corals *Montipora* (Stobart & Benzie, 1994); *Montastraea* (Lopez & Knowlton, 1997; Lopez et al., 1999; Medina et al., 1999), and *Platygyra* (Miller & Benzie, 1997); the octocorals *Briareum* (Brazeau & Harvell, 1994) and *Plexaura* (Lasker et al., 1996); the soft coral *Alcyonium* (McFadden, 1999); and five genera of zoanthids (Burnett et al., 1997). Genetic analyses in other cnidarian classes are largely lacking, with the hydrozoan *Millepora* (Manchenko et al., 1993; Amaral et al., 1997) and the scyphozoan *Aurelia* (Greenberg et al., 1996) being the primary examples.

In apparent contrast to the situation with sponges, where genetic differences between sibling species are usually high, there are a number of well-studied cases of sympatric taxa that show very low levels of genetic differentiation, including the absence of fixed allozyme differences (Solé-Cava et al., 1985; Solé-Cava & Thorpe, 1992; Brazeau & Harvell, 1994; Weil & Knowlton, 1994; Edmands, 1996; McFadden et al., 1997; Monteiro et al., 1997). The low level of differentiation observed between some sympatric species makes the interpretation of negative evidence in other groups (e.g. Burnett et al., 1997; Miller & Benzie, 1997) especially difficult. Moreover, in the *Montastraea annularis* species complex, the three taxa show no differentiation in ITS and COI sequences (Lopez and Knowlton, 1997; Medina et al., 1999), although one of these taxa has recently been clearly distinguished after screening of numerous AFLP primers (Lopez et al., 1999).

One possible reason for the small degree of genetic differentiation between what appear to be reproductively isolated taxa is that rates of molecular evolution in this group may be unusually slow (Romano & Palumbi, 1997). In corals, for example, rates of divergence of the mtDNA COI and cyt b genes appear to be less than 0.1% per million years (Medina et al., 1999; Van Oppen et al., 1999), compared to 2% per million years in many vertebrates (Avise et al., 1998), and 1.5–0.5% for crustaceans and fish separated by the Isthmus of Panama (Knowlton & Weigt, 1998; Schubart et al., 1998). Pont-Kingdon et al. (1998) report the existence of a homologue to the bacterial MutS gene (part of a mismatch repair pathway) in the mitochondria of an octocoral; although in principle this could contribute to low rates of molecular evolution, its absence in the sea anemone *Metridium* makes its general significance for Cnidaria unclear.

Alternatively, the lack of genetic differentiation could reflect the effects of regular hybridization. This perspective has been strongly argued by Veron (1995) for corals, many of which reproduce in mass spawning events at approximately the same time. Some genetic and chromosomal data have been interpreted from this point of view (Kenyon, 1997; Odorico & Miller, 1997).

This controversy is likely to persist for some time for several reasons. First, it is intrinsically difficult to distinguish between reproductively isolated taxa that have very few fixed genetic differences and taxa that are only imperfectly isolated. In both cases, most loci will show either no significant differences or only modest frequency differences. The weakness of negative evidence is particularly evident in this situation, as has been well illustrated by the effort required to find genetic differences between some plants (Howard et al., 1997). Second, the existence of puzzling patterns in existing data sets [e.g. highly divergent nuclear ITS-1 and mitochondrial 16S rDNA sequences within single individuals or single species (Odorico & Miller, 1997; Romano & Palumbi, 1997; Takabayashi et al., 1998)] suggests that conventional genetic assumptions may be inappropriate for cnidarians, further complicating interpretation of results.

Polychaeta

Most of the previously reviewed studies of polychaete sibling species did not include genetic analyses (Knowlton, 1993). The list of genera analyzed genetically is now longer, and additions include *Polydora* (Mustaquim, 1988; Manchenko & Radashevsky, 1993, 1994, 1998), *Capitella* (Baoling et al., 1991; Gamenick et al., 1998), *Marenzellaria* (Bastrop et al., 1998), *Arenicola* (Cadman & Nelson-Smith, 1993), *Eulalia* (Bonse et al., 1996), *Petitia* (Van Soosten et al., 1998), *Nereis* (Rodriguez-Trelles et al., 1996), *Nerilla* (Schmidt & Westheide, 1997), and *Hediste* (Fong & Garthwaite, 1994; Abbiati & Maltagliati, 1996; Rohner et al., 1997; Sato & Masuda, 1997). Nevertheless, genetic information remains scanty, even for the genus *Capitella*, despite the highly influential early paper by Grassle and Grassle (1976) and the importance of this group for pollution studies.

Most genetic studies of polychaetes are based on protein electrophoresis, and very large genetic distances are not uncommon. *Petitia* and *Nerilla* have been analyzed by RAPDs and *Marenzelleria* by mitochondrial 16S rDNA sequencing. In many cases, morphological differences are very subtle, although sympatric forms often have clear differences in life history or ecology. Studies of allopatric forms often leave many unanswered questions, particularly when studied with techniques like RAPDs for which there is no established relationship between genetic divergence and reproductive isolation or ecological distinctiveness. However, the finding of reproductive incompatibility associated with very small levels of allozyme divergence between populations (e.g. Nei's $D < 0.01$; Rice & Simon, 1980) suggests that one cannot simply ignore subtle levels of differentiation. More breeding studies (e.g. as in Gamenick et al., 1998) would be particularly informative in this regard. The many small polychaetes that lack a dispersive phase may be genetically structured like the copepod *Tigriopus*, making the cohesion species concept particularly appropriate (see above).

Mollusca

Mollusks are arguably the best genetically studied phylum of marine invertebrates. Hence, the following summary of analyses of species boundaries is far from complete and highlights only the most recent work where multiple studies exist. Mollusks have also provided a superb model system for understanding the evolution of the gamete recognition proteins that may underlie the origin of reproductive barriers between some species (Metz et al., 1998b; Swanson and Vacquier, 1998; Hellberg & Vacquier, 1999). Finally, mollusks have an excellent fossil record, so that there is enormous potential for combining genetic and paleontological approaches (e.g. Collins et al., 1996; Reid et al., 1996).

Among the bivalves, species in the genera *Crassostrea* (e.g. Boudry et al., 1998; Hare & Avise, 1998) and particularly *Mytilus* (e.g. Beynon & Skibinksi, 1996; Gardner, 1996; Quesada et al., 1998; Comesaña et al., 1999) have been especially well studied. Other recently analyzed groups of bivalves include *Geukensia* (Sarver et al., 1992), *Cardium* (Hummel et al., 1994), *Lithophaga* (Mokady et al., 1994), *Chamelea* (Backeljau et al., 1994), mytilids and three vesicomyid genera from the deep sea (Craddock et al., 1995; Peek et al., 1997), and *Donax* (Adamkewicz & Harasewych, 1996). Species of gastropods have been even more extensively investigated. The genus *Littorina*, for example, has a long history of study [see recent works by Reid et al. (1996), Grahame et al. (1997), Tatarenkov & Johannesson (1998), and Kyle & Boulding (1998)]. Other genera that have been recently analyzed include *Hydrobia* (Ponder & Clark, 1988), *Stramonita* (Liu et al., 1991), *Alviniconcha* (Denis et al., 1993), *Trochus* (Borsa & Benzie, 1993), *Tectus* (Borsa & Benzie, 1993), *Austrocochlea* (Parsons & Ward, 1994), *Drupella* (Johnson & Cumming, 1995), *Columbella* (Oliverio, 1995), *Patella* (Côrte-Real et al., 1996a,b; Ridgway et al., 1998), *Nassarius* (Sanjuan et al., 1997), *Nucella* (Kirby et al., 1997; Marko, 1998), and *Dendronotus* (Thollesson, 1998). Cephalopods are comparatively less studied, but genera where species boundaries have recently been revised or identified based on genetic analyses include *Photololigo* (Yeatman & Benzie, 1994), *Nautilus* (Wray et al., 1995), *Sepioteuthis* (Izuka et al., 1996) and *Illex* (Carvalho & Nigmatullin, 1998).

Molluscan alpha taxonomy has long attracted the attention of professional and amateurs alike because shells are readily collected and studied. However, differences in shell morphology are often the result of phenotypic plasticity, so that many past descriptions of species were unjustified. Consequently, the number of genetic studies that have resulted in the synonymization of species is probably higher for mollusks than most other marine invertebrate phyla. The above-mentioned studies of *Donax* (Adamkewicz & Harasewych, 1996) and *Nautilus* (Wray et al., 1995) are but two examples of numerous cases where distinctions between sympatric taxa based on shell morphology were not supported by genetic data. On the other hand, genetic data have also demonstrated that minor differences in shell morphology (or other characters) can sometimes be taxonomically important (e.g. Borsa & Benzie, 1993; Mokady et al., 1994; Parsons & Ward, 1994; Johnson & Cumming, 1995;

Oliverio, 1995; Izuka et al., 1996; Sanjuan et al., 1997; Thollesson, 1998). The unexpected discovery of highly divergent sympatric taxa within morphologically uniform groups is uncommon in mollusks, but it does occur; for example, in the deep sea (Craddock et al., 1995; Peek et al., 1997) and in cephalopods (Yeatman & Benzie, 1994).

The problem of the uncertain basis for shell variability is exacerbated in allopatry, where genetic and environmental differences are often confounded. Moreover, the same geographic comparison can reveal little or no genetic differentiation in some taxa and substantial genetic differentiation in others, sometimes even within the same genus (Adamkewicz & Harasewych, 1996; Côrte-Real et al., 1996a,b). Along continuous coastlines, the existence of narrow zones of overlap with little evidence of hybridization provides definitive evidence for species boundaries (Liu et al., 1991; Ridgway et al., 1998), but far more complex patterns of hybridization are sometimes observed (e.g. *Mytilus*, see below), and in other cases no zone of overlap exists (see Discussion).

Several molluscan genera have been extensively studied using a variety of approaches, and the complexity of the resulting data sets is sobering. Studies of *Mytilus* have revealed a variety of patterns of hybridization depending on the taxa and the geographic locale, as well as a bizarre sex-linked mitochondrial transmission system (Beynon & Skibinksi, 1996; Gardner, 1996; Quesada et al., 1998; Comesaña et al., 1999). The multipronged attack on the population structure of *Crassostrea* across the Gulf of Mexico and Atlantic coast of North America (see most recently Hare & Avise, 1998) has shown that different techniques can give fundamentally different answers for reasons that still remain unclear. In *Littorina*, closely related pairs of taxa that occur in different habitats have in some studies been shown to exhibit reproductive barriers across environmental gradients within a site but, in other cases, samples from different environments within sites are more similar genetically than are those samples which are geographically isolated but environmentally similar (Kyle & Boulding, 1998; Tatarenkov & Johannesson, 1998). Some of this complexity is probably due to the proximity of the divergences being studied in all three genera (see Discussion), but is such complexity lying below the surface in the many other groups that have yet to be investigated so intensively?

Crustacea

Recent genetic analyses of species boundaries in decapod crustaceans include studies of the lobsters *Homarus* (Kornfield et al., 1995), *Jasus* (Ovenden et al., 1997) and *Panulirus* (Sarver et al., 1998); the crabs *Pinnotheres* (Stevens, 1990), *Liopetrolisthes* (Weber and Galleguillos, 1991), *Helice* (Irawan et al., 1993), *Sesarma* (Felder & Staton, 1994), *Uca* (Felder and Staton, 1994), *Emerita* (Tam et al., 1996), *Scylla* (Fuseya & Watanabe, 1996), *Carcinus* (Geller et al., 1997) and *Menippe* (Bert et al., 1996; Schneider-Broussard et al., 1998); and the shrimps *Alpheus* (McClure & Greenbaum, 1994; Knowlton & Weigt, 1997, 1998), *Synalpheus* (Duffy, 1996), and *Rimicaris* (Shank et al., 1998). Studies of other crustaceans include those of four genera of calanoid copepods (Bucklin et al., 1995, 1996, 1998); the isopods *Cyathura* (Brown et al., 1988), *Sphaeroma* (Laulier, 1989) and *Idotea* (Charfi-Cheikhrouha et al., 1998); the amphipods *Themisto* (Schneppenheim & Weigmann-Haass, 1986), *Abyssorchomene* (France, 1994) and *Eurythenes* (France & Kocher, 1996) and the mysid *Mysis* (Väinölä & Vainio, 1998).

Most genetic analyses of species boundaries in Crustacea confirm or reveal the existence of cryptic species, some of which are distinguished by surprisingly large genetic differences given their morphological similarity (e.g. Palumbi & Benzie, 1991; Bucklin et al., 1995; Knowlton & Weigt, 1998; Sarver et al., 1998). However, some taxa have smaller genetic differences than expected (Kornfield et al., 1995), and others have been shown to be conspecific based on the absence of genetic differentiation between size classes (Shank et al., 1998) or spatially disjunct populations (Ovenden et al., 1997). Sympatric or parapatric taxa typically exhibit ecological differences, suggesting that species are less generalized than previously assumed; good recent examples of this include zonation by depth within the deep sea (France, 1994; France & Kocher, 1996) and the existence of species complexes of symbiotic crustaceans associated with different hosts (Stevens, 1990; Weber & Galleguillos, 1991; Duffy, 1996). Genetic analyses have been used to document species introductions (e.g. Geller et al., 1997) and can provide rapid methods for characterizing planktonic copepod samples (Bucklin et al., 1998).

Some groups of crustaceans are well studied taxonomically and genetically, and several genetic studies include explicit calibrations of rates of molecular evolution for the cytochrome oxidase I and rRNA mitochondrial genes (Cunningham et al., 1992; Sturmbauer et al., 1996; Knowlton & Weigt, 1998; Schubart et al., 1998). Thus there is considerable potential not only for the recognition of species boundaries but also for estimating the timing of their divergences (see Discussion). Genetic studies of crustaceans can also draw on a large base of genetic data available for other arthropods (e.g. *Drosophila*). Analyses of Crustacea are consequently more advanced than those of many other marine invertebrates in some respects. As with the molluscan examples discussed above, these more detailed analyses have brought an appreciation of potential complexities in the interpretation of genetic data. Examples include the problem of pseudogenes (Schneider-Broussard & Neigel, 1997), coadaptation between nuclear and mitochondrial genomes (Burton et al., 1999) and inconsistency of outcomes when different genes are compared (Bert et al., 1996; Tam et al., 1996; Schneider-Broussard et al., 1998).

Echinodermata

Sea urchins are a model system for developmental biologists, and have also been extensively studied using molecular genetic techniques from the perspectives of population genetics and dispersal (e.g. Palumbi et al., 1997; Lessios et al., 1998) and the nature of fertilization barriers (e.g. Metz & Palumbi, 1996; Metz et al., 1998a). Echinoderms have also figured prominently in attempts to calibrate rates of molecular evolution based on the Isthmus of Panama (Lessios, 1998), thereby facilitating estimates of the times of species divergences (see Discussion).

The use of genetic techniques to discriminate species is, however, comparatively limited; recent studies of potential species complexes not previously summarized include those of the starfish *Leptasterias* (Hrincevich & Foltz, 1996; Foltz, 1997) and the sea cucumber *Cucumaria* (Arndt et al., 1996). In the latter case molecular data were used to synonymize taxa as well as to separate them. Results from *Leptasterias*, as well as an earlier study of *Echinaster* (Tuttle & Lindahl, 1980), suggest that sibling species of starfish may often lack fixed differences in allozymes.

Other marine invertebrates

Cryptic species have been investigated genetically in only a few bryozoan genera, including studies of *Bugula* (Davidson & Haygood, 1999) and *Stylopoma* (Jackson & Cheetham, 1994) not previously reviewed. *Stylopoma* is one of the few cases where genetic divergence and morphological divergence have been compared in detail; within this genus the correspondence is quite strong, even where differences are subtle, and lends support to the use of high-resolution morphometrics in recognizing species in the fossil record. The case of *Bugula* has applied importance: two cryptic taxa that differ by about 9% in the mtDNA COI gene harbor bacterial symbionts that differ in 4 of 1024 bp of the small subunit rRNA gene, and differ as well in the production of anticancer bryostatin compounds (Davidson & Haygood, 1999). The existence of cryptic species that differ in their bioactive compounds is likely to be a general phenomenon for which genetic analyses are ideally suited (King et al., 1995).

Genetic studies of species boundaries in ascidians are similarly limited, with only a few genera emerging from bibliographic searches: *Botryllus* (Aron & Solé-Cava, 1991), *Pyura* (Dalby, 1997) and *Herdmania* (Degnan & Lavin, 1995). In each of these cases, the species could be readily distinguished by color or other features, and genetic differences were substantial. This suggests that the alpha taxonomy of ascidians is plagued by excessive lumping, with many more cryptic taxa yet to be discovered.

Cryptic species of nermerteans, often distinguishable by color, are also characterized by large genetic differences (taxa from Table 1 of Knowlton, 1993; Rogers et al., 1995; Manchenko & Kulikova, 1996a). However, genetic analyses of nemerteans have also shown that some forms that differ in color are in fact conspecific (Sundberg & Andersson, 1995; Manchenko & Kulikova, 1996b), indicating that color is a potentially useful cue, but one that demands genetic follow-up.

For most other groups, recent genetic analyses of species boundaries are even more limited (e.g. for hemichordates, see King et al., 1995; for gastrotrichs, see Todaro et al., 1996; for priapulids, see Schrieber et al., 1996; for vestimentiferans, see Kojima et al., 1997). General conclusions about the extent to which taxa are over- or under-split are impossible to draw from such a scattered record.

Fish

Sharks and bony fishes have been extensively studied genetically, largely from the perspective of elucidating stock structure, but also for recognizing species boundaries; Ward & Grewe (1994) and Shaklee &

Bentzen (1998) provide general reviews of this literature. Additional analyses of species boundaries not cited by these authors include studies of *Myliobatis* (Levy & Conceição, 1989), *Siganus* (Lacson & Nelson, 1993), *Naso* (Dayton et al., 1994), *Pleuronichthys* (Watanabe et al., 1994), *Osmerus* (Taylor & Dodson, 1994), several damselfish genera (Lacson, 1994; Lacson & Clark, 1995; Lessios et al., 1995), *Lutjanus* (Lee & Cheng, 1996), *Trematomus* (Bernardi & Goswami, 1997), *Cyclothone* (Miya & Nishida, 1997), *Mugil* (Rossi et al., 1998), and *Chaetodon* (McMillan et al., 1999).

Many of these studies exhibit the same kinds of pattern as those shown by marine invertebrates, namely the recognition that similar forms in either sympatry or allopatry are clearly distinct on a genetic basis. What is somewhat surprising, however, is the considerable inconsistency among authors in their assignment of species level status to such forms. For example, allopatric populations of damselfish species belonging to several genera and *Mugil cephalus* are in each case distinguished by several fixed allozyme differences (Lacson, 1994; Lacson & Clark, 1995; Lessios et al., 1995; Rossi et al., 1998), but only in one publication is species level distinction recommended (Lessios et al., 1995). At the opposite end of the spectrum, allopatric butterflyfishes that can only be distinguished genetically by diagnostic sequences in the rapidly evolving control region of mtDNA are nevertheless recognized as distinct at the species level (McMillan et al., 1999).

Many fishes are brightly colored, and fish systematists have often drawn on these color patterns to distinguish closely related species. Several recent studies (Planes & Doherty, 1997; McMillan et al., 1999) show how complex the relationship can be between color pattern and genetic divergence. In the tropical damselfish and butterflyfish studied by these authors, some evidence for assortative mating was found, but patterns were often inconsistent across regions. In both cases partial reproductive boundaries associated with color pattern appear to have been established very recently (see Discussion).

Marine turtles and mammals

Slow rates of molecular evolution have been suggested for both marine turtles (Bowen et al., 1993) and marine mammals (Garcia-Rodriguez et al., 1998). As a consequence, even sympatric sibling species in these groups may be hard to distinguish genetically, and the interpretation of small differences in allopatry is problematic. Nevertheless, genetic analyses have confirmed or suggested cryptic species in a number of groups. In the turtles, genetic data were critical for the recognition of the distinctiveness of the Kemp's and olive ridley turtles (*Lepidochelys*) and for showing that Atlantic and eastern Pacific green turtles (*Chelonia mydas*) were more distinct than the eastern Pacific green turtle and the so-called black turtle from the western Pacific (*C. agassizi*) (Bowen et al., 1993; Dutton et al., 1996). In marine mammals, examples of sibling species that have been recognized or confirmed using molecular approaches include spotted vs. harbor seals (*Phoca*), cryptic species within both Bryde's and minke whales (*Balaenoptera*), the shortbeaked vs. longbeaked common dolphin (*Delphinus*), and several taxa of bottlenose dolphins (*Tursiops*) (Rice, 1998). Killer whales (*Orcinus*) exhibit slight but fixed mtDNA differences between northern 'residents', southern 'residents' and 'transients' (Hoelzel et al., 1998); transients and residents differ ecologically as well as genetically, but the role of cultural transmission of behaviors coupled with very limited dispersal between social groups complicates the analysis of these data. Recent mitochondrial DNA analyses of salt-water manatees (*Trichechus manatus*) have revealed three deep, allopatric clades that are as genetically distinct from each other as they are from the freshwater manatee (*T. inunguis*) and comparable to the differences observed between genera of whales (Garcia-Rodrigues et al., 1998).

Many marine mammals have several geographically defined subspecies (Rice, 1998). Genetic techniques may lead to the discovery of unique genetic features that would permit their recognition on the grounds of the phylogenetic species concept. Regardless of one's opinion of the various species concepts, some form of recognition of such taxa may facilitate conservation of these often endangered groups.

Common themes and problems

Many of the patterns that emerged in this review, based primarily on the more recent literature, are similar to those I outlined before (Knowlton, 1993): many sibling species continue to be discovered, and in sympatry they often have characteristic differences in ecology or life history, and can, in retrospect, be identified by subtle differences in morphology or color pattern. The importance of identifying species boundaries for both basic and applied science is widely acknowledged.

The biggest change in the time between these two reviews is the increasing prominence of DNA analyses. For the 91 marine invertebrate genera listed in Knowlton (1993, Table 1), protein data had been published for 68 (75%) while DNA analyses were available for only 6 (7%). In contrast, in the papers surveyed for this review, analyses of species boundaries used DNA-based techniques in 36 of 102 genera.

DNA and allozyme data sets have different strengths and weaknesses which have been well reviewed by others (e.g. see Shaklee & Bentzen, 1998); my purpose in focusing on the former is not to disparage the latter, but rather to consider where newly available DNA studies have important contributions or implications for studies of species boundaries. Two areas merit special attention: (1) the greater resolving power of DNA analyses and the consequent implications for analyses of allopatric taxa, and (2) the ability to estimate the time of origin of divergent taxa using the molecular 'clock'.

Lumping vs. splitting of allopatric taxa

Can the degree of genetic differentiation be used to determine whether allopatric taxa merit recognition at the species level? For Cracraft's (1989) version of the phylogenetic species concept the answer is simple: any fixed difference will do. When genetic research was dominated by allozyme analysis, this was a reasonable and often used approach, since fixed allozyme differences do not rapidly evolve. Thorpe's (1983) summary of levels of genetic divergence (e.g. Nei's D) associated with clearly distinct species provided a related criterion that has also been widely adopted by systematists attempting to make taxonomic decisions for allopatric taxa.

As Avise & Wollenberg (1997) point out, however, in an age of ready DNA analysis, finding fixed differences becomes far simpler, with the real possibility of needing to recognize trivially divergent taxa at the species level if Cracraft's (1989) version of the phylogenetic species concept is applied. They persuasively argue that multiple concordant differences are a far better criterion for recognizing species boundaries and, that moreover, much of the conflict between the biological and phylogenetic species concepts evaporates when this approach is used.

An alternative, but not unrelated, approach would be to use estimates of time since divergence to make assessments of likelihood of interbreeding. Studies of the Isthmus of Panama, for example, suggest that 3 million years is long enough for reproductive barriers to form in a variety of marine taxa (Knowlton et al., 1993; Lessios, 1998), although additional taxa should be studied to determine how general is this result. Data from the Isthmus are of particular relevance because they are based on allopatric taxa that have not been influenced evolutionarily by secondary contact since the connection between the Caribbean and eastern Pacific closed. This absence of secondary contact is critical, because currently sympatric and allopatric species do not exhibit the same relationships between reproductive isolation and genetic differentiation. In *Drosophila*, the only group for which we have extensive data, prezygotic barriers are three times stronger in sympatric taxa than in allopatric taxa for closely related species of similar age, presumably due to the effects of reinforcement (Coyne & Orr, 1997). Thus it is appropriate, even within a single genus, that some sympatric species are more similar to each other genetically than are allopatric populations of other species (e.g. Russo et al., 1994).

The age of recently diverged species

The ability to estimate the divergence times of species by comparing DNA sequences, even if only approximate, offers the opportunity to examine the relationship between speciation and the broader paleontological record of climatic change. In particular, did many of the closely related taxa we are discovering with genetic techniques emerge during the enormous and cyclical changes in temperature and sea level associated with the waxing and waning of the ice ages over the last 2.5 million years? Alternatively, are they older or much younger?

Few attempts have been made to answer this question for marine organisms generally, and indeed the same question has generated debate for the much better studied terrestrial realm (Avise & Walker, 1998; Avise et al., 1998). Benzie (1999) provides the only general summary of which I am aware which is based on genetic data for marine organisms, and he focuses on coral reef taxa. He argues that both intraspecific population structure and interspecific species divergences often date to the Pleistocene (less than 2 *mya*). The importance of lowered sea level in generating divergences between conventionally conspecific populations in the Indian Ocean and western Pacific seems robust as it is based on a number of analyses; the patterns observed are similar to those summarized on land by Avise et al. However, species-level divergences calibrated by mtDNA clocks are based on a very limited sample in Benzie's (1999) analysis and are dominated by Indo-West Pacific butterflyfishes, and by taxa separated by the Isthmus of Panama (which cannot have diverged more recently than 3 *mya*, see below).

Taxa separated by the Isthmus of Panama provide other important data on this subject, however. If the most similar transisthmian sister species taxa are assigned a divergence time of 3 million years ago, one can then ask: how many non-transisthmian taxa are younger? For snapping shrimp, the answer is none: all other sibling species analyzed to date exhibit divergences suggesting times of origin that precede the closure of the Isthmus (Knowlton et al., 1993). A number of other genetic studies also suggest older divergences between sibling species (e.g. Bakker et al., 1995a,b; Bucklin et al., 1995; Lessios et al., 1995).

There do, however, seem to be situations where more recent divergences are not uncommon. Marine snails and fish provide some of the best documented examples of speciation events which are almost certainly comparatively recent (Reid et al., 1996; Marko, 1998; Metz et al., 1998b; McMillan et al., 1999). Interestingly, many snails do not seem to conform to the classical allopatric model, in that sister species often come from the same biogeographic province (Hellberg & Vacquier, 1999). It is difficult to know whether these cases of speciation during the past 2 *my* reflect some special attribute of the groups involved [e.g. sexual selection acting on color patterns in fishes (see Galis & Metz, 1998 for freshwater counterparts) and divergent selection in gamete recognition proteins in gastropods (Swanson & Vacquier, 1998; Hellberg & Vacquier, 1999)], or whether similar situations will emerge elsewhere when other groups are more intensively studied. However, marine taxa provide no good cases of instant speciation or speciation in historical times; the only proposed example (a founder event in a laboratory population of polychaetes) has since been withdrawn based on new genetic data (Rodrígues-Trelles et al., 1996).

Final comments

As part of a symposium on the genetics of marine organisms, this paper has focused almost exclusively on the application of molecular analyses to the problem of understanding species in the sea. There is no denying the power of these techniques, but they have far more potential when combined with insights based on other approaches. Many have made this point, but perhaps none so eloquently as Evelyn Hutchinson (1975). His tribute to Robert MacArthur (best known today for his mathematical theory) closed with the following: "... a wide and quite deep understanding of organisms, past and present, is as basic a requirement as anything else ... it may best be self-taught, but how often is this difficult process made harder by a misplaced emphasis on a quite specious modernity. Robert MacArthur really knew his warblers."

Unfortunately, we simply do not know our squids, starfish and shrimp as well as ornithologists know their birds, in part because of long-established taxonomic traditions that depend on preserved material. If in our enthusiasm for molecular genetics we simply replace formalin with liquid nitrogen or DMSO, we will make only limited progress in elucidating the pattern and process of speciation in the sea.

Acknowledgments

I would especially like to thank the organizing and editorial committees of the International Workshop on Marine Genetics – Rio 98 for giving me the opportunity to participate in the conference and to revisit the problem of species boundaries. W. Toller, D. Kline and the editors made many helpful comments on the manuscript. I am also very grateful to the people who have over the years shared with me their manuscripts, knowledge and thoughts on many groups of marine organisms.

References

Abbiata, M. & F. Maltagliati, 1996. Allozyme evidence of genetic differentiation between populations of *Hediste diversicolor* (Polychaeta: Nereididae) from the western Mediterranean. J. mar. biol. Ass. U.K. 76: 637–647.

Adamkewicz, S. L. & M. G. Harasewych, 1996. Systematics and biogeography of the genus *Donax* (Bivalvia: Donacidae) in eastern North America. Am. malac. Bull. 13: 97–103.

Amaral, F. D., R. S. Silva, L. Mauricio-da-Silva & A. M. Solé-Cava, 1997. Molecular systematics of *Millepora alcicornis* Linnaeus, 1758 and *M. braziliensis* Verrill, 1868 (Hydrozoa: Milleporidae) from Brazil. Proc. 8th Intl. Coral Reef Symp. 2: 1577–1580.

Arndt, A., C. Marquez, P. Lambert & M. J. Smith, 1996. Molecular phylogeny of eastern Pacific sea cucumbers (Echinodermata: Holothuroidea) based on mitochondrial DNA sequence. Mol. Phylogenet. Evol. 6: 425–437.

Aron, S. & A. M. Solé-Cava, 1991. Genetic evaluation of the taxonomic status of two varieties of the cosmopolitan ascidian

Botryllus niger (Ascidiaceae: Botryllidae). Biochem. Syst. Ecol. 19: 271–276.

Avise, J. C. & R. M. Ball Jr., 1990. Principles of genealogical concordance in species concepts and biological taxonomy. Oxford Surv. evol. Biol. 7: 45–67.

Avise, J. C. & D. Walker, 1998. Pleistocene phylogeographic effects on avian populations and the speciation process. Proc. r. Soc. Lond. B 265: 457–463.

Avise, J. C. & K. Wollenberg, 1997. Phylogenetics and the origin of species. Proc. natn. Acad. Sci. U.S.A. 94: 7748–7755.

Avise, J. C., D. Walker & G. C. Johns, 1998. Speciation durations and Pleistocene effects on vertebrate phylogeography. Proc. r. Soc. Lond. B 265: 1707–1712.

Backeljau, T., P. Bouchet, S. Gofas & L. de Bruyn, 1994. Genetic variation, systematics and distribution of the venerid clam *Chamelea gallina*. J. mar. biol. Ass. U.K. 74: 211–223.

Bakker, F. T., J. L. Olsen & W. T. Stam, 1995a. Evolution of nuclear rDNA ITS sequences in the *Cladophora albida/sericea* clade (Chlorophyta). J. mol. Evol. 40: 640–651.

Bakker, F. T., J. L. Olsen & W. T. Stam, 1995b. Global phylogeography in the cosmopolitan species *Cladophora vagabunda* (Chlorophyta) based on nuclear rDNA internal transcribed spacer sequences. Europ. J. Phycol. 30: 197–208.

Baoling, W., Q. Pei-Yuan & Z. Sonling, 1991. Morphology, reproduction, ecology and allozyme electrophoresis of three *Capitella* sibling species in Qingdao (Polychaeta: Capitellidae). Ophelia (Suppl.) 5: 391–400.

Barbieri, M., G. Bavestrello & M. Sarà, 1995. Morphological and ecological differences in two electrophoretically detected species of *Cliona* (Porifera, Demospongiae). Biol. J. linn. Soc. 54: 193–200.

Bastrop, R., K. Jurss & C. Sturmbauer, 1998. Cryptic species in a marine polychaete and their independent introduction from North America to Europe. Mol. Biol. Evol. 15: 97–103.

Bavestrello, G. & M. Sarà, 1992. Morphological and genetic differences in ecologically distinct populations of *Petrosia* (Porifera, Demospongiae). Biol. J. linn. Soc. 47: 49–60.

Benzie, J. A. H., 1999. Genetic structure of coral reef organisms: ghosts of dispersal past. Am. Zool. 39: 131–145.

Benzie, J. A. H., I. R. Price & E. Ballment, 1997. Population genetics and taxonomy of *Caulerpa* (Chlorophyta) from the Great Barrier Reef, Australia. J. Phycol. 33: 491–504.

Bernardi, G. & U. Goswami, 1997. Molecular evidence for cryptic species among the Antarctic fish *Trematomus bernacchii* and *Trematomus hansoni*. Antarct. Sci. 9: 381–385.

Bert, T. M. & W. S. Arnold, 1995. An empirical test of predictions of two competing models for the maintenance and fate of hybrid zones: both models are supported in a hard-clam hybrid zone. Evolution 49: 276–289.

Bert, T. M., K. J. McCarthy, H. Cruz-Lopez & S. Bogdanowicz, 1996. Character discriminatory power, character-set congruence, and the classification of individuals from hybrid zones: an example using stone crabs (*Menippe*). Evolution 50: 655–671.

Beynon, C. M. & D. O. F. Skibinski, 1996. The evolutionary relationships between three species of mussel (*Mytilus*) based on anonymous DNA polymorphisms. J. exp. mar. Biol. Ecol. 203: 1–10.

Blomster, J., C. A. Maggs & M. J. Stanhope, 1998. Molecular and morphological analysis of *Enteromorpha intestinalis* and *E. compressa* (Chlorophyta) in the British Isles. J. Phycol. 34: 319–340.

Bonse, S., H. Schmidt, D. Eibye-Jacobsen & W. Westheide, 1996. *Eulalaia viridis* (Polychaeta: Phyllodocidae) is a complex of two species in northern Europe: results from biochemical and morphological analyses. Cah. Biol. mar. 37: 33–48.

Borsa, P. & J. A. H. Benzie, 1993. Genetic relationships among the topshells *Trochus* and *Tectus* (Prosobranchia: Trochidae) from the Great Barrier Reef. J. moll. Stud. 59: 275–284.

Boudry, P., S. Heurtebise, B. Collet, F. Cornette & A. Gérard, 1998. Differentiation between populations of the Portuguese oyster, *Crassostrea angulata* (Lamark) and the Pacific oyster, *Crassostrea gigas* (Thunberg), revealed by mtDNA RFLP analysis. J. exp. mar. Biol. Ecol. 226: 279–291.

Boury-Esnault, N., M. Klautau, C. Bezac, J. Wulff & A. M. Solé-Cava, 1999. Comparative study of putative conspecific sponge populations from both sides of the Isthmus of Panama. J. mar. biol. Assoc. U.K. 79: 39–50.

Bowen, B. W., W. S. Nelson & J. C. Avise, 1993. A molecular phylogeny for marine turtles: trait mapping, rate assessment, and conservation relevance. Proc. natn. Acad. Sci. U.S.A. 90: 5574–5577.

Brazeau, D. A. & C. D. Harvell, 1994. Genetic structure of local populations and divergence between growth forms in a clonal invertebrate, the Caribbean octocoral *Briareum asbestinum*. Mar. Biol. 119: 53–60.

Brodie, J., P. K. Hayes, G. L. Barker & L. M. Irvine, 1996. Molecular and morphological characters distinguishing two *Porphyra* species (Rhodophyta: Bangiophycidae). Europ. J. Phycol. 31: 303–308.

Brown, L. D., 1995. Genetic evidence for hybridisation between *Haliotis rubra* and *H. laevigata*. Mar. Biol. 123: 89–93.

Brown, C. J., W. W. Anderson, W. D. Burbanck & C. T. Hackney, 1988. Genetic transition between northern and southern populations of the estuarine isopod, *Cyathura polita*, and the discovery of a new species of *Cyathura*. Estuaries 11: 96–98.

Bucklin, A., B. W. Frost & T. D. Kocher, 1995. Molecular systematics of six *Calanus* and three *Metridia* species (Calanoida: Copepoda). Mar. Biol. 121: 655–664.

Bucklin, A., A. M. Bentley & S. P. Franzen, 1998. Distribution and relative abundance of *Pseudocalanus moultoni* and *P. newmani* (Copepoda: Calanoida) on Georges Bank using molecular identification of sibling species. Mar. Biol. 132: 97–106.

Bucklin, A., T. C. LaJeunesse, E. Curry, J. Wallinga & K. Garrison, 1996. Molecular diversity of the copepod *Nannocalanus minor*: genetic evidence of species and population structure in the North Atlantic Ocean. J. mar. Res. 54: 285–310.

Burnett, W. J., J. A. H. Benzie, J. A. Beardmore & J. S. Ryland, 1997. Zoanthids (Anthozoa, Hexacorallia) from the Great Barrier Reef and Torres Strait, Australia: systematics, evolution and a key to species. Coral Reefs 16: 55–68.

Burton, R. S., P. D. Rawson & S. Edmands, 1999. Genetic architecture of physiological phenotypes: empirical evidence for coadapted gene complexes. Am. Zool. 39: 451–462.

Cadman, P. S. & A. Nelson-Smith, 1993. A new species of lugworm: *Arenicola defodiens* sp. nov. J. mar. biol. Ass. U.K. 73: 213–223.

Carvalho, G. R. & Ch. M. Nigmatullin, 1998. Stock structure analysis and species identification. FAO Fish. tech. Pap 376: 199–232.

Charfi-Cheikhrouha, F., M. Laulier, E. Hamelin & J.-P. Mocquard, 1998. Genetic differentiation and evolutionary process of speciation in the *Idotea chelipes* complex (Crustacea, Isopoda). Genet. Sel. Evol. 30: 289–303.

Collins, T. M., K. Frazer, A. R. Palmer, G. J. Vermeij & W. M. Brown, 1996. Evolutionary history of northern hemisphere *Nucella* (Gastropoda, Muricidae): molecular, morphological, ecological, and paleontological evidence. Evolution 50: 2287–2304.

Comesaña, A. S., J. E. Toro, D. J. Innes & R. J. Thompson, 1999. A molecular approach to the ecology of a mussel (*Mytilus edulis – M. trossulus*) hybrid zone on the east coast of Newfoundland, Canada. Mar. Biol. 133: 213–221.

Côrte-Real, H. B. S. M., S. J. Hawkins & J. P. Thorpe, 1996a. An interpretation of the taxonomic relationship between the limpets *Patella rustica* and *P. piperata*. J. mar. biol. Ass. U.K. 76: 717–732.

Côrte-Real, H. B. S. M., S. J. Hawkins & J. P. Thorpe, 1996b. Population differentiation and taxonomic status of the exploited limpet *Patella candei* in the Macaronesian islands (Azores, Madeira, Canaries). Mar. Biol. 125: 141–152.

Coyne, J. A. & H. A. Orr, 1997. 'Patterns of speciation in *Drosophila*' revisited. Evolution 51: 295–303.

Cracraft, J., 1989. Speciation and its ontology: The empirical consequences of alternative species concepts for understanding patterns and processes of differentiation. In Otte, D. & J. A. Endler (eds), Speciation and its Consequences. Sinauer, Sunderland, Massachusetts: 28–59.

Craddock, C., W. R. Hoeh, R. G. Gustafson, R. A. Lutz, J. Hashimoto & R. J. Vrijenhoek, 1995. Evolutionary relationships among deep-sea mytilids (Bivalvia: Mytilidae) from hydrothermal vents and cold-water methane/sulfide seeps. Mar. Biol. 121: 477–485.

Cunningham, C. W., N. W. Blackstone & L. W. Buss, 1992. Evolution of king crabs from hermit crab ancestors. Nature 355: 539–542.

Dalby, J. E. Jr., 1997. Reproductive and electrophoretic evidence for genetic maintenance of dimorphism in the ascidian *Pyura stolonifera* near Melbourne, Australia. Ophelia 47: 227–243.

Davidson, S. K. & M. G. Haygood, 1999. Identification of sibling species of the bryozoan *Bugula neritina* that produce different anticancer bryostatins and harbor distinct strains of the bacterial symbiont '*Candidatus* Endobugula sertula.' Biol. Bull. 196: 273–280.

Dayton, C., A. C. Santayana & J. M. Lacson, 1994. Genetic evidence for reproductive isolation of the recently described unicornfish *Naso caesius* and its sibling *N. hexacanthus*. Mar. Biol. 118: 551–554.

De Queiroz, K., 1998. The general lineage concept of species, species criteria, and the process of speciation. In Howard, D. J. & S. H. Berlocher (eds), Endless Forms: Species and Speciation. Oxford University Press, New York: 57–75.

Degnan, B. M. & M. F. Lavin, 1995. Highly repetitive DNA sequences provide evidence for a lack of gene flow between two morphological forms of *Herdmania momus* (Ascidiacea: Stolidobranchia). Mar. Biol. 124: 293–299.

Denis, F., D. Jollivet & D. Moraga, 1993. Genetic separation of two allopatric populations of hydrothermal snails *Alviniconcha* spp. (Gastropoda) from two south-western Pacific back-arc basins. Biochem. Syst. Ecol. 21: 431–440.

Duffy, J. E., 1996. Species boundaries, specialization, and the radiation of sponge-dwelling alpheid shrimp. Biol. J. linn. Soc. 58: 307–324.

Dutton, P. H., S. K. Davis, T. Guerra & D. Owens, 1996. Molecular phylogeny for marine turtles based on sequences of the ND4-leucine tRNA and control regions of mitochondrial DNA. Mol. Phylogenet. Evol. 5: 511–521.

Edmands, S., 1996. The evolution of mating systems in a group of brooding sea anemones (*Epiactis*). Invert. Reprod. Devel. 30: 227–237.

Embley, T. M. & E. Stackebrandt, 1997. Species in practice: exploring uncultured prokaryote diversity in natural samples. In Claridge, M. F., H. A. Dawah & M. R. Wilson (eds), Species: the Units of Biodiversity. Chapman & Hall, London: 61–81.

Felder, D. L. & J. L. Staton, 1994. Genetic differentiation in trans-Floridian species complexes of *Sesarma* and *Uca* (Decapoda: Brachyura). J. Crust. Biol. 14: 191–209.

Ferris, M. J. & B. Palenik, 1998. Niche adaptation in ocean cyanobacteria. Nature 396: 226–228.

Foltz, D. W., 1997. Hybridization frequency is negatively correlated with divergence time of mitochondrial DNA haplotypes in a sea star (*Leptasterias* spp.) species complex. Evolution 51: 283–288.

Fong, P. P. & R. L. Garthwaite, 1994. Allozyme electrophoretic analysis of the *Hediste limnicola – H. diversicolor – H. japonica* species complex (Polychaeta: Nereididae). Mar. Biol. 118: 463–470.

France, S. C., 1994. Genetic population structure and gene flow among deep-sea amphipods, *Abyssorchomene* spp., from six California continental borderland basins. Mar. Biol. 118: 67–77.

France, S. C. & T. D. Kocher, 1996. Geographic and bathymetric patterns of mitochondrial 16S rRNA sequence divergence among deep-sea amphipods, *Eurythenes gryllus*. Mar. Biol. 126: 633–643.

Fuseya, R. & S. Watanabe, 1996. Genetic variability in the mud crab genus *Scylla* (Brachyura: Portunidae). Fish. Sci. 62: 705–709.

Galis, F. & J. A. J. Metz, 1998. Why are there so many cichlid species? Trends Ecol. Evol. 13: 1–2.

Gamenick, I., M. Abbiati & O. Giere, 1998. Field distribution and sulphide tolerance of *Capitella capitata* (Annelida: Polychaeta) around shallow water hydrothermal vents off Milos (Aegean Sea). A new sibling species? Mar. Biol. 130: 447–453.

Garcia-Rodriguez, A. I., B. W. Bowen, D. Domning, A. A. Mignucci-Giannoni, M. Marmontel, R. A. Montoya-Opina, B. Morales-Vela, M. Rudin, R. K. Bonde & P. M. McGuire, 1998. Phylogeography of the West Indian manatee (*Tirchechus manatus*): how many populations and how many taxa? Mol. Ecol. 7: 1137–1149.

Gardner, J. P. A., 1996. The *Mytilus edulis* species complex in southwest England: effects of hybridization and introgression upon interlocus associations and morphometric variation. Mar. Biol. 125: 385–399.

Gardner, J. P. A., 1997. Hybridization in the sea. Adv. mar. Biol. 31: 1–78.

Geller, J. B., E. D. Walton, E. D. Grosholz & G. M. Ruiz, 1997. Cryptic invasions of the crab *Carcinus* detected by molecular phylogeography. Mol. Ecol. 6: 901–906.

Goff, L. J., D. A. Moon & A. W. Coleman, 1994. Molecular delineation of species and species relationships in the red algal agarophytes *Gracilariopsis* and *Gracilaria* (Gracilariales). J. Phycol. 30: 521–537.

Grahame, J., S. L. Hull, P. J. Mill & R. Hemingway, 1997. Discovering unrecognized diversity among marine molluscs. In Ormond, R. F. G., J. D. Gage & M. V. Angel (eds), Marine Diodiversity: Patterns and Processes. Cambridge University Press, Cambridge: 293–318.

Grassle, J. P. & J. F. Grassle, 1976. Sibling species in the marine pollution indicator *Capitella* (Polychaeta). Science 192: 567–569.

Greenberg, N., R. L. Garthwaite & D. C. Potts, 1996. Allozyme and morphological evidence for a newly introduced species of *Aurelia* in San Francisco Bay, California. Mar. Biol. 125: 401–410.

Hare, M. P. & J. C. Avise, 1998. Population structure in the American oyster as inferred by nuclear gene genealogies. Mol. Biol. Evol. 15: 119–128.

Hellberg, M. E. & V. D. Vacquier, 1999. Rapid evolution of fertilization selectivity and lysin cDNA sequences in teguline gastropods. Mol. Biol. Evol. 16: 839–848.

Hinrichs, K.-U., J. M. Hayes, S. P. Sylva, P. G. Brewer & E. F. DeLong, 1999. Methane-consuming archaebacteria in marine sediments. Nature 398: 802–805.

Hoelzel, A. R., M. Dalheim & S. J Stern, 1998. Low genetic variation among killer whales (*Orcinus orca*) in the eastern North Pacific and genetic differentiation between foraging specialists. J. Heredity 89: 121–128.

Howard, D. J., R.W. Preszler, J. Williams, S. Fenchel & W. J. Boecklen, 1997. How discrete are oak species: Insights from a hybrid zone between *Quercus grisea* and *Quercus gambelii*. Evolution 51: 747–755.

Hrincevich, A. W. & D. W. Foltz, 1996. Mitochondrial DNA sequence variation in a sea star (*Leptasterias* spp.) species complex. Mol. Phylogenet. Evol. 6: 408–415.

Huber, B. T., J. Bijma & K. Darling, 1997. Cryptic speciation in the living planktonic foraminifer *Globigerinella siphonifera* (d'Orbigny). Paleobiology 23: 33–62.

Hughes, R. N., 1989. A Functional Biology of Clonal Animals. Chapman & Hall, London.

Hummel, H., M. Wolowicz & R. H. Bogaards, 1994. Genetic variability and relationships for populations of *Cerastoderma edule* and of the *C. glaucum* complex. Neth. J. Sea Res. 33: 81–89.

Hutchinson, G. E., 1975. Variations on a theme by Robert MacArthur. In Cody, M. L. & J. M. Diamond (eds), Ecology and Evolution of Communities. Belknap Press of Harvard University Press, Cambridge: 492–521.

Irawan, B., A. Kijima & Y. Fujio, 1993. Genetic divergence among the three species of estuarine crab, *Helice tridens*, *H. japonica* and *Chiromantes dehaani* (Sesarminae, Grapsidae; Decapoda). Tohoku J. agric. Res. 43: 101–110.

Izuka, T., S. Segawa & T. Okutani, 1996. Biochemical study of the population heterogeneity and distribution of the oval squid *Sepioteuthis lessoniana* complex in southwestern Japan. Am. malacol. Bull. 12: 129–135.

Jackson, J. B. C. & A. H. Cheetham, 1994. Phylogeny reconstruction and the tempo and mode of speciation in cheilostome Bryozoa. Paleobiology 20: 407–423.

Johnson, M. S. & R. L. Cumming, 1995. Genetic distinctness of three widespread and morphologically variable species of *Drupella* (Gastropoda, Muricidae). Coral Reefs 14: 71–78.

Kenyon, J. C., 1997. Models of reticulate evolution in the coral genus *Acropora* based on chromosome numbers: parallels with plants. Evolution 51: 756–767.

King, G. M., C. Giray & I. Kornfield, 1995. Biogeographical, biochemical and genetic differentiation among North American saccoglossids (Hemichordata; Enteropneusta; Harrimaniidae). Mar. Biol. 123: 369–377.

Kirby, R. R., R. J. Berry & D. A. Powers, 1997. Variation in mitochondrial DNA in a cline of allele frequencies and shell phenotype in the dog-whelk *Nucella lapillus* (L.). Biol. J. linn. Soc. 62: 299–312.

Kirsten, J. H., C. J. Dawes & B. J. Cochrane, 1998. Randomly amplified polymorphism detection (RAPD) reveals high genetic diversity in *Thalassia testudinum* Banks ex Konig (Turtlegrass). Aquat. Bot. 61: 269–287.

Klautau, M., A. M. Solé-Cava & R. Borojevic, 1994. Biochemical systematics of sibling sympatric species of *Clathrina* (Porifera: Calcarea). Biochem. Syst. Ecol. 22: 367–375.

Klautau M., C. A. M. Russo, C. Lazoski, N. Boury-Esnault, J. P. Thorpe & A. M. Solé-Cava, 1999. Does cosmopolitanism result from overconservative systematics? A case study using the marine sponge *Chondrilla nucula*. Evolution 53: 1414–1422.

Knowlton, N., 1993. Sibling species in the sea. Ann. Rev. Ecol. Syst. 24: 189–216.

Knowlton, N. & J. B. C. Jackson, 1993. Inbreeding and outbreeding in marine invertebrates. In Thornhill, N. W. (ed.), The Natural History of Inbreeding and Outbreeding. University of Chicago Press, Chicago: 200–249.

Knowlton, N. & L. A. Weigt, 1997. Species of marine invertebrates: a comparison of the biological and phylogenetic species concepts. In Claridge, M. F., H. A. Dawah & M. R. Wilson (eds), Species: the Units of Biodiversity. Chapman & Hall, London: 199–219.

Knowlton, N. & L. A. Weigt, 1998. New dates and new rates for divergence across the Isthmus of Panama. Proc. r. Soc. Lond. B 265: 2257–2263.

Knowlton, N., L. A. Weigt, L. A. Solórzano, D. E. K. Mills & E. Bermingham, 1993. Divergence in proteins, mitochondrial DNA, and reproductive compatibility across the Isthmus of Panama. Science 260: 1629–1632.

Kojima, S., R. Segawa, J. Hashimoto & S. Ohta, 1997. Molecular phylogeny of vestimentiferans collected around Japan, revealed by the nucleotide sequences of mitochondrial DNA. Mar. Biol. 127: 507–513.

Kornfield, I., A. B. Williams & R. S. Steneck, 1995. Assignment of *Homarus capensis* (Herbst, 1792), the Cape lobster of South Africa, to the new genus *Homarinus* (Decapoda: Nephropidae). U.S. natn. Mar. Fish. Serv. Fish. Bull. 93: 97–102.

Kyle, C. J. & E. G. Boulding, 1998. Molecular genetic evidence for parallel evolution in a marine gastropod, *Littorina subrotundata*. Proc. r. Soc. Lond. B 265: 303–308.

Lacson, J. M., 1994. Fixed allele frequency differences among Palauan and Okinawan populations of the damselfishes *Chrysiptera cyanea* and *Pomacentrus coelestis*. Mar. Biol. 118: 359–365.

Lacson, J. M. & S. Clark, 1995. Genetic divergence of Maldivian and Micronesian demes of the damselfishes *Stegastes nigricans*, *Chrysiptera biocellata*, *C. glauca* and *C. leucopoma*. Mar. Biol. 121: 585–590.

Lacson, J. M. & S. G. Nelson, 1993. Genetic distances among fishes of the genus *Siganus* (Siganidae) from the western Pacific Ocean. Mar. Biol. 116: 187–192.

Lasker, H. R., K. Kim & M. A. Coffroth, 1996. Reproductive and genetic variation among Caribbean gorgonians: the differentiation of *Plexaura kuna*, new species. Bull. mar. Sci. 58: 277–288.

Laulier, M., 1989. Génétique et systématique évolutives du complexe d'espèces *Spaeroma hookeri* Leach, *Sphaeroma levii* Argano et *Sphaeroma rugicauda* Leach (Crustacés, Isopodes Flabellifères). 2. Variabilité génétique, distances génétiques et allèles diagnostiques. Gen. Sel. Evol. 21: 131–145.

Lazoski, C., S. Peixinho, C. A. M. Russo & A. M. Solé-Cava, 1999. Genetic confirmation of the specific status of two sponges of the genus *Cinachyrella* (Porifera: Spirophorida: Demospongiae) in the southwest Atlantic. Mem. Queensland Mus. 44: 299–306.

Lee, S.-C. & H.-L. Cheng, 1996. Genetic difference between two snappers, *Lutjanus ophuysenii* and *L. vitta* (Teleostei: Lutjanidae). Ichthyol. Res. 43: 340–344.

Lessios, H. A., 1998. The first stage of speciation as seen in organisms separated by the Isthmus of Panama. In Howard, D. J. & S. H. Berlocher (eds), Endless Forms: Species and Speciation. Oxford University Press, New York: 186–201.

Lessios, H. A. & J. S. Pearse, 1996. Hybridization and introgression between Indo-Pacific species of *Diadema*. Mar. Biol. 126: 715–723.

Lessios, H. A. & J. R. Weinberg, 1994. Genetic and morphological divergence among morphotypes of the isopod *Excirolana* on the two sides of the Isthmus of Panama. Evolution 48: 530–548.

Lessios, H. A., B. D. Kessing & D. R. Robertson, 1998. Massive gene flow across the world's most potent marine biogeographic barrier. Proc. r. Soc. Lond. B 265: 583–588.

Lessios, H. A., J. R. Weinberg & V. R. Starczak, 1994. Temporal variation in populations of the marine isopod *Excirolana*: how stable are gene frequencies and morphology? Evolution 48: 549–563.

Lessios, H. A., G. R. Allen, G. M. Wellington & E. Bermingham, 1995. Genetic and morphological evidence that the eastern Pacific damselfish *Abudefduf declivifrons* is distinct from *A. concolor* (Pomacentridae). Copeia 1995: 277–288.

Levy, J. A. & M. B. Conceição, 1989. Biochemical evidences for two sibling species of genus *Myliobatis* (Chondrichthyes: Myliobatidae) in south Brazil. Comp. Biochem. Physiol. 94B: 687–690.

Lindstrom, S. C. & K. M. Cole, 1992. A revision of the species of *Porphyra* (Rhodophyta: Bangiales) occurring in British Columbia and adjacent waters. Can. J. Bot. 70: 2066–2075.

Liu, L. L., D. W. Foltz & W. B. Stickle, 1991. Genetic population structure of the southern oyster drill *Stramonita* (= *Thais*) *haemostoma*. Mar. Biol. 111: 71–79.

Lopez, J. V. & N. Knowlton, 1997. Discrimination of species in the *Montastraea annularis* complex using multiple genetic loci. Proc. 8th Intl. Coral Reef Symp. 2: 1613–1618.

Lopez, J. V., R. Kersanach, S. A. Rehner & N. Knowlton, 1999. Molecular determination of species boundaries in corals: genetic analysis of the *Montastraea annularis* complex using amplified fragment length polymorphisms and a microsatellite marker. Biol. Bull. 196: 80–93.

Maggs, C. A. & B. A. Ward, 1996. The genus *Pikea* (Dumontiaceae, Rhodophyta) in England and the north Pacific: comparative morphological, life history, and molecular studies. J. Phycol. 32: 176–193.

Maggs, C. A., S. E. Douglas, J. Fenety & C. J. Bird, 1992. A molecular and morphological analysis of the *Gymnogongrus devoniensis* (Rhodophyta) complex in the north Atlantic. J. Phycol. 28: 214–232.

Manchenko, G. P. & V. I. Kulikova, 1996a. Allozyme and colour differences between two sibling species of the heteronemertean *Lineus torquatus* from the Sea of Japan. Mar. Biol. 125: 687–691.

Manchenko, G. P. & V. I. Kulikova, 1996b. Enzyme and colour variation in the hoplonemertean *Tetrastemma nigrifrons* from the Sea of Japan. Hydrobiologia 337: 69–76.

Manchenko, G. P & V. I. Radashevsky, 1993. Genetic differences between two sibling species of the *Polydora ciliata* complex (Polychaeta: Spionidae). Biochem. Syst. Ecol. 21: 543–548.

Manchenko, G. P & V. I. Radashevsky, 1994. Genetic differences between two allopatric sibling species of the genus *Polydora* (Polychaeta: Spionidae) from the west Pacific. Biochem. Syst. Ecol. 22: 767–773.

Manchenko, G. P & V. I. Radashevsky, 1998. Genetic evidence for two sibling species within *Polydora* cf. *ciliata* (Polychaeta: Spionidae) from the Sea of Japan. Mar. Biol. 131: 489–495.

Manchenko, G. P., A. V. Moschenko & V. S. Odintsov, 1993. Biochemical genetics and systematics of *Millepora* (Coelenterata: Hydrozoa) from the shore of South Vietnam. Biochem. Syst. Ecol. 21: 729–735.

Marko, P. B., 1998. Historical allopatry and the biogeography of speciation in the prosobranch snail genus *Nucella*. Evolution 52: 757–774.

Mayden, R. L., 1997. A hierarchy of species concepts: the denouement in the saga of the species problem. In Claridge, M. F., H. A. Dawah & M. R. Wilson (eds), Species: the Units of Biodiversity. Chapman & Hall, London: 381–424.

Mayr, E., 1963. Animal Species and Evolution. Harvard University Press, Cambridge, Massachusetts.

Mayr, E. & P. D. Ashlock, 1991. Principles of Systematic Zoology, 2nd edn. McGraw Hill, New York.

McClure, M. R. & I. F. Greenbaum, 1994. Biochemical variation in *Alpheus* (Decapoda, Caridea, Alpheidae) from the coast of Texas: evidence for cryptic species. Southwest. Nat. 39: 63–66.

McFadden, C. S., 1999. Genetic and taxonomic relationships among Northeastern Atlantic and Mediterranean populations of the soft coral *Alcyonium coralloides*. Mar. Biol. 133: 171–184.

McFadden, C. S., R. K. Grosberg, B. B. Cameron, D. P. Karlton & D. Secord, 1997. Genetic relationships within and between clonal and solitary forms of the sea anemone *Anthopleura elegantissima* revisited: evidence for the existence of two species. Mar. Biol. 128: 127–139.

McMillan, W. O., L. A. Weigt & S. R. Palumbi, 1999. Color pattern evolution, assortative mating, and genetic differentiation in brightly colored butterflyfishes (Chaetodontidae). Evolution 53: 247–260.

Medina, M., E. Weil & A. M. Szmant, 1999. Examination of the *Montastraea annularis* species complex (Cnidaria: Scleractinia) using ITS and COI sequences. Mar. Biotechnol. 1: 89–97.

Metz, E. C. & S. R. Palumbi, 1996. Positive selection and sequence rearrangements generate extensive polymorphism in the gamete recognition protein bindin. Mol. Biol. Evol. 13: 397–406.

Metz, E. C., G. Gomez-Gutierrez & V. D. Vacquier, 1998a. Mitochondrial DNA and bindin gene sequence evolution among allopatric species of the sea urchin genus *Arbacia*. Mol. Biol. Evol. 15: 185–195.

Metz, E. C., R. Robles-Sikisaka & V. D. Vacquier, 1998b. Nonsynonymous substitution in abalone sperm fertilization genes exceeds substitution in introns and mitochondrial DNA. Proc. natn. Acad. Sci. U.S.A. 95: 10676–10681.

Miller, K. J. & J. A. H. Benzie, 1997. No clear genetic distinction between morphological species within the coral genus *Platygyra*. Bull. mar. Sci. 61: 907–917.

Miya, M. & M. Nishida, 1997. Speciation in the open ocean. Nature 389: 803–804.

Mokady, O., S. Rozenblatt, D. Graur & Y. Loya, 1994. Coral-host specificity of Red Sea *Lithophaga* bivalves: interspecific and intraspecific variation in 12S mitochondrial ribosomal RNA. Mol. mar. Biol. Biotechnol. 3: 158–164.

Monteiro, F. A., A. M. Solé-Cava & J. P. Thorpe, 1997. Extensive genetic divergence between populations of the common intertidal sea anemone *Actinia equina* from Britain, the Mediterranean and the Cape Verde Islands. Mar. Biol. 129: 425–433.

Moore, L. R., G. Rocap & S. W. Chisolm, 1998. Physiology and molecular phylogeny of coexisting *Prochlorococcus* ecotypes. Nature 393: 464–467.

Muricy, G., A. M. Solé-Cava, J. P. Thorpe & N. Boury-Esnault, 1996. Genetic evidence for extensive cryptic speciation in the subtidal sponge *Plakina trilopha* (Porifera: Demospongiae: Homoscleromorpha) from the western Mediterranean. Mar. Ecol. Prog. Ser. 138: 181–187.

Mustaquim, J., 1988. Electrophoretic variation of isozymes in *Polydora ciliata* complex (Polychaeta: Spionidae). Comp. Biochem. Physiol. 91B: 197–205.

Odorico, D. M. & D. J. Miller, 1997. Variation in the ribosomal internal transcribed spacers and 5.8 S rDNA among five species of *Acropora* (Cnidaria; Scleractinia): patterns of variation consistent with reticulate evolution. Mol. Biol. Evol. 14: 465–473.

Ó Foighil, D. & M. J. Smith. 1996. Phylogeography of an asexual marine clam complex, *Lasaea*, in the northeastern Pacific based on Cytochrome Oxidase III sequence variation. Mol. Phylogenet. Evol. 6: 134–142.

Oliverio, M., 1995. Larval development and allozyme variation in east Atlantic *Columbella* (Gastropoda: Prosobranchia: Columbellidae). Sci. mar. 59: 77–86.

Olsen, J. L., W. T. Stam, P. V. M. Bot & C. Van den Hoek, 1987. scDNA-DNA hybridization studies in Pacific and Caribbean isolates of *Dictyosphaeria cavernosa* (Chlorophyta) indicate a long divergence. Helgoländer Wiss. Meeresunters 41: 377–383.

Olsen, J. L., M. Valero, I. Meusnier, S. Boele-Bos & W. T. Stam, 1998. Mediterranean *Caulerpa taxifolia* and *C. mexicana* (Chlorophyta) are not conspecific. J. Phycol. 34: 850–856.

Ovenden, J. R., J. D. Booth & A. J. Smolenski, 1997. Mitochondrial DNA phylogeny of red and green rock lobsters (genus *Jasus*). Mar. freshwat. Res. 48: 1131–1136.

Palumbi, S. R. 1994. Genetic divergence, reproductive isolation, and marine speciation. Ann. Rev. Ecol. Syst. 25: 547–572.

Palumbi, S. R. & J. A. H. Benzie, 1991. Large mitochondrial DNA differences between morphologically similar penaeid shrimp. Mol. mar. Biol. Biotech. 1: 27–34.

Palumbi, S. R., G. Grabowsky, T. Duda , L. Geyer & N. Tachino, 1997. Speciation and population genetic structure of tropical Pacific sea urchins. Evolution 51: 1506–1517.

Parsons, K. E. & R. D. Ward, 1994. Electrophoretic and morphological examination of *Austrocochlea constricta* (Gastropoda: Trochidae): a species complex. Aust. J. mar. Freshwat. Res. 45: 1065–1085.

Peek, A. S., R. G. Gustafson, R. A. Lutz & R. C. Vrijenhoek, 1997. Evolutionary relationships of deep-sea hydrothermal vent and cold-water seep clams (Bivalvia: Vesicomyidae): results from mitochondrial cytochrome oxidase subunit I. Mar. Biol. 130: 151–161.

Pillmann, A., G. W. Woolcott, J. L. Olsen, W. T. Stam & R. J. King, 1997. Inter- and intraspecific genetic variation in *Caulerpa* (Chlorophyta) based on nuclear rDNA ITS sequences. Europ. J. Phycol. 32: 379–386.

Planes, S. & P. J. Doherty, 1997. Genetic relationships of the colour morphs of *Acanthochromis polyacanthus* (Pomacentridae) on the northern Great Barrier Reef. Mar. Biol. 130: 109–117.

Ponder, W. F. & G. A. Clark, 1988. A morphological and electrophoretic examination of '*Hydrobia buccinoides*', a variable brackish-water gastropod from temperate Australia (Mollusca: Hydrobiidae). Aust. J. Zool. 36: 661–690.

Pont-Kingdon, G., N. A. Okada, J. L. Macfarlane, C. T. Beagley, C. D. Watkins-Sims, T. Cavelier-Smith, G. D. Clark-Walker & D. R. Wolstenholme, 1998. Mitochondrial DNA of the coral *Sarcophyton glaucum* contains a gene for a homologue of bacterial MutS: a possible case of gene transfer from the nucleus to the mitochondrion. J. mol. Evol. 46: 419–431.

Potter, D., T. C. Lajeunesse, G. W. Saunders & R. A. Anderson, 1997. Convergent evolution masks extensive biodiversity among marine coccoid picoplankton. Biodiv. Conserv. 6: 99–107.

Quesada, H., C. Gallagher, D. A. G. Skibinski & D. O. F. Skibinski, 1998. Patterns of polymorphism and gene flow of gender-associated mitochondrial DNA lineages in European mussel populations. Mol. Ecol. 7: 1041–1051.

Rappé, M. S., M. T. Suzuki, K. L. Vergin & S. J. Giovannoni, 1998. Phylogenetic diversity of ultraplankton plastid small-subunit rRNA genes recovered in environmental nucleic acid samples from the Pacific and Atlantic coasts of the United States. Apl. envir. Microbiol. 64: 294–303.

Reid, D. G., E. Rumbak & R. H. Thomas, 1996. DNA, morphology and fossils: phylogeny and evolutionary rates of the gastropod genus *Littorina*. Phil. Trans. r. Soc., Lond. B 351: 877–895.

Rice, D. W., 1998. Marine mammals of the world: systematics and distribution. Soc. mar. Mamm. Spec. Publ. 4.

Rice, S. A. & J. L. Simon, 1980. Intraspecific variation in the pollution indicator polychaete *Polydora ligni* (Spionidae). Ophelia 19: 79–115.

Ridgway, T. M., B. A. Stewart, G. M. Branch & A. N. Hodgson, 1998. Morphological and genetic differentiation of *Patella granularis* (Gastropoda: Patellidae): recognition of two sibling species along the coast of southern Africa. J. Zool., Lond. 245: 317–333.

Rieseberg, L. H. & C. R. Linder, 1999. Hybrid classification: insights from genetic map-based studies of experimental hybrids. Ecology 80: 361–370.

Rodríguez-Trelles, F., J. R. Weinberg & F. J. Ayala, 1996. Presumptive rapid speciation after a founder event in a laboratory population of *Nereis*: allozyme electrophoretic evidence does not support the hypothesis. Evolution 50: 457–461.

Rogers, A. D., J. P. Thorpe & R. Gibson, 1995. Genetic evidence for the occurrence of a cryptic species with the littoral nemerteans *Lineus ruber* and *L. viridis* (Nemertea: Anopla). Mar. Biol. 122: 305–316.

Rohner, M., R. Bastrop & K. Jurss, 1997. Genetic differentiation in *Hediste diversicolor* (Polychaeta: Nereididae) for the North Sea and Baltic Sea. Mar. Biol. 130: 171–180.

Romano, S. L. & S. R. Palumbi, 1997. Molecular evolution of a portion of the mitochondrial 16S ribosomal gene region in scleractinian corals. J. mol. Evol. 45: 397–411.

Rossi, A. R., M. Capula, D. Crosetti, L. Sola & D. E. Campton, 1998. Allozyme variation in global populations of striped mullet, *Mugil cephalus* (Pisces: Mugilidae). Mar. Biol. 131: 203–212.

Rowan, R., 1998. Diversity and ecology of zooxanthellae on coral reefs. J. Phycol. 34: 407–417.

Russo, C. A. M., A. M. Solé-Cava & J. P. Thorpe, 1994. Population structure and genetic variation in two tropical sea anemones (Cnidaria, Actinidae) with different reproductive strategies. Mar. Biol. 119: 267–276.

Sanjuan, A., M. Pérez-Losada & E. Rolán, 1997. Allozyme evidence for cryptic speciation in sympatric populations of *Nassarius* spp. (Mollusca: Gastropoda). J. mar. biol. Ass. U.K. 77: 773–784.

Sarà, M., G. Corriero & G. Bavestrello, 1993. *Tethya* (Porifera, Demospongiae) species coexisting in a Maldivian coral reef lagoon: taxonomical, genetic and ecological data. Mar. Ecol. 14: 341–355.

Sarver, S. K., M. C. Landrum & D. M. Foltz, 1992. Genetics and taxonomy of ribbed mussels (*Geukensia* spp.). Mar. Biol. 113: 385–390.

Sarver, S. K., J. D. Silberman & P. J. Walsh, 1998. Mitochondrial DNA sequence evidence supporting the recognition of two subspecies or species of the Florida spiny lobster *Panulirus argus*. J. Crust. Biol. 18: 177–186.

Sato, M. & Y. Masuda, 1997. Genetic differentiation in two sibling species of the brackish-water polychaete *Hediste japonica* complex (Nereididae). Mar. Biol. 130: 163–170.

Schmidt, H. & W. Westheide, 1997. RAPD-PCR experiments confirm the distinction between three morphologically similar species of *Nerilla* (Polychaeta: Nerillidae). Zool. Anz. 236: 277–285.

Schneider-Broussard, R. & J. E. Neigel, 1997. A large subunit mitochondrial ribosomal DNA sequence translocated to the nuclear genomes of two stone crabs. Mol. Biol. Evol. 14: 156–165.

Schneider-Broussard, R., D. L. Felder, C. A. Chlan & J. E. Neigel, 1998. Tests of phylogeographic models with nuclear and mitochondrial DNA sequence variation in the stone crabs, *Menippe adina* and *Menippe mercenaria*. Evolution 52: 1671–1678.

Schneppenheim, R. & R. Weigmann-Haass, 1986. Morphological and electrophoretic studies of the genus *Themisto* (Amphipoda: Hyperiidea) from the South and North Atlantic. Polar Biol. 6: 215–226.

Schreiber, A., M. Eisinger & V. Storch, 1996. Allozymes characterize sibling species of bipolar Priapulida (*Priapulus*, *Priapulopsis*). Polar Biol. 16: 521–526.

Schubart, C. D., R. Diesel & S. B. Hedges, 1998. Rapid evolution to terrestrial life in Jamaican crabs. Nature 393: 363–365.

Shaklee, J. B. & P. Bentzen, 1998. Genetic identification of stocks of marine fish and shellfish. Bull. mar. Sci. 62: 589–621.

Shank, T. M., R. A. Lutz & R. C. Vrijenhoek, 1998. Molecular systematics of shrimps (Decapoda: Bresiliidae) from deep-sea hydrothermal vents, I: Enigmatic 'small orange' shrimp from the Mid-Atlantic Ridge are juvenile *Rimicaris exoculata*. Mol. mar. Biol. Biotechnol. 7: 88–96.

Solé-Cava, A. M. & N. Boury-Esnault, 1999. Patterns of intra and interspecific divergence in marine sponges. Mem. Queensland. Mus. 44: 591–602.

Solé-Cava, A. M. & J. P. Thorpe, 1992. Genetic divergence between colour morphs in populations of the common intertidal sea anemones *Actinia equina* and *A. prasina* (Anthozoa: Actiniaria) in the Isle of Man. Mar. Biol. 112: 243–252.

Solé-Cava, A. M., J. P. Thorpe & J. G. Kaye, 1985. Reproductive isolation with little genetic divergence between *Urticina* (= *Tealia*) *felina* and *U. eques* (Anthozoa: Actiniaria). Mar. Biol. 85: 279–284.

Stevens, P. M., 1990. A genetic analysis of the pea crabs (Decapoda: Pinnotheridae) of New Zealand. I. Patterns of spatial and host-associated genetic structuring in *Pinnotheres novaezelandiae* Filhol. J. exp. mar. Biol. Ecol. 141: 195–212.

Stiller, J. W. & J. R. Waaland, 1993. Molecular analysis reveals cryptic diversity in *Porphyra* (Rhodophyta). J. Phycol. 29: 506–517.

Stobart, B. & J. A. H. Benzie, 1994. Allozyme electrophoresis demonstrates that the scleractinian coral *Montipora digitata* is two species. Mar. Biol. 118: 183–190.

Sturmbauer, C., J. S. Levinton & J. Christy, 1996. Molecular phylogeny analysis of fiddler crabs: test of the hypothesis of increasing behavioral complexity in evolution. Proc. natn. Acad. Sci. U.S.A. 93: 10855–10857.

Sundberg, P. & S. Andersson, 1995. Random amplified polymorphic DNA (RAPD) and intraspecific variation in *Oerstedia dorsalis* (Hoplonemertea, Nemertea). J. mar. biol. Ass. U.K. 75: 483–490.

Swanson, W. J. & V. D. Vacquier, 1998. Concerted evolution in an egg receptor for a rapidly evolving abalone sperm protein. Science 281: 710–712.

Takabayashi, M., D. Carter, S. Ward & O. Hoegh-Guldberg, 1998. Inter- and intra-specific variability in ribosomal DNA sequence in the internal transcribed spacer region of corals. In Greenwood, J. G. & N. J. Hall (eds), Proceedings of the Australian Coral Reef Society 75 Anniversary Conference. University of Queensland, Brisbane: 237–244.

Tam, Y. K., I. Kornfield & F. P. Ojeda, 1996. Divergence and zoogeography of mole crabs, *Emerita* spp. (Decapoda: Hippidae), in the Americas. Mar. Biol. 125: 489–497.

Tatarenkov, A. & K. Johannesson, 1998. Evidence of a reproductive barrier between two forms of the marine periwinkle *Littorina fabalis* (Gastropoda). Biol. J. linn. Soc. 63: 349–365.

Taylor, E. B. & J. J. Dodson, 1994. A molecular analysis of relationships and biogeography within a species complex of Holarctic fish (genus *Osmerus*). Mol. Ecol. 3: 235–248.

Templeton, A. R., 1989. The meaning of species and speciation: a genetic perspective. In Otte, D. & J. A. Endler (eds), Speciation and its Consequences. Sinauer, Sunderland, Massachusetts: 3–27.

Thollesson, M., 1998. Discrimination of two *Dendronotus* species by allozyme electrophoresis and the reinstatement of *Dendronotus lacteus* (Thompson, 1840) (Nudibranchia, Dendronotoidea). Zool. Scr. 27: 189–195.

Thorpe, J. P., 1983. Enzyme variation, genetic distance and evolutionary divergence in relation to levels of taxonomic separation. In Oxford, G. S. & D. Rollison (eds), Protein Polymorphism: Adaptive and Taxonomic Significance. Academic Press, London: 131–152.

Thorpe, J. P. & A. M. Solé-Cava, 1994. The use of allozyme electrophoresis in invertebrate systematics. Zool. Scr. 23: 3–18.

Todaro, M. A., J. W. Fleeger, Y. P. Hu, A. W. Hrincevich & D. W. Foltz, 1996. Are meiofaunal species cosmopolitan? Morphological and molecular analysis *Xenotrichula intermedia* (Gastrotricha: Chaetonotida). Mar. Biol. 125: 735–742.

Tuttle, R. D. & R. Lindahl, 1980. Genetic variability in 3 cooccurring forms of the starfish genus *Othilia* (= *Echinaster*). Experientia 36: 923–925.

Väinölä, R. & J. K. Vainio, 1998. Distributions, life cycles and hybridization of two *Mysis relicta* group species (Crustacea: Mysida) in the northern Baltic Sea and Lake Båven. Hydrobiologia 368: 137–148.

Van Oppen, M. J. H., J. L. Olsen & W. T. Stam, 1995a. Genetic variation within and among North Atlantic and Baltic populations of the benthic alga *Phycodrys rubens* (Rhodophyta). Europ. J. Phycol. 30: 251–260.

Van Oppen, M. J. H., B. L. Willis & D. J. Miller, 1999. Atypically low rate of cytochrome *b* evolution in the scleractinian coral genus *Acropora*. Proc. r. Soc. Lond. B 266: 179–183.

Van Oppen, M. J. H., S. G. A. Draisma, J. L. Olsen & W. T. Stam, 1995b. Multiple trans-Arctic passages in the red alga *Phycodrys rubens*: evidence from nuclear rDNA ITS sequences. Mar. Biol. 123: 179–188.

Van Oppen, M. J. H., H. Klerk, J. L. Olsen & W. T. Stam, 1996. Hidden diversity in marine algae: some examples of genetic variation below the species level. J. mar. biol. Assoc. U.K. 76: 239–242.

Van Soosten, C., H. Schmidt & W. Westheide, 1998. Genetic variability and relationships among geographically widely separated populations of *Petitia amphophthalma* (Polychaeta: Syllidae). Results from RAPD-PCR investigations. Mar. Biol. 131: 659–669.

Veron, J. E. N., 1995. Corals in space and time: The Biogeography and Evolution of the Scleractinia. University of New South Wales Press, Sydney.

Ward, R. D. & P. M. Grewe, 1994. Appraisal of molecular genetic techniques in fisheries. Rev. Fish Biol. Fish. 4: 300–325.

Watanabe, K., K. Numachi, M. Goto, S. Wada, T. Kobayashi & T. Kamaishi, 1994. Genetic differentiation of two forms of flatfishes, genus *Pleuronichthys* in Kiisuido channel waters off the Pacific coast of central Japan, by restriction analysis of mitochondrial DNA. Nippon Suisan Gakkaishi 60: 515–520.

Weber, L. I. & R. Galleguillos, 1991. Morphometric and electrophoretic evidences for two species of the genus *Liopetrolisthes*

(Crustacea: Decapoda: Porcellanidae) and some aspects of their variability. Comp. Biochem. Physiol. 100B: 201–207.

Weil, E. & N. Knowlton, 1994. A multi-character analysis of the Caribbean coral *Montastraea annularis* (Ellis and Solander, 1786) and its two sibling species, *M. faveolata* (Ellis and Solander, 1786) and *M. franksi* (Gregory, 1895). Bull. mar. Sci. 55: 151–175.

Wray, C. G., N. H. Landman, W. B. Saunders & J. Bonacum, 1995. Genetic divergence and geographic diversification in *Nautilus*. Paleobiology 21: 220–228.

Yeatman, J. & J. A. H. Benzie, 1994. Genetic structure and distribution of *Photololigo* spp. in Australia. Mar. Biol. 118: 79–87.

Hydrobiologia **420:** 91–94, 2000.
A. M. Solé-Cava, C. A. M. Russo & J. P. Thorpe (eds), Marine Genetics.

Gene transfer technology in aquaculture

J. A. Levy[1], L. F. Marins[1] & A. Sanchez[2]
[1]*Fundação Universidade Federal de Rio Grande, Departamento de Química, Laboratório de Bioquímica Marinha, C.P. 474, Rio Grande, RS, 96201-900 Brazil*
E-mail: dqmlevy@super.furg.br
[2]*Universitat Autônoma de Barcelona, Facultat de Veterinária, Unitat de Genética, Bellaterrra, 08193 Barcelona, Spain*
E-mail: sanchez@uab.es

Key words: genetically modified organisms (GMOs), transgenic fish, food safety

Abstract

The gene transfer technique, transgenesis, has permitted the transfer of genes from one organism to another to create new lineages of organisms with improvement in traits important to aquaculture. Genetically modified organisms (GMOs), therefore, hold promise for producing genetic improvements, such as enhanced growth rate, increased production and efficiency, disease resistance and expanded ecological ranges. The basic procedure to generate transgenic fish for aquaculture includes: (1) design and construction of transgenic DNA; (2) transfer of the gene construct into fish germ cells; (3) screening for transgenic fish; (4) determination of transgene expression and phenotype; (5) study of inheritance; and (6) selection of stable lines of transgenics.

GMOs offer economic benefits, but also pose environmental threats. Optimising the mix of benefits and risks is of fundamental importance. The potential economic benefits of transgenic technology to aquaculture are obvious. Transgenic fish production has the goal of producing food for human consumption; thus the design of genetic constructs must take into consideration the potential risks to consumer health, as well as marketing strategies and product acceptance in the market.

Introduction

Aquaculture is one of the fastest growing food-producing sectors, providing an acceptable supplement to and substitute for wild fish and other aquatic products. Regional, cultural and historic attributes have played major roles influencing both the production base and rate of expansion of aquaculture. The historic tradition of growing fish in Asia, which is well documented in countries such as China, India, Cambodia and Indonesia, has played a significant role in maintaining Asia's dominant role in aquaculture. Production of finfish, shellfish and plants from culture in Asia increased from 8.4 million metric tons (mt) in 1984 at a growth rate (APR) of 10.4 to over 25 million mt in 1995 and accounted for over 90% of world output. Much of this growth, however, specifically relates to China. When China is excluded, the APR of the sector in Asia is only 4.4, not too dissimilar to that in Europe (3.9) and North America (3.6). In Africa and Latin America, the aquaculture production base is considerably lower and although there is no tradition of aquaculture, the sector has expanded at higher rates. Data from Brazil show that, between 1996 and 1997, the aquaculture production increased from 40.5 mt yr^{-1} to 60.7 mt yr^{-1} (CNPq, 1998).

The potential of aquaculture to meet the challenges of food demand and to generate employment and foreign exchange is clearly demonstrated by the rapid expansion of this sector which has grown at an average annual rate of almost 10% since 1984 compared with 3% for livestock meat and 1.6% for capture fisheries production. In contrast, with the exception of the increase in 1993–1994 which is probably attributable to small pelagic species, data reported for capture fisheries show near zero or negative growth (Rana, 1997). In fact, in several parts of the world this activity has declined seriously. FAO (1997) evaluated 17 world fisheries and determined that four have been commercially depleted and nine are in serious decline. Off the coast of New England, catches of cod, haddock and yellowtail flounder in 1993 were 85% lower than in the mid-1960s. In eastern South America some coastal fish stocks seem to have diminished in recent years (Haimovici, 1997). Such data suggest that the fish food supply will be affected in the near future, because of the progressive global demand stimulated by the increasing human population.

However, whilst aquaculture shows potential for expansion to meet global demand for fish for food, this activity has a slow growth rate within the low income food deficit countries (LIFDCs). This may be for a combination of reasons, including countries being landlocked, low national priority of aquaculture, small and unsuitable coastline, inadequate water supply, poor infrastructure and limited capacity of institutions and technical and financial constraints. A solution to this problem lies in increasing aquaculture production through genetic improvement, thus leading to more efficient use of facilities.

The traditional methods of genetic improvement such as selective breeding, have not always been successful in increasing the capacity and diversity of farmed fish. Genetic gains through selective breeding have been reduced because of the difficulty of separating beneficial from undesirable traits. There is a need to adopt new technology to further stimulate the growth of this industry by overcoming several of its major obstacles, including high feed costs, susceptibility of the farmed fish to diseases, as well as the limitations of culture facilities and the environment.

Today, the search for increased productivity follows two trends:

1. the search for new species with better yield; and
2. a further development and application of methods for genetic improvement.

The introduction of new methodologies based on molecular biology has lead to new research fields in this sector and new markets for aquaculture industries.

Gene transfer techniques (transgenesis) have permitted the transfer of traits from one organism to another to create new lineages of organisms with improved aquaculture performances. Genetically modified organisms (GMOs), therefore, hold promise for producing genetic improvements, such as enhanced growth rate, increased production and efficiency, disease resistance and expanded ecological ranges.

According to Beardmore (1997), in the strictest sense, 'transgenic' should be used to denote a individual or a cell with a DNA sequence which has been transferred and integrated, into that individual or cell, by genetic engineering techniques. The first transgenic fish was produced by Zhu et al. (1985) and considerable progress has been made since that time. The application of gene transfer technology is obviously not limited to the topics maintained above. It should be possible to generate other beneficial traits useful in aquaculture such as adaptation to salinity, pH, stress and the development of sterile fish. With the development of fish molecular biology, other candidate genes should be available for transfer.

The basic procedure to generate transgenic fish for aquaculture includes:

1. design and construction of transgenic DNA;
2. transfer of the gene construct into fish germ cells;
3. screening for transgenic fish;
4. determination of transgene expression and phenotype;
5. study of inheritance; and
6. selection of stable lines of transgenics.

The transgene is constructed, essentially, by associating an ubiquitous promoter, a structural gene coding for a desirable trait, and a termination sequence (polyadenylation site). One of the principal problems to generate transgenic fish is the choice of promoter/enhancer. The first promoters that have been used include mouse metallothionein-1 (mMt-1), Rous Sarcoma Virus and SV40. Constructs made with fish promoters, for example, anti-freeze protein (AFP) and carp B-actin are expressed more efficiently in fish than are mammalian constructs.

Microinjection is a well established method for gene transfer in fish. This method was fully described in the zebrafish by Khoo et al. (1993). Essentially the process is straightforward, but technically tedious, and there are numerous variations (review by Khoo, 1995). Other procedures of gene transfer can take advantage of the large number of eggs produced by most fish species. Electroporation, which uses short electric pulses to permeate cell membranes, has been used as an effective means of introducing foreign DNA in zebrafish, salmon, loach and others. More recently, gene transfer via electroporated sperm, subsequently used to fertilise eggs, has been reported in marine molluscs, but its efficacy is still a matter of debate in other species.

Alternative methods based on the use of pantropic retroviral vectors, which direct DNA integration, will probably be of minor impact in aquaculture, even if they can provide a convenient stable approach to introduce and express foreign genes in fish.

Transgenesis, therefore, can be applied to genetic improvement in fish production. The use of genetically enhanced organisms may increase production efficiency through improvements in growth rates, food conversion, disease resistance and product quality. The application of biotechnology to aquaculture also may help conserve wild species and genetic resources and provide unique models for biomedical research.

Food safety of transgenic fish

The development and use of genetically modified organisms (GMOs) pose both economic benefits and risks to human health and the environment. Optimising the mix of benefits and risks is fundamental. The potential economic benefits of transgenic technology to aquaculture are obvious. Transgenic fish production has as a goal the production of food for human consumption and the design of genetic constructs must take into consideration the potential risks to consumer health, as well as marketing strategies and product acceptance in the market.

For evaluating the food safety of transgenic fish derived from common food fish stock consideration should be given to:

1. the transgene; and
2. the gene product.

The DNA used for transgenesis in fish, at the present, include genes such as the growth hormone gene associated with ubiquitous promoters (Beta-actin, Histone H3). New alternative structural genes such as insulin-like growth factor (IGF) are being tested. These genes code for mitogenic peptide hormones that play an important function in the growth and differentiation of vertebrates.

The tendency is that both structural genes and promoters will be derived from the target species (in Beardmore's 1997 terminology 'autotransgenic'), or from fish species as closely related as possible, thus decreasing considerably the health risks and facilitating commercialisation. Furthermore, coding sequences derived from distantly related species may pose problems, not only at a transcriptional or translational level, but in terms of physiological effectiveness (Maclean, 1998).

Autotransgenic gene constructs have given impressive increases in growth rate. Hew et al. (1991) and Du et al. (1992) used growth hormone gene from the chinook salmon linked to the ocean pout antifreeze protein promoter; second generation transgenic salmon. These fish are fast growing and fully capable of entering full strength seawater as smolts almost a year earlier than non-transgenic siblings (Hew & Fletcher, 1997). The most dramatic increase of growth rate was obtained by Devlin et al. (1994) who inserted into coho salmon a gene construct containing the metallothionein promoter and the type 1 growth hormone gene, both from sockeye salmon. This gave an 11 fold average increase in growth of the transgenic coho salmon.

The evaluation of gene products and their secondary effects parallel the safety evaluations of chemicals or drugs introduced into animal tissues. Observations of the health of transgenic animals and monitoring for unexpected effects will yield critical information on safety. Genetically modified organisms or products derived from them are likely to be accepted by the public only after the demonstration, by governmental agencies, that they are safe to eat. It will be necessary to educate the public about the safety of the products. At present, the commercial use of transgenic fish has only just started, although transgenic salmon are being grown in Scotland (Martinez, 1997) and New Zealand and applications have been filed in the United States (Dunham, 1999).

The food safety of transgenic animals has been addressed internationally by a joint consultation of the FAO and WHO (World Health Organisation, 1991) and by OECD (Organisation for Economic Cooperation and Development, 1992). These groups have concluded that modern techniques of molecular biology and biotechnology do not inherently result in foods that are less safe and that these foods should be assessed by comparing them with their closest conventional counterparts. Chatakondi et al. (1995) compared the body composition of transgenic common carp containing rainbow trout growth hormone gene with non-transgenic full-siblings. The transgenic carp muscle had an increase protein content of about 7.5% and a decreased fat content of 13%. The results showed that significant, but relatively subtle, changes occurred in the flesh of transgenic common carp and were not detrimental to human health and safety and may have been beneficial.

Public acceptance of genetically modified marine organisms or products derived from them is very much influenced by the public perception of safety. The exclusive use of fish DNA in the constructs for gene transfer will probably facilitate the approval by regulatory agencies and overcome the considerable resistance of public opinion to transgenic food and products.

Perspectives

Transgenic technology has proved successful landmarks in several aquatic species, in particular in salmon and trout, in contrast to the relative inefficiency of the technology in the generation of 'improved' transgenic farm animals. This difference can be attributed to the relative ease of availability of fish embryos for microinjection of foreign DNA, the traditional technology of gene transfer used in animals. However, embryo mortality in some cases is high, rates of integration are often low and the process of random gene insertion is subject to positional effects for transgene expression.

Despite large efforts, totipotent cell lines for use in techniques of homologous recombination have been available only to generate transgenic mice, but progress in the establishment of pluripotent cells in livestock species and preliminary results on ES-like cell cultures in the Japanese medaka and in zebrafish have been reported (Sun et al., 1995; Hong et al., 1996).

There are numerous problems to be resolved in the successful development of transgenic broodstock for aquaculture. Ecological, socio-economic as well as ethical issues have to be considered. New techniques need to be developed to control sterility and broodstock management. Nevertheless, after the arrival of 'Dolly', the first mammal to be cloned from differentiated adult cells (Wilmut et al., 1997), the advent of new transfection-mediated homologous recombination methodology in animal transgenesis, by cloning, seems more feasible and opens new horizons for research and practical applications in animal production.

Genomes of aquatic species can be analysed and characterised and quantitative trait loci identified. This can be applied to improve our understanding of the molecular basis of gene regulation and expression as well as sex determination and thereby create methods for defining species, stocks and populations. Approaches that will play a key role in the aquaculture industry in the near future include developing marker-assisted selection technologies, improving the precision and efficiency of transgenic techniques, and improving technologies for the cryo-preservation of gametes and embryos.

References

Beardmore, J. A., 1997. Transgenics: autotransgenics and allotransgenics. Transgenic Res. 6: 107–108.

Chatakondi, N., R. T. Lovell, P. L. Duncan, M. Hayat, T. T. Chen, D. A. Powers, J. D. Weete, K. Cummins & R. A. Dunham, 1995. Body composition of transgenic common carp, *Cyprinus carpio*, containing rainbow trout growth hormone gene. Aquaculture 138: 1–4.

CNPq, 1998. Apoio ao crescimento da aqüicultura no Brasil. Conselho Nacional de Desenvolvimento Científico e Tecnológico, Boletim Informativo no. 4, Brasilia.

Devlin, R. H., T. Y. Yesaki, C. A. Biagl, E. M. Donaldson, P. Swanson & W.-K. Chan, 1994. Extraordinary salmon growth. Nature 371: 209–210.

Du, S.-J., Z. Gong, G. L. Fletcher, M. A. Shears, M. J. King, D. R. Idler & C. L. Hew, 1992. Growth enhancement in transgenic Atlantic salmon by the use of an 'all fish' chimeric growth hormone gene construct. Biotechnology 10: 176–181.

Dunham, R. A., 1999. Utilization of transgenic fish in developing countries: potential benefits and risks. J. World Aquacult. Soc. 30: 1–11.

FAO, 1997. Review of the state of world fishery resources: marine fisheries. FAO Fisheries Circular No. 920 FIRM/C920, FAO, Rome.

Haimovici, M., 1997. Recursos pesqueiros demersais da região sul. Fundação de Estudos do Mar, Rio de Janeiro.

Hew, C. L., G. Fletcher, S. Du, Z. Gong, M. Shears & P. Davies, 1991. Biotechnology in aquatic sciences: improved freezing tolerance and enhanced growth in Atlantic salmon by gene transfer. Bull. Inst. Zool. 16: 341–356.

Hew, C. L. & G. Fletcher, 1997. Transgenic fish for aquaculture. Chem. Indust. 1997: 311–314.

Hong, Y., C. Winkler & M. Schartl, 1996. Pluripotency and differentiation of embrionic stem cell lines from the medakafish (*Oryzias latipes*). Mech. Dev. 60: 33–44.

Khoo, H. W., L. H. Ang & H. B. Lim, 1993. Gene transfer by microinjection in the zebrafish *Brachydanio rerio*. In Murphy, D. & D. A. Carter (eds), Transgenesis Techniques, Principles and Protocols. Methods in Molecular Biology. Humana Press, Totowa, vol. 18: 87–94.

Khoo, H. W., 1995. Transgenesis and its applications in aquaculture. Asian Fish. Sci. 8: 1–25.

Maclean, N., 1998. Regulation and exploitation of transgenes in fish. Mutat. Res. 399: 255–266.

Martinez, R., 1997. Engineering the blue revolution. Seedling 14 20–30.

Organization for Economic Cooperation and Development, 1992 The proceedings of the OECD symposium on aquatic biotechnology and food safety. O. E. C. D., Paris.

Rana, K., 1997. Recent trends in global aquaculture production 1984–1995. FAO Aquacult. Newslt. 16: 14–19.

Sun, L., C. S. Bradford, C. Ghosh, P. Collodi & D. W. Barnes, 1995 ES-like cell cultures derived from early zebrafish embryos. Mol mar. Biol. Biotechnol. 4: 193–199.

Wilmut, I., A. E. Schnieke, J. McWhir, A. J. Kind & K. H. S Campbell, 1997. Viable offspring derived from fetal and adul mammalian cells. Nature 385: 810–813.

World Health Organization, 1991. Strategies for assessing th safety of foods produced by biotechnology. Report of a Joir FAO/WHO Consultation, Geneva.

Zhu, Z., G. Li, L. He & S. Chen, 1985. Novel gene transfer into th fertilized eggs of gold fish (*Carassius auratus* L. 1758). Zeitsch f. angewandt Ichthyol. 1: 31–34.

Hydrobiologia **420:** 95–98, 2000.
A. M. Solé-Cava, C. A. M. Russo & J. P. Thorpe (eds), Marine Genetics.

Seven *Hox* gene sequences from the asterinid starfish *Patiriella exigua* (Echinodermata: Asteroidea)

S. C. Long[1,*], V. B. Morris[2] & M. Byrne[1]
[1]*Department of Anatomy & Histology F13, University of Sydney, NSW 2006, Australia*
[2]*School of Biological Sciences A12, University of Sydney, NSW 2006, Australia*

Key words: echinoderms, asteroids, *Hox* genes, evolution of development, *Patiriella exigua*

Abstract

Partial homeobox gene fragments were amplified from *Patiriella exigua* genomic DNA by PCR using degenerate primers. The primers spanned a region of 82 nucleotides within the homeobox, encompassing amino acids 21–47 of the homeodomain. The deduced amino acid sequences of the seven *Hox* gene sequences plus one *Xlox*-type gene sequence from *P. exigua* are presented and compared with similar sequences from other organisms. This work is a preliminary step in the analysis of the evolution of development in the *Patiriella* species group and the roles of *Hox* gene expression therein.

Introduction

Hox genes are a subset of the homeobox genes whose defining character is a conserved 180 base pair homeobox encoding a 60 amino acid DNA-binding homeodomain (Duboule, 1994). Homeobox genes are transcription factors. For the purposes of determining evolutionary relationships between the different genes they are classified into classes and families based on sequence similarity in the homeodomain (Bürglin, 1994). Members of the *Hox* gene class lie in a cluster on the chromosome and the order of the different genes in the cluster is evolutionarily conserved. *Hox* genes in the vertebrate cluster are grouped on sequence similarity into 13 paralogous groups named from Hox1 to Hox13 (Scott, 1992). *Hox* genes from other organisms are aligned with paralogous groups of the vertebrate *Hox* cluster using sequence similarities in the homeodomain, where the order in which they lie on the chromosome is unknown.

Here we report partial homeodomain sequences from seven *Hox* genes in the asterinid starfish *Patiriella exigua*. This is the second report of *Hox* sequences in sea stars (asteroids), following the report by Mito & Endo (1997) of *Hox* genes in another asterinid starfish *Asterina minor*. The other reports of *Hox* genes in echinoderms are for sea urchins (echinoids) (Dolecki et al., 1986, 1988; Wang et al., 1990; Pfeffer & von Holt, 1991; Zhao et al., 1991; Di Bernardo et al., 1994; Dobias et al., 1996; Popodi et al., 1996; Morris et al., 1997; Arenas-Mena et al., 1998; Wu et al., 1998; Martinez et al., 1999). In their recent work Martinez et al. (1999) identified the *Hox* gene cluster in the echinoid *Strongylocentrotus purpuratus* and, by establishing the order of the genes along the chromosome, aligned them with vertebrate *Hox* paralogous groups. We have therefore aligned the partial homeodomains from *P. exigua* with those of *S. purpuratus* and so now refer to them as *Hox* genes rather than *Hox*-type genes as they were previously (Popodi et al., 1996).

Hox genes are important in patterning the anterior-posterior axis in both insects and vertebrates and are expressed in the embryo from anterior to posterior colinearly with their 3′–5′ order in the gene cluster on the chromosome (McGinnis & Krumlauf, 1992). Of late their role in shaping body plans in development and evolution is being addressed (Gellon & McGinnis, 1998). Echinoderms are of particular interest in this regard given their bilaterally symmetrical larval and radially symmetrical adult body plans. It is in this context that our work on *Hox* genes in *Patiriella* is being undertaken.

The *Patiriella* genus displays the greatest developmental diversity known in asteroids (Hart et al., 1997).

* Author for correspondence

Although the adults are phenotypically similar, the larvae of the various species range from typical feeding bipinnaria and brachiolaria (*P. regularis*), through modified brachiolaria undergoing benthic development (*P. exigua*), to viviparous species in which developmental stages are simplified and occur within the parental gonad (*P. vivipara, P. parvivipara*) (Byrne, 1995, 1996). The *Patiriella* genus thus offers a range of developmental modes in which the role of *Hox* genes in body plan patterning can be studied. Our report of *Hox* sequences in *P. exigua* is the first step in this investigation.

Materials and methods

Adult *P. exigua* were collected from rocky shores at Clovelly, New South Wales, Australia. Genomic DNA was prepared from gonad tissue: 1–5 mg of dissected testis and ovotestis was added to 1 ml aliquots of Chelex solution (5% v/v Chelex 100 resin (Bio-Rad) in 10 mM Tris-HCl, 0.1 mM EDTA at pH 7.6) that had been heated to 95°C. The mixtures were incubated at 95°C for 15 min, during which time they were mixed three times by inversion.

Partial homeobox nucleotide sequences were amplified from the genomic DNA by PCR using the degenerate primers HoxE (5′-AAAGGATCCTGCAGARYT IGARAARGARTT-3′) and HoxF (5′-ACAAGCTTGA ATTCATICKICKRTTYTGRAACCA-3′) designed by Pendleton et al. (1993), where R=A or G, Y=C or T, K=G or T and I=inosine. These primers spanned a region of 82 base pairs, encompassing amino acids 21–47 of the homeodomain. Three μl of the DNA-containing supernatant from the Chelex preparations was used in a 25 μl reaction mixture, which also contained 0.05 U/μl Taq DNA polymerase (Boehringer Mannheim) in the supplier's buffer, 200 μM each dNTP, 1.5 mM $MgCl_2$, and HoxE and HoxF at 1 μM concentrations. After an initial 97°C 5 min melt in a Perkin-Elmer 480 thermal cycler, 35 cycles of 94°C for 1 min, 37°C for 1 min and 72°C for 1 min were followed by a final extension at 72°C for 7 min.

The PCR products were purified by polyacrylamide gel electrophoresis using conventional protocols (Sambrook et al., 1989) and ligated into a pBluescript vector (Stratagene) and cloned. Recombinant clones were selected by blue-white screening on LAXI agar plates and purified plasmid DNA was sequenced across the insert in an ABI 377 (Applied Biosystems) automated sequencer.

Forty cloned amplification products were sequenced. They were identified by comparing the deduced amino acid sequences with amino acid sequences from the homeodomains of other organisms, using BLASTp to search GenBank databases. The genes were named following the established convention for naming homeobox genes in echinoids (Popodi et al., 1996). This is that an abbreviated species name (Pe) precedes the shortened gene family name (Hbox), followed by a number that represents the chronological order in which the genes were reported. Thus, a *P. exigua* gene that was similar to a previously reported echinoid gene was given the same number as that gene, but where a *P. exigua* gene had not been reported for echinoids, it was given the next highest unused number. For example, *PeHbox15* means it is the fifteenth homeobox gene reported for echinoderms.

Results

Seven *Hox* gene sequences plus an *Xlox*-type homeobox gene sequence were identified amongst those characterized from *P. exigua*. The deduced amino acid sequences (partial homeodomains) from the seven *Hox* gene sequences are shown in Figure 1, where they are aligned with the partial homeodomains of the echinoid *Strongylocentrotus purpuratus* in the paralogous groups identified by Martinez et al. (1999). Most of the partial homeodomains identified from *P. exigua* align with echinoid paralogous groups, with the exception of *PeHbox7* and *PeHbox15*, whose paralogy is less clear. The partial homeodomain of the *Xlox*-type gene is also shown (Figure 1).

Discussion

The most 3′ of the *Hox* gene sequences we report is *PeHbox16*. The partial homeodomain is 100% similar to that of *AM-1* reported for the asteroid *A. minor* (Mito & Endo, 1997) and 93% similar to *SpHox1* of the echinoid *S. purpuratus* (Martinez et al., 1999) of paralogous group 1. It is thus putatively a homologue of the most anteriorly-expressed *Hox* gene in insects and vertebrates (McGinnis & Krumlauf, 1992).

The next three *Hox* sequences, *PeHbox9*, *PeHbox3* and *PeHbox6* (Figure 1), show sequence similarities in their partial homeodomains with those of paralogous groups 4/5, 6 and 8 respectively from *S. purpuratus*. The partial homeodomains of *PeHbox9* and *PeHbox6*

	21				25					30					35					40					45		47	%
3' END																												
PG1																												
PeHbox16	H	F	N	K	Y	L	T	R	A	R	R	V	E	I	A	A	M	L	G	L	N	E	T	Q	V	K	I	
SpHox1	-	-	-	-	-	-	-	-	-	-	-	-	-	-	-	-	S	-	Q	-	-	-	-	-	-	-	-	93
PG4+PG5																												
PeHbox9	H	F	N	R	Y	L	T	R	R	R	R	I	E	I	A	H	A	L	G	L	T	E	R	Q	I	K	I	
SpHox4/5	-	-	-	-	-	-	-	-	-	-	-	-	-	-	-	-	-	-	-	-	-	-	-	-	-	-	-	100
PG6																												
PeHbox3	H	F	N	R	Y	L	T	R	R	R	R	I	E	I	A	Q	S	L	G	L	S	E	R	Q	I	K	I	
SpHox6	-	-	S	-	-	V	-	-	-	-	-	F	-	-	-	-	-	-	-	-	-	-	-	-	-	-	-	89
PG8																												
PeHbox6	H	F	N	R	Y	L	T	R	K	R	R	I	E	I	A	Q	S	V	C	L	T	E	R	Q	I	K	I	
SpHox8	-	Y	-	-	-	-	-	-	-	-	-	-	-	-	-	-	A	-	-	-	S	-	-	-	-	-	-	89
PG9+PG10																												
PeHbox4	L	F	N	M	Y	L	T	R	D	R	R	V	D	I	A	R	M	L	N	L	T	E	R	Q	V	K	I	
SpHox9/10	-	-	-	-	-	-	-	-	-	-	-	L	E	-	-	-	L	-	S	-	-	-	-	-	-	-	-	85
PG11+PG12+PG13																												
PeHbox7	Q	A	N	M	Y	L	T	R	D	R	R	S	K	L	S	H	S	L	S	L	S	E	R	Q	I	K	I	
SpHox11/13b	T	T	-	-	-	-	-	-	-	-	-	-	-	-	-	Q	A	-	D	-	T	-	-	-	V	-	-	74
PeHbox15	Q	A	H	Q	Y	L	T	R	D	R	R	A	R	L	A	Q	T	L	S	L	T	E	R	Q	V	K	I	
SpHox11/13b	T	T	N	M	-	-	-	-	-	-	-	S	K	-	S	-	A	-	D	-	-	-	-	-	-	-	-	70
5' END																												
NON-*HOX* HOMEOBOX SEQUENCE																												
PeXlox	H	F	N	K	Y	I	S	R	P	R	R	I	E	L	A	A	M	L	N	L	T	E	R	H	I	K	I	
AM-XlH8	-	-	-	-	-	-	-	-	-	-	-	-	-	-	-	-	-	-	-	-	-	-	-	-	-	-	-	100

Figure 1. Deduced partial homeodomain sequences for *Hox* and *Xlox*-type homeobox sequences, amino acids 21–47 of the homeodomain, from the asteroid *P. exigua*. The sequences are listed in 3′–5′ order in the *Hox* gene cluster according to the findings of Martinez et al. (1999). Echinoid homeobox naming conventions are followed where precedents exist (see 'Materials and methods'). Each *P. exigua* partial Hox homeodomain is aligned with the most similar Hox homeodomain from the echinoid *S. purpuratus* (Martinez et al., 1999), with the percent similarity (%) indicated. Dashes indicate amino acid identity. Paralogy is less clear amongst the more 5′ sequences. The sequence of the non-Hox Xlox-type partial homeodomain of *P. exigua* is aligned with that of the Xlox-type partial homeodomain from the asteroid *A. minor, AM-XlH8* (Mito & Endo, 1997).

are identical with the partial homeodomains of *AM-5* and *AM-6* respectively reported for *A. minor* (Mito & Endo, 1997), while *PeHbox3* is 92% similar to *AM-4* (Mito & Endo, 1997).

The last three *Hox* sequences in 3′–5′ order, *PeHbox4*, *PeHbox7* and *PeHbox15* (Figure 1), probably belong with the most 5′ paralogous groups, based on the sequence similarities of their partial homeodomains. *PeHbox4* appears to be paralogous to *SpHox9/10*, while *PeHbox7* is 74% similar to *SpHox11/13b* (Figure 1). *PeHbox15* is the most divergent, with no unequivocal echinoid paralogue. It does, however, have strong sequence similarity with the partial homeodomain of *AM-9* of *A. minor* (92%) (Mito & Endo, 1997) and *AtHbox15* from the asteroid *Archaster typicus* (96%) (J.-Y. Wu & C.-P. Chen, pers. comm.).

PeXlox is a member of the highly conserved family of *Xlox* genes (Bürglin, 1994). It displays 100% amino acid similarity in the partial homeodomain with *Xlox*-type sequences from an echinoid (Morris et al., 1997), amphioxus (Garcia-Fernandez & Holland, 1994) and an acorn worm (Pendleton et al., 1993). In insects and vertebrates, *Xlox* is an endodermally expressed *Hox*-like gene that is not physically located within the *Hox* cluster (Bürglin, 1994).

Acknowledgements

We would like to thank J-Y Wu and C-P Chen of Academica Sinica, Taiwan for their unpublished *Archaster typicus* sequences, Pedro Martinez for useful discussions and Marianne Frommer and colleagues for providing facilities and technical advice. This work is supported by an Australian Postgraduate Award and a grant from the Australian Research Council.

References

Arenas-Mena, C., P. Martinez, R. A. Cameron & E. H. Davidson, 1998. Expression of the *Hox* gene complex in the indirect development of a sea urchin. Proc. natnl. Acad. Sci. U.S.A. 95: 13062–13067.

Bürglin, T. R., 1994. A comprehensive classification of the homeobox genes. In Duboule, D. (ed.), Guidebook to the Homeobox Genes. Sambrook & Tooze Publications, London: 27–71.

Byrne, M., 1995. Changes in larval morphology in the evolution of benthic development by *Patiriella exigua* (Asteroidea: Asterinidae), a comparison with the larvae of *Patiriella* species with planktonic development. Biol. Bull. 188: 293–305.

Byrne, M., 1996. Viviparity and intragonadal cannibalism in the dimunitive sea stars *Patiriella vivipara* and *P. parvivipara* (family Asterinidae). Mar. Biol. 125: 551–567.

Di Bernardo, M., R. Russo, P. Oliveri, R. Melfi & G. Spinelli, 1994. Expression of homeobox-containing genes in the sea urchin (*Paracentrotus lividus*) embryo. Genetica 94: 141–150.

Dobias, S. L., A. Z. Zhao, H.-Y. Tan, J. R. Bell & R. Maxson, 1996. *SpHbox7*, a new *Abd-B* class homeobox gene from the sea urchin *Strongylocentrotus purpuratus*: insights into the evolution of *Hox* gene expression and function. Devl. Dynam. 207: 450–460.

Dolecki, G. J., S. Wannakrairoj, R. Lum, G. Wang, H. D. Riley, R. Carlos, A. Wang & T. Humphreys, 1986. Stage-specific expression of a homeo box-containing gene in the non-segmented sea urchin embryo. EMBO J. 5: 925–930.

Dolecki, G. J., G. Wang & T. Humphreys, 1988. Stage- and tissue-specific expression of two homeobox genes in sea urchin embryos and adults. Nucl. Acids Res. 16: 11543–11558.

Duboule, D., 1994. Guidebook to the Homeobox Genes. Sambrook & Tooze Publications, London.

Garcia-Fernandez, J. & P. J. Holland, 1994. Archetypal organization of the amphioxus *Hox* gene cluster. Nature 370: 563–566.

Gellon, G. & W. McGinnis, 1998. Shaping animal body plans in development and evolution by modulation of *Hox* expression patterns. BioEssays 20: 116–125.

Hart, M. W., M. Byrne & M. J. Smith, 1997. Molecular phylogenetic analysis of life history evolution in asterinid starfish. Evolution 51: 1848–1861.

Martinez P., J. P. Rast, C. Arenas-Mena & E. H. Davidson, 1999. Organization of an echinoderm *Hox* gene cluster. Proc. natnl. Acad. Sci. U.S.A. 96: 1469–1474.

McGinnis, W. & R. Krumlauf, 1992. Homeobox genes and axial patterning. Cell 68: 283–302.

Mito, T. & K. Endo, 1997. A PCR survey of *Hox* genes in the sea star *Asterina minor*. Mol. Phylogenet. Evol. 8: 218–224.

Morris, V. B., J. Brammall, M. Byrne & M. Frommer, 1997. *Hox*-type and non-*Hox* homeobox gene sequences in genomic DNA of the sea urchin *Holopneustes purpurescens*. Gene 201: 107–110.

Pendleton, J. W., B. K. Nagai, M. T. Murtha & F. H. Ruddle, 1993. Expansion of the *Hox* gene family and the evolution of chordates. Proc. natnl. Acad. Sci. U.S.A. 90: 6300–6304.

Pfeffer, P. L. & C. von Holt, 1991. Stage- and adult tissue-specific expression of a homeobox gene in embryo and adult *Parechinus angulosus* sea urchins. Gene 108: 219–226.

Popodi, E., J. C. Kissinger, M. E. Andrews & R. A. Raff, 1996. Sea urchin Hox genes: insights into the ancestral Hox cluster. Mol. Biol. Evol. 13: 1078–1086.

Sambrook J., E. F. Fritsch E. & T. Maniatis, 1989. Molecular Cloning – a Laboratory Manual (2nd edn). Cold Spring Harbor Laboratory Press, Cold Spring Harbor.

Scott, M. P., 1992. Vertebrate homeobox gene nomenclature. Cell 71: 551–553.

Wang, G. V. L., G. J. Dolecki, R. Carlos & T. Humphreys, 1990. Characterization and expression of two sea urchin homeobox gene sequences. Devl. Genet. 11: 77–87.

Wu, J.-Y., C.-P. Chen, C.-F. Hui, S.-P. L. Hwang & D. J. Miller, 1998. Preliminary studies on *Hox*-type homeoboxes in the echinoids *Stomopneustes variolaris* and *Tripneustes gratilla*. In Mooi, R. & M. Telford (eds), Echinoderms: San Francisco. Balkema, Rotterdam: 893 pp.

Zhao, A. Z., G. Vansant, J. Bell, T. Humphreys & R. Maxson, 1991. Activation of the L1 late H2B histone gene in blastula-stage sea urchin embryos by an *Antennapedia*-class homeoprotein. Mech. Devel. 34: 21–28.

Hydrobiologia **420:** 99–102, 2000.
A. M. Solé-Cava, C. A. M. Russo & J. P. Thorpe (eds), Marine Genetics.

The karyotype of *Nodipecten nodosus* (Bivalvia: Pectinidae)

E. Pauls & P. R. A. M. Affonso
Lab. de Genética em Organismos Aquáticos, Departamento de Zootecnia, Universidade Federal Fluminense, Niterói, RJ, Brazil E-mail: mzokika@vm.uff.br

Key words: Bivalvia, *Nodipecten nodosus*, chromosomes, NOR, C-banding, meiosis

Abstract

Earlier karyotypical work on *Nodipecten nodosus* embryos indicated that this species has a diploid number of 38, with six pairs respectively of metacentric and submetacentric chromosomes and seven pairs of subtelocentric chromosomes, although there were some difficulties in obtaining complete metaphases. The present work provides additional results on specific regions of the chromosomes in *N. nodosus* and, by meiotic studies, confirms the chromosome number with more reliability. Active nucleolar organizer regions (NOR), detected in mitotic metaphases from embryos, can be characterized in *N. nodosus* by a high level of heteromorphism of NOR-sites, indicating that these regions are not appropriate as chromosomal markers in this species. The procedure for detecting constitutive heterochromatin of chromosomes allowed us to observe most of the heterochromatic blocks at a pericentromeric position and some at telomeric and interstitial positions. The analysis of meiotic chromosomes from gonad tissue revealed the presence of 19 bivalents during metaphase I, all homomorphic and isopicnotic, confirming the previously described diploid chromosomal number of 38 for *N. nodosus*. From these results, some evolutionary aspects of the Pectinidae are briefly discussed.

Introduction

Genetic and cytogenetic studies on bivalves have allowed the characterization of species, subspecies and populations, besides the identification of important aspects of variability in this group. These data have provided information on evolution and phylogeny and have been applied to the improvement of bivalves of commercial interest (Beaumont, 1991; Rodriguez-Juiz et al., 1996). However, when compared to the diversity of such animals, these studies seem to be still underrepresented. This is the case in the family Pectinidae, which includes a large number of species (about 350; Wada, 1978). Most scallops are of economic importance and their intensive culture is developing worldwide (Gjedren, 1997), although fewer than 7% of species have been described cytogenetically (Brand et al., 1990), with chromosome banding studies even fewer.

In the present work we analyzed mitotic and meiotic chromosomes of *Nodipecten nodosus* (Pectinidae), a survey initiated by Pauls et al. (1996), with a view to gaining increased genetic information and karyological definition for a native bivalve of aquaculture potential.

Materials and methods

Cytogenetic studies were carried out on embryos of *Nodipecten nodosus* from the Ilha Grande Bay (Rio de Janeiro, Brazil). For this, mature individuals were selected from the broodstock and the release of eggs and spermatozoa was induced by thermal shocks. Fertilization was performed preventing self-fertilization (this species, like most pectinids, is a functional hermaphrodite). Mitotic chromosomes were obtained from embryos 8 h after fertilization (see Pauls et al., 1996). Slides were conventionally stained with 5% Giemsa. Active nucleolar organizer regions (NOR) were detected by silver nitrate (Howell & Black, 1980) and constitutive heterochromatin by C-banding (Sumner, 1972). Meiotic chromosomes were obtained from testicular tissue of mature individuals (Kligermann & Bloom, 1977).

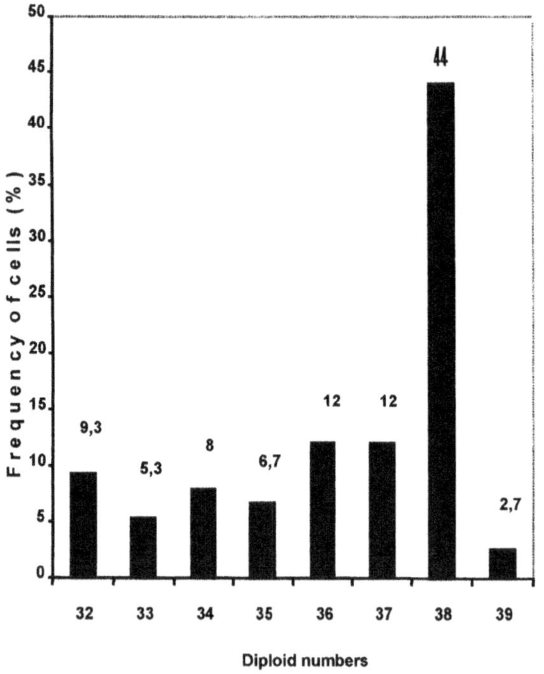

Figure 1. Frequency of chromosome number taken from 75 somatic metaphases of embryos cells of *Nodipecten nodosus*.

Results and discussion

From metaphase analysis, carried out on the 8 h old embryos, it was possible to observe a diploid chromosome number of $2n$=38 in 44% of 75 counted spreads. This number has been suggested as the modal chromosomal number for the species by Pauls et al. (1996). However, a large number of cells with other numbers were found, ranging from 32 to 39 chromosomes (Figure 1).

Normally, when using embryos, chromosomal number variations, particularly reductions are generally considered to be artefactual (Beaumont & Gruffydd, 1974; Thiriot-Quiévreux, 1986). However, it should be considered that the occurrence of aneuploid embryo cells may be a consequence of pollution (Dixon, 1982) or correlated with high levels of homozygosity and growth rate decrease within commercial stocks (Thiriot-Quiévreux et al., 1988; Zouros, 1998). Ilha Grande Bay has suffered the effects of contamination by hydrocarbons and cadmium from a harbor (A.D. Marques, pers. comm.). Otherwise, as *N. nodosus* has recently been collected from the wild and inbreeding has been avoided, problems with homozygosity seem to be unlikely. In addition, aneuploidy in embryos (as compared to adults) could reflect a minor role of natural selection, as pointed out by earlier authors (Longwell, 1976; Dixon, 1982). Further specific studies are required to elucidate whether aneuploidy is present in *N. nodosus* and for what reason(s).

Six pairs of metacentric, six pairs of submetacentric and seven pairs of subtelocentric chromosomes were identified by conventional staining confirming the results of Pauls et al. (1996). Also, we detected the presence of secondary constrictions at an interstitial location in two subtelocentric chromosomes. Similar constrictions were previously found in another pectinid, *Argopecten purpuratus*, by Brand et al. (1990).

Active NOR sites in *N. nodosus* embryos were characterized by a high level of polymorphism in their expression pattern (from two to four marks), the exception being the interstitial site in one chromosomal pair, which probably equated with the location of a secondary constriction revealed using Giemsa stain. This is to be expected, since such constrictions usually include active rDNA regions (Hsu et al., 1975).

Although NORs can be used as a tool for species identification and hence as important genetic markers in several species, this does not seem to be true for all bivalves. Thiriot-Quiévreux & Insua (1992) and Insua & Thiriot-Quiévreux (1991) have detected, by silver nitrate, intraindividual variation of number and size of NOR sites in *Ostrea edulis*, *O. denselamellosa* and *Crassostrea gigas*, while NOR polymorphisms were found between and within individuals of various species of Mytilidae (Dixon et al., 1986; Martinéz-Expósito et al., 1994). As silver nitrate reveals only NORs active during early interphase (Warburton & Henderson, 1979), it is suggested that bivalves may show different levels of gene expression of rDNA among individuals and cells, giving rise to this variable NOR pattern. However, it should be borne in mind that this technique has been applied to very few species of bivalves. Furthermore, Martinéz-Expósito et al. (1994) demonstrated that embryos and young individuals of bivalves have a greater and variable number of active NORs if compared to older ones. This may reflect metabolic requirements and physiological stage and may explain the present result in *N. nodosus*.

In most animals, heterochromatin is detected at the pericentromeric region (Sumner, 1990), and its pattern can be very useful in karyotypical definition and comparison (Martinéz-Lage et al., 1994). C-banding analyses in *O. denselamellosa* (Insua & Thiriot-Quiévreux, 1991), *M. edulis* (Dixon et al.,

1986), *M. galloprovincialis* (Martinéz-Lage et al., 1994) and *Perna perna* (Jacobi et al., 1990) indicate that pericentromeric heterochromatin is not common in bivalves, although telomeric and interstitial heterochromatin are. On the other hand, most chromosomes in *N. nodosus* showed C-positive regions near to centromeres and some revealed terminal and interstitial heterochromatic blocks. Normally, as the NOR sites are associated with constitutive heterochromatin (Dixon et al., 1986), the interstitial heterochromatin is probably located in the same chromosomal pair as the interstitial secondary constrictions and NORs.

The meiotic studies were performed primarily to define the diploid number described for *N. nodosus*, since, as previously discussed, intact mitotic metaphases were not easy to obtain. Analysis of meiotic chromosomes during metaphase I revealed the presence of 19 bivalents, all homomorphic and isopicnotic, thus confirming the chromosome number ($2n$=38) for *N. nodosus*. This appears to be the most common number within pectinids, particularly in the genus *Argopecten* (Beaumont & Gruffydd, 1974; Wada, 1978; Brand et al., 1990).

Our data seem to corroborate the divergence and karyotypic diversity of the Pectinidae from other bivalves of the subclass Pteriomorpha. This conclusion stems from the higher diploid number in most oysters and mussels ($2n\approx20$); the presence of subtelocentric chromosomal pairs (rare in non-pectinid species) and secondary constrictions and pattern of constitutive heterochromatin distribution (see e.g. Thiriot-Quiévreux, 1984; Komaru & Wada, 1985; Brand et al., 1990). However, further studies using different banding techniques, including *in situ* hybridization, are essential for a better understanding of chromosomal evolution and species identification in this quite diverse group of aquatic animals.

Acknowledgements

This research was supported by grants from Universidade Federal Fluminense, Institute for the Ecodevelopment of the Ilha Grande Bay (IED-BIG), Furnas, Petrobrás and Technical and Scientific Development of National Council of Brazil.

References

Beaumont, A. R., 1991. Genetic distances between some scallop species. In Shumway, S. E. & P. A. Sandifer (eds), Scallop Biology and Culture. World Aquaculture Society, Baton Rouge (Louisiana): 151–155.

Beaumont, A. R. & L. D. Gruffydd, 1974. Studies on the chromosomes of the scallop *Pecten maximus* (L.) and related species. J. mar. biol. Ass. U.K. 54: 713–718.

Brand, E. V., G. Bellolio & K. Lohrmann, 1990. Chromosome number of the Chilean scallop *Argopecten purpuratus*. Tohoku J. agric. Res. 40: 91–95.

Dixon, D. R., 1982. Aneuploidy in mussel embryos (*Mytilus edulis* L.) originating from a polluted dock. Mar. Biol. Let. 3: 155–161.

Dixon, D. R., I. R. B. McFadzen & K. Sisley, 1986. Heterochromatic marker regions (nucleolar organizers) in the chromosomes of the common mussel, *Mytilus edulis* (Mollusca: Pelecypoda). J. exp. mar. Biol. Ecol. 97: 205–212.

Gjedren, T., 1997. Selective breeding to improve aquaculture production. World Aquacul. 28: 33–45.

Howell, W. M. & D. A. Black, 1980. Controlled silver staining of nucleolus organizer region with protective colloidal developer: a 1-step method. Experientia 36: 1014–1015.

Hsu, T. C., S. E. Spirito & M. L. Pardue, 1975. Distribution of 18+28S ribosomal genes in mammalian genomes. Chromosoma 53: 25–36.

Insua, A. & C. Thiriot-Quiévreux, 1991. The characterization of *Ostrea denselamellosa* (Mollusca, Bivalvia) chromosomes: karyotype, constitutive heterochromatin and nucleolus organizer regions. Aquaculture 97: 317–325.

Jacobi, C. M., C. Rosenberg & A. M. Vianna-Morgante, 1990. The karyotype of the brown mussel *Perna perna* (L.) (Bivalvia: Mytilidae). Brazil. J. Gen. 13: 669–673.

Kligermann, A. D. & S. E. Bloom, 1977. Rapid chromosome preparations from solid tissues of fishes. J. Fish. Res. Bd. Can. 34: 266–269.

Komaru, A. & K. T. Wada, 1985. Karyotypes of four species in the Pectinidae (Bivalvia, Pteriomorphia). Jap. J. Malac. 44: 249–259.

Longwell, A. C., 1976. Review of genetic and related studies on commercial oysters and other pelecypod mollusks. J. Fish. Res. Bd. Can. 33: 1100–1107.

Martinéz-Expósito, M. J., J. J. Pasantes & J. Méndez, 1994. NOR activity in larval and juvenile mussels (*Mytilus galloprovincialis* Lmk.). J. exp. mar. Biol. Ecol. 175: 155–165.

Martinéz-Lage, A., A. González-Tizón & J. Méndez, 1994. Characterization of different chromatin types in *Mytilus galloprovincialis* L. after C-banding, fluorochrome and restriction endonuclease treatments. Heredity 72: 242–249.

Pauls, E., M. L. Pacheco & M. P. Cabeza, 1996. Cytogenetic characterization of *Nodipecten nodosus* embryos (Bivalvia: Pectinidae) from South Atlantic Coast of Brazil. Arch. Zootecnia 45: 225–228.

Rodriguez-Juiz, A., M. Torrado & J. Méndez, 1996. Genome-size variation in bivalve mollusks determined by flow cytometry. Mar. Biol. 126: 489–497.

Sumner, A. T., 1972. A simple technique for demonstrating centromeric heterochromatin. Exp. Cell Res. 75: 304–306.

Sumner, A. T., 1990. Chromosome banding. Unwin Hyman, London.

Thiriot-Quiévreux, C. 1984. Analyse comparée des caryotypes d'Ostreidae (Bivalvia). Cah. Biol. mar. 25: 407–418.

Thiriot-Quiévreux, C. 1986. Etude de l'aneuploïdie dans différents naissains d'Ostreidae (Bivalvia). Genetica 70: 225–231.

Thiriot-Quiévreux, C. & A. Insua, 1992. Nucleolar organizer region variation in the chromosomes of three oysters species. J. exp. mar. Biol. Ecol. 157: 33–40.

Thiriot-Quiévreux, C., T. Noël, S. Bougrier & S. Dallot, 1988. Relationships between aneuploidy and growth rate in pair matings of the oyster *Crassostrea gigas*. Aquaculture 75: 89–96.

Wada, K., 1978. Chromosome karyotypes of three bivalves: the oysters, *Isognomon alatus* and *Pinctada imbricata* and the Bay scallop *Argopecten irradians irradians*. Biol. Bull. 155: 235–245.

Warburton, D. & A. S. Henderson, 1979. Sequential silver staining and hybridization *in situ* on nucleolus organizing regions in human cells. Cytogen. Cell Gen. 24: 168–175.

Zouros, E., 1998. Gene-dose as the underlying cause of variation in growth rate in wild and cultured populations of bivalves. In Aquaculture'98 – book of abstracts. Las Vegas, World Aquaculture Society: 602 pp.

Hydrobiologia **420:** 103–118, 2000.
A. M. Solé-Cava, C. A. M. Russo & J. P. Thorpe (eds), Marine Genetics.

Taxonomic status of the '*Patella miniata*' species complex (Mollusca: Gastropoda) in southern Africa

T. M. Ridgway[1,4], G. M. Branch[1], B. A. Stewart[2] & A. N. Hodgson[3]
[1]*Marine Biology Research Institute, Zoology Department, University of Cape Town, Rondebosch 7701, South Africa*
[2]*Department of Zoology, University of Stellenbosch, Private Bag X1, Matieland 7602, South Africa*
[3]*Department of Zoology and Entomology, Rhodes University, Grahamstown 6140, South Africa*
[4]*Present address: The Centre for Marine Studies, University of Queensland, Brisbane, QLD 4072, Australia*
E-mail: tridgway@bio.usyd.edu.au

Key words: patellid, limpet, allozymes, genetics, cladistics, morphology

Abstract

The species boundaries between members of the southern African '*Patella miniata*' species complex (*Patella miniata miniata*, *P. miniata sanguinans* and *P. compressa*) are poorly defined on morphological grounds. Doubt also exists whether all populations of *P. miniata miniata* constitute a single taxon, with '*P.* cf. *miniata*' having uncertain affinities. The status of *P. adansonii* (known only from shells) is also obscure. Protein gel electrophoresis was employed to resolve these taxonomic uncertainties. In addition, a population of *P. safiana* from southern Angola was included in the study as a reference population. Electrophoretic analysis of 16 enzyme loci revealed four distinctive groupings. All populations of *P. miniata miniata* and *P.* cf. *miniata* proved closely related and separated from *P. compressa* populations at *I*=0.85; the *P. miniata sanguinans* populations clustered out at *I*=0.79; and *P. safiana* separated at *I*=0.69. Phylogenetic analyses, using *P. safiana* as an outgroup, indicated that *P. miniata miniata*, *P. miniata sanguinans* and *P. compressa* form a closely related monophyletic group in which *P. miniata miniata* and *P. compressa* are more closely related to each other than either one of them is to *P. miniata sanguinans*. Morphological analyses (shells, radula, sperm microstructure) revealed differences that mirrored the differences detected by the allozymes. Thus, it was concluded that *P. miniata sanguinans* and *P. miniata miniata* warrant full specific status and we therefore raise the former to *P. sanguinans*. *Patella miniata miniata* and *P.* cf. *miniata* are geographic variants of the same species, *P. miniata*, and *P. adansonii* is sunk within this species. Furthermore, the suggestion that *P. miniata* and *P. compressa* may be ecomorphs of the same species are not supported and both warrant specific status. Thus, the '*Patella miniata*' species complex in southern Africa is a closely related monophyletic group of three species, with *P. miniata* and *P. compressa* being more closely related to each other than either one is to *P. sanguinans*.

Introduction

The genus *Patella* contains some of the most common intertidal mollusc species on rocky shores. As a consequence of their abundance, availability and ecological significance, much is known about the biology and behaviour of this genus (see Branch, 1981), yet the systematics of the genus *Patella* is less well known. Consensus has yet to be reached on the number of species within the genus *Patella*, with the two most recent revisions (Christiaens, 1973; Powell, 1973) differing by as many as eight species. Moreover, a recent review of the world's patellids argues that the genus *Patella* is polyphyletic, comprising three clades (*Patella*, *Cymbula* and *Scutellastra*) in addition to *Helcion* (S. Ridgway et al., 1998). We support this, but have retained the older broader use of the name *Patella sensu lato* here to facilitate comparisons of species known until now under this name.

The problems of limpet taxonomy are largely attributable to the considerable morphological plasticity of individuals (Côrte-Real et al., 1996a) and the fact

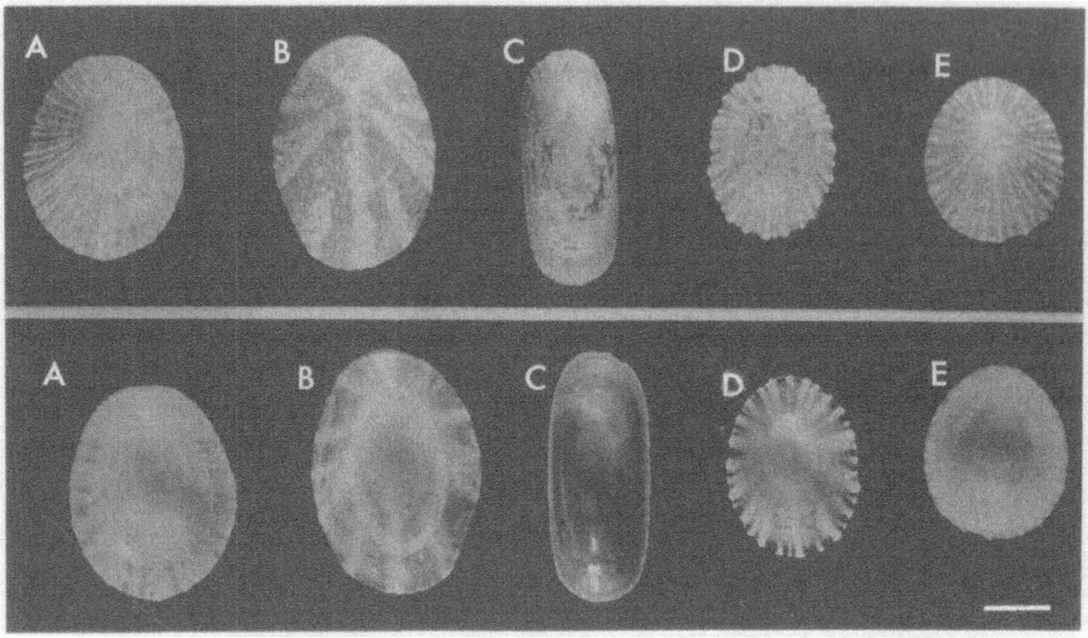

Figure 1. Dorsal and ventral views of the shells of (a) *P. miniata miniata*, (b) *P. miniata sanguinans*, (c), *P. compressa*, (d) *P. safiana* and (e) the type specimen of *P. adansonii*. Scale bar: 17.5 mm.

that early taxonomy was based purely on morphological characters, with shell form and radular structure being the most commonly used characters. Nevertheless, despite the confusion surrounding the old genus *Patella*, in comparison with other regions, the southern African representatives of the group are fairly well characterised. There remains, however, some taxonomic uncertainty for a suite of closely related species within the *Cymbula* clade in southern Africa. In their cladistic phylogeny of the family Patellidae, S. Ridgway et al. (1998) recognise *Patella miniata* Born, 1778, *P. compressa* Linnaeus, 1758, and *P. adansonii* Dunker, 1853, as members of a monophyletic southern African clade (Figure 1).

Patella miniata, although showing marked colour gradations around the southern African coast (Koch, 1949), is currently recognised as a valid species on South African shores. *Patella sanguinans* Reeve, 1854, on the other hand, was recognised as a valid species (Bartsch, 1915; Tomlin & Stephenson, 1942), but since then it has been dismissed and included as a synonym of *P. miniata* (Koch, 1949; Christiaens, 1973; Powell, 1973). However, Kilburn & Rippey (1982) and Branch et al. (1994), although not recognising *P. sanguinans* as a full species, divide *P. miniata* into two subspecies, *P. miniata miniata* and *P. miniata sanguinans* (Figure 1). Nevertheless, S. Ridgway et al. (1998) exclude the possibility of *P. sanguinans* as a species, since they consider it identical to *P. miniata* in the morphological characters examined in their cladistic study, differing only in the size and coloration of the shell.

The west coast variants of *P. miniata* have been given various names, including *P. safiana* Lamarck, 1819 (R. Kilburn, pers. comm.) and *P. adansonii* (S. Ridgway et al., 1998) (Figure 1). We refer to them here as *P.* cf. *miniata*. Thus, the status of these west-coast individuals also need resolution.

The kelp limpet, *P. compressa*, is usually recognised as a valid species on South African shores, but several authors have suggested that *P. miniata* and *P. compressa* may be ecomorphs of the same species (living on rocks and kelp respectively) (Pilsbry, 1891; Kilburn & Rippey, 1982; S. Ridgway et al., 1998).

The aim of this paper is to examine the species boundaries between *P.* cf. *miniata*, *P. miniata miniata*, *P. miniata sanguinans* and *P. compressa*, and thus provide a revision of the taxonomic status of the '*Patella miniata*' 'clade' (S. Ridgway et al., 1998) in southern Africa. A further species, *P. safiana* Lamarck, 1819, from Angola, which according to S. Ridgway et al. (1998) is closely related to *P. miniata miniata* and *P. compressa*, is included in the study to provide a suitable reference population. Both

Table 1. Enzymes, locus abbreviations, buffer systems and tissue types used. See text for details of buffers

Enzyme (abbreviation)	E.C. number	Locus	Buffer system	Tissue
Arginine kinase (*Ark*)	2.7.3.3	*Ark-2*	(1)	Foot
Glucose-6-phosphate (*Gpi*)	5.3.1.9	*Gpi-1*	(1)	Foot
Sorbitol dehydrogenase (*Sdh*)	1.1.1.14	*Sdh-1*	(1)	Foot
Superoxide dismutase (*Sod*)	1.15.1.1	*Sod-1*	(2)	Mantle
Malic enzyme (*Me*)	1.1.1.40	*Me-1*	(2)	Mantle
		Me-2	(2)	Mantle
Peptidase - Glycyl-leucine (*Gl*) as substrate	3.4.-.-	*Gl-2*	(2)	Foot
Peptidase - Leucyl-glycyl-glycine (*Lgg*) as substrate	3.4.-.-	*Lgg-1*	(2)	Foot
Peptidase - Phenylalanine-proline (*Php*) as substrate	3.4.-.-	*Php-1*	(2)	Foot
Hexokinase (*Hex*)	2.7.1.1	*Hex-1*	(2)	Foot
Malate dehydrogenase (*Mdh*)	1.1.1.37	*Mdh-1*	(3)	Mantle
Isocitrate dehydrogenase (*Idh*)	1.1.1.42	*Idh-1*	(3)	Mantle
		Idh-2	(3)	Mantle
Aspartate amino transferase (*Got*)	2.6.1.1	*Got-1*	(3)	Mantle
Phosphoglucomutase (*Pgm*)	2.7.5.1	*Pgm-1*	(4)	Foot
Aldolase (*Ald*)	4.1.2.13	*Ald-1*	(4)	Foot

morphological and biochemical techniques are integrated, as a combined approach has helped to elucidate the taxonomic status of a variety of molluscan species (Skibinski et al., 1980; Hedgecock & Okazaki, 1984; Heller & Dempster, 1991; Cretella et al., 1994; Côrte-Real et al., 1996a,b). Thus, soft-part morphology, shell morphometrics, radular morphology and sperm microstructure, as well as protein electrophoretic patterns were examined to assess the taxonomic status and relationships of the species in question. *Patella adansonii* is known only from shells, so that an integrated approach could not be applied to it, but the type specimens were examined to resolve its status.

Materials and methods

Collection

Specimens of *Patella miniata miniata* (abbreviated herein as *min*), *P. miniata sanguinans* (*sang*), *P. adansonii* (here referred to as *P.* cf. *miniata* (cf. *min*)) and *P. compressa* (*comp*) were collected from South African and Namibian waters, and *P. safiana* (*safi*) from southern Angola (see Figure 2). Foot and mantle tissues were generally dissected from each individual and stored in liquid nitrogen. Upon returning to the laboratory, the dissected tissue samples were transferred to an ultra deep-freeze (−80 °C) until required for genetic analysis. The Kommetije and Clovelly samples were transported live to the laboratory, where they were placed directly in the ultra deep-freeze. Fifteen animals from each site, together with the remains of the dissected animals, were preserved in 70% alcohol.

Electrophoretic analyses

Genetic variation at 16 allozyme loci was examined using standard horizontal starch gel electrophoresis (see Harris & Hopkinson, 1976). Foot and mantle tissue samples were roughly minced using a scalpel and then homogenised in 0.01 M Tris buffer (pH 8.0) using a glass rod attached to a variable speed motor. Prior to electrophoresis, the homogenate was centrifuged at 2500 **g** for 5 min. Gels were prepared with 12.5% hydrolysed potato starch. The buffer systems used were: 1. Tris-citrate-lithium hydroxide-borate buffer, pH 8.0 (Ridgeway et al., 1970), 2. Tris-borate-EDTA buffer, pH 8.6 (Markert & Faulhaber, 1965), 3. Tris-citrate buffer, pH 6.9 (Whitt, 1970), and 4. Tris-citrate buffer, pH 8.0 (Ward & Beardmore, 1977). Gels were run for 3.5–4.5 h at a constant current of 40 mA. Gels were sliced and the staining methods followed those of Shaw & Prasad (1970) and Harris & Hopkinson (1976). The enzymes assayed, the buffer systems and the tissues used are listed in Table 1.

The electrophoretic data were analysed using the programme BIOSYS-1 Release 1.7 (Swofford & Selander, 1981). Allele frequencies and levels of polymorphism and heterozygosity were calculated for each population. Deviations of genotype frequencies from Hardy-Weinberg equilibrium were measured using Levene's (1949) test, which makes allowance for small sample sizes. Average heterozygosity (H) per

Table 2. Patella cf. *miniata* (cf. *min*), *P. miniata miniata* (*min*), *P. miniata sanguinans* (*sang*), *P. compressa* (comp) and *P. safiana* (*safi*). Distribution of allele frequencies at 11 loci in 13 populations. (*N*)=sample size

Locus	Swakopmund (cf. *min*)	Groen River (cf. *min*)	Kommetjie (*min*)	Clovelly (*min*)	Still Bay (*min*)	Knysna (*min*)	St Francis Bay (*min*)	Dwesa (*min*)	Dwesa (*sang*)	Ballito (*sang*)	Kommetjie (*comp*)	Clovelly (*comp*)	Angola (*safi*)
Gpi-1													
(*N*)	20	16	12	25	19	12	20	8	28	9	20	20	25
A	0.300	0.781	0.667	0.600	0.789	0.542	0.525	0.938	0.839	0.833	0.925	1.00	0.060
B	0.700	0.219	0.333	0.400	0.211	0.458	0.475	0.063	0.161	0.467	–	–	0.940
C	–	–	–	–	–	–	–	–	–	–	0.075	–	–
Pgm-1													
(*N*)	20	16	12	25	19	12	20	8	28	9	18	20	25
A	0.850	0.656	0.667	0.780	1.000	0.917	0.725	0.625	0.339	0.444	0.806	0.750	0.820
B	0.150	0.313	0.333	0.040	–	0.042	0.050	0.250	0.018	–	0.083	–	–
C	–	0.031	–	0.180	–	0.042	0.225	0.125	0.643	0.556	0.111	0.250	0.180
Gl-2													
(*N*)	20	16	12	25	19	12	20	8	28	9	20	20	25
A	0.275	0.500	0.250	0.420	0.763	0.458	0.500	0.188	–	–	0.500	0.200	–
B	0.075	0.031	0.042	0.300	0.079	0.458	0.400	0.250	–	–	0.075	0.100	–
C	0.650	0.469	0.708	0.280	0.158	0.083	0.100	0.563	1.000	1.000	0.425	0.700	–
D	–	–	–	–	–	–	–	–	–	–	–	–	1.000
Lgg-1													
(*N*)	20	16	12	25	19	12	20	8	28	9	20	20	25
A	1.000	1.000	1.000	1.000	1.000	1.000	0.875	1.000	0.179	0.833	0.950	0.925	–
B	–	–	–	–	–	–	0.100	–	0.821	0.167	0.025	0.050	–
C	–	–	–	–	–	–	0.025	–	–	–	0.025	0.025	–
D	–	–	–	–	–	–	–	–	–	–	–	–	1.000
Me-1													
(*N*)	20	16	12	25	19	12	20	8	28	9	20	20	25
A	0.950	0.875	0.625	0.780	0.526	0.750	0.500	1.000	1.000	1.000	0.550	0.500	1.000
B	0.050	0.125	0.375	0.220	0.474	0.250	0.500	–	–	–	0.450	0.500	–
Got-1													
(*N*)	20	16	12	25	19	12	20	8	28	9	20	20	25
A	1.000	1.000	1.000	0.920	0.921	0.917	0.950	0.625	0.036	–	0.100	–	–
B	–	–	–	0.080	0.079	0.083	0.050	0.375	0.964	1.000	0.900	1.000	–
C	–	–	–	–	–	–	–	–	–	–	–	–	1.000
Idh-1													
(*N*)	20	16	12	25	19	12	20	8	28	9	20	20	25
A	0.925	0.969	1.000	0.960	0.947	1.000	1.000	1.000	0.946	1.000	0.925	0.925	–
B	–	0.031	–	–	–	–	–	–	0.036	–	–	0.025	–
C	0.075	–	–	0.040	0.053	–	–	–	0.018	–	0.075	0.050	–
D	–	–	–	–	–	–	–	–	–	–	–	–	1.000
Mdh-1													
(*N*)	20	16	12	25	19	12	20	8	28	9	20	20	25
A	0.900	0.969	0.958	0.980	0.947	1.000	1.000	0.938	0.929	0.722	1.000	0.900	1.000
B	0.100	0.031	0.042	0.020	0.053	–	–	0.063	0.071	0.278	–	0.100	–

Continued on p. 107

Table 2. Continued

Locus	Swakopmund (cf. *min*)	Groen River (cf. *min*)	Kommetjie (*min*)	Clovelly (*min*)	Still Bay (*min*)	Knysna (*min*)	St Francis Bay (*min*)	Dwesa (*min*)	Dwesa (*sang*)	Ballito (*sang*)	Kommetjie (*comp*)	Clovelly (*comp*)	Angola (*safi*)
Mdh-2													
(*N*)	20	16	12	25	19	12	20	8	28	9	20	20	25
A	1.000	1.000	0.917	1.000	1.000	1.000	1.000	1.000	1.000	1.000	1.000	1.000	1.000
B	–	–	0.083	–	–	–	–	–	–	–	–	–	–
Ald-1													
(*N*)	20	16	12	25	19	12	20	8	28	9	20	20	25
A	1.000	1.000	1.000	1.000	1.000	1.000	1.000	1.000	–	–	1.000	1.000	1.000
B	–	–	–	–	–	–	–	–	1.000	1.000	–	–	–
Sod-1													
(*N*)	20	16	12	25	19	12	20	8	28	9	20	20	25
A	1.000	1.000	1.000	1.000	1.000	1.000	1.000	1.000	1.000	1.000	–	–	1.000
B	–	–	–	–	–	–	–	–	–	–	1.000	1.000	–

locus for each population were calculated using Nei's (1978) unbiased estimator. The percentage of polymorphic loci in each population was determined, with a locus being considered polymorphic if the frequency of the most common allele did not exceed 0.99. Allele frequencies were used to calculate a matrix of Nei's (1978) unbiased genetic identity (*I*) and genetic distance (*D*), and these were used to construct a dendrogram using the UPGMA clustering algorithm (Sokal & Sneath, 1973).

The taxonomic relationships among the species was assessed by a phylogenetic analysis carried out on the allozyme data. We used two methods to code the electrophoretic data. The first was the *independent alleles* model (Mickevich & Mitter, 1981), treating the allele as the character, with the states being either present or absent. This method has been criticised (Buth, 1984, Swofford & Olsen, 1990), because the alleles at a single locus are not independent and because hypothetical ancestors may contain no alleles for some loci. Thus, we also used an alternative approach, with the locus as the character, and the alleles or allelic combinations as the states. We ordered states using the *minimum allele turnover* model (Mickevich & Mitter, 1983).

According to S. Ridgway et al. (1998), *P. safiana* is closely related to *P. miniata* and *P. compressa*, and it therefore provides a suitable outgroup taxon. All the cladistic analyses were performed using the implicit enumeration ('ie') command in the HENNIG86 programme (Farris, 1988). The consistency (CI) and retention (RI) indices were used to measure the fit of the data to the phylogenetic trees. All characters used were unweighted.

Morphometric analysis

Quantitative measurements of shell length (SL, greatest distance between anterior and posterior end), shell width (SW, greatest distance perpendicular to the anterior-posterior axis), shell height (SH, greatest vertical distance from the apex of the shell to the plane of the aperture) of the shells of 231 individuals from the 13 sample sites were measured to the nearest 0.05 mm using vernier callipers. In addition, shell dry weight (WT) was measured and the internal shell volume determined from the weight of 70% alcohol which filled an upturned shell. The morphometric measurements were analysed by Principal Component Analysis (PCA) (Statistica for Windows Release 5.1, Statsoft Inc., 1996) to test whether there was any visible separation of the 13 populations based on shell morphometrics alone. The PCA was based on log-transformed variables and the significance of the separation of the groups was tested by one-way analysis of variance (ANOVA) of the overall principal-component scores (Statistica for Windows Release 5.1, Statsoft Inc., 1996).

The shell morphometric data from the 13 populations used in the PCA were further analysed by discriminant function analysis. The data were grouped by genetic characters, log transformed and analysed in Statistica for Windows Release 5.1, Statsoft Inc. (1996). A set of discriminant function coefficients was

Table 3. *Patella* cf. *miniata* (cf. *min*), *P. miniata miniata* (*min*), *P. miniata sanguinans* (*sang*), *P. compressa* (*comp*) and *P. safiana* (*safi*). Matrix of Nei's (1978) genetic identity (above diagonal) and genetic distance (below diagonal) averaged over 16 loci in 13 populations

Population	1	2	3	4	5	6	7	8	9	10	11	12	13
1. Swakopmund (cf. *min*)	*****	0.981	0.984	0.985	0.955	0.977	0.964	0.962	0.771	0.817	0.828	0.810	0.750
2. Groen River (cf. *min*)	0.019	*****	0.995	0.992	0.981	0.982	0.975	0.987	0.783	0.825	0.862	0.838	0.714
3. Kommetjie (*min*)	0.016	0.005	*****	0.987	0.974	0.976	0.976	0.987	0.783	0.825	0.862	0.838	0.714
4. Clovelly (*min*)	0.016	0.008	0.013	*****	0.987	1.000	0.995	0.980	0.781	0.824	0.866	0.844	0.738
5. Still Bay (*min*)	0.046	0.019	0.026	0.013	*****	0.987	0.987	0.956	0.743	0.789	0.876	0.844	0.704
6. Knysna (*min*)	0.023	0.018	0.025	0.000	0.013	*****	0.997	0.968	0.755	0.798	0.860	0.831	0.738
7. St Francis Bay (*min*)	0.036	0.026	0.025	0.006	0.013	0.003	*****	0.953	0.758	0.790	0.856	0.830	0.728
8. Dwesa (*min*)	0.038	0.013	0.022	0.020	0.045	0.032	0.048	*****	0.839	0.884	0.892	0.884	0.715
9. Dwesa (*sang*)	0.260	0.244	0.246	0.248	0.297	0.282	0.277	0.175	*****	0.970	0.766	0.788	0.626
10. Ballito (*sang*)	0.202	0.192	0.192	0.194	0.237	0.225	0.236	0.123	0.030	*****	0.809	0.831	0.631
11. Kommetjie (*comp*)	0.189	0.148	0.153	0.143	0.133	0.151	0.156	0.114	0.267	0.212	*****	0.994	0.629
12. Clovelly (*comp*)	0.210	0.177	0.168	0.170	0.169	0.185	0.186	0.123	0.238	0.185	0.006	*****	0.614
13. Angola (*safi*)	0.288	0.337	0.347	0.304	0.351	0.303	0.317	0.335	0.461	0.469	0.464	0.488	*****

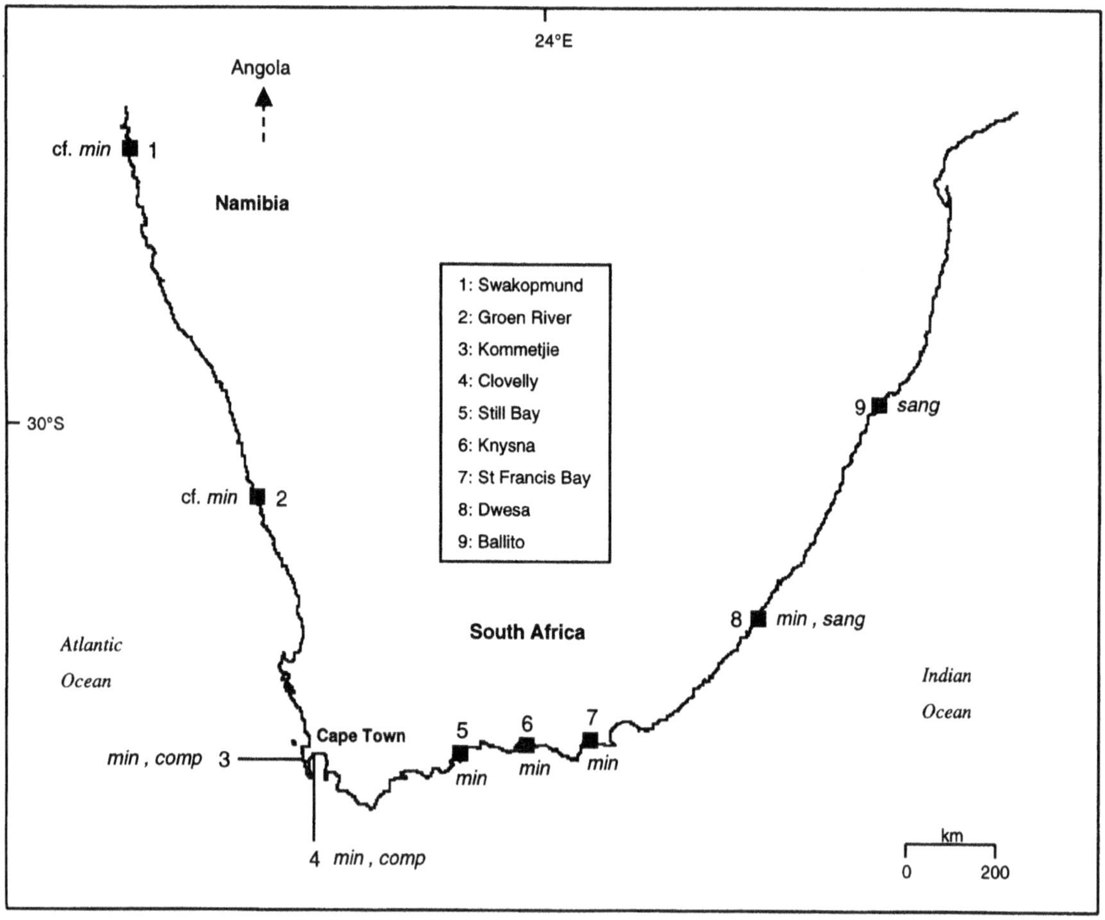

Figure 2. Map of southern Africa showing the locations of 12 of the 13 sites (■) at which limpets were sampled. See 'Methods' for species abbreviations.

Table 4. Data matrix for using the allele as the character, showing the presence or absence of the characters foor each species. All autapomorphs have been removed. See 'Methods' for species abbreviations

Locus/allele		Character number	*min*	*sang*	*comp*	*safi*
Gl-2	A	1	1	0	1	0
	B	2	1	0	1	0
	C	3	1	1	1	0
	D	4	0	0	0	1
Got-1	A	5	1	1	1	0
	B	6	1	1	1	0
	C	7	0	0	0	1
Idh-1	A	8	1	1	1	0
	B	9	1	1	1	0
	C	10	1	1	1	0
	D	11	0	0	0	1
Lgg-1	A	12	1	1	1	0
	B	13	1	1	1	0
	C	14	1	0	1	0
	D	15	0	0	0	1
Mdh-1	B	16	1	1	1	0
Me-1	B	17	1	0	1	0
Pgm-1	B	18	1	1	1	0

calculated whereby a specimen is assigned to one of the groups on which the analysis was based (Côrte-Real et al., 1996a). Group differences were evaluated by the proportion of correctly re-classified specimens in each group. A two-dimensional plot of the individuals was composed based on the first two canonical (discriminant) variables.

Morphological analyses

Shell and soft-part morphology

Visual observations were made on the external and internal appearance of the shell, as well as the pigmentation of the head, tentacles, underside and side of the foot.

Radular morphology

Radulae were dissected from five individuals from each of the populations. Due to abrasion and variation

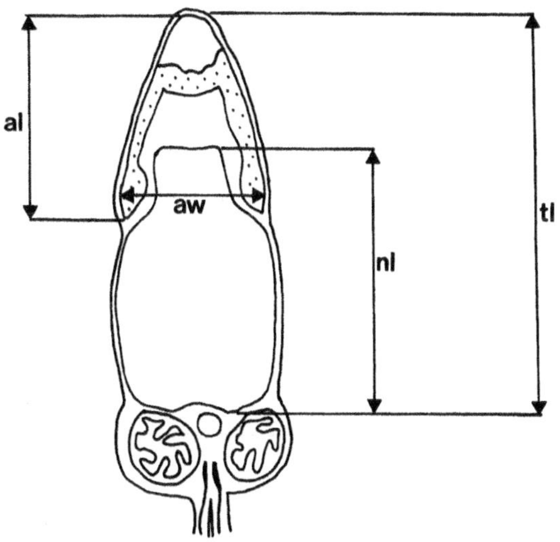

Figure 3. Diagrammatic representation of a mid-longitudinal section through a sperm, showing the dimensions measured. a, acrosome; al, length of acrosome; aw, basal width of acrosome; n, nucleus; nl, length of nucleus; tl, total length of nucleus and acrosome.

in the degree of mineralisation, only the anterior midsection of the radula was examined. The extracted radulae were processed following the methods outlined in T. Ridgway et al. (1998) and examined with a Cambridge S-200 Electron microscope. For each radula, the presence or absence of the rachidian tooth, and the number and arrangement of the lateral and marginal teeth were recorded.

Sperm microstructure

Sperm was examined from *P. miniata miniata* from Clovelly and St Francis Bay, from *P. miniata sanguinans* from Dwesa and Ballito, from *P.* cf. *miniata* from Groen River and Swakopmund, and from *P. compressa* from Kommetjie and Clovelly, following the methods in T. Ridgway et al. (1998). Using the Image Measuring System (IMS) of the JEOL 1210 TEM, the measurements shown in Figure 3 were taken from 10 mid-longitudinal sections per individual per population from the TEM images: 1. total length of the nucleus and acrosome, 2. length of the nucleus, 3. length of the acrosome, and 4. basal width of the acrosome. Differences between the populations were tested using nested ANOVA and Tukey's honestly significant difference test.

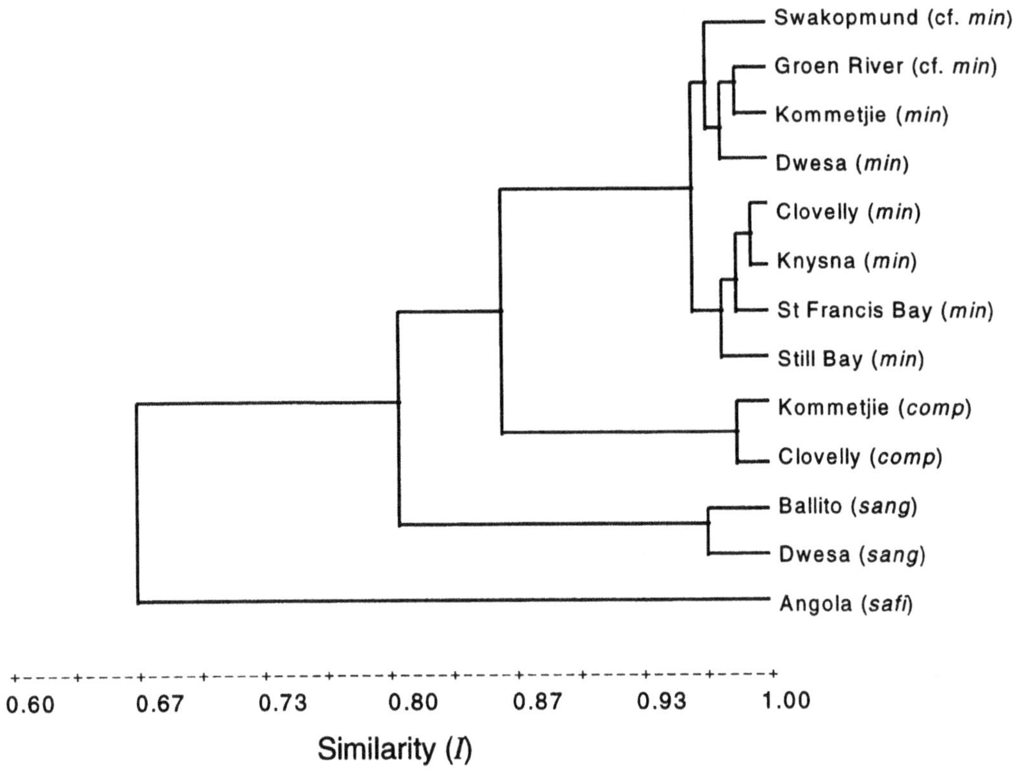

Figure 4. Patella cf. *miniata, P. miniata miniata, P. miniata sanguinans, P. compressa* and *P. safiana*. UPGMA dendrogram derived from Nei's (1978) genetic identity based on 16 loci. See 'Methods' for species abbreviations.

Table 5. Data matrix for using the locus as the character, with the states ordered using the *minimum turnover model*, showing the character states for each species. See 'Methods' for species abbreviations

Locus	Character number	*min*	*sang*	*comp*	*safi*
Gl-2	1	2	1	2	0
Got-1	2	1	1	1	0
Idh-1	3	1	1	1	0
Lgg-1	4	2	1	2	0
Mdh-1	5	1	1	1	0
Me-1	6	1	0	1	0
Pgm-1	7	1	1	1	0

Results

Electrophoretic analyses

Five loci were consistently monomorphic across all populations (*Ark*-2, *Sdh*-1, *Hex*-1, *Me*-2, *Idh*-2). The allele frequencies for the nine polymorphic loci and the two loci showing fixed allele differences are shown in Table 2. The locus *Ald*-1 was fixed for allele *Ald*-1^B for *P. miniata sanguinans*, and for allele *Ald*-1^A in the other taxa. The locus *Sod*-1 was fixed for allele *Sod*-1^B for *P. compressa* and for allele *Sod*-1^A for the other taxa. *P. miniata sanguinans* was fixed for allele *Gl*-2^C, and *P. safiana* was uniquely fixed for alleles *Gl*-2^D, *Lgg*-1^D, *Idh*-1^D and *Got*-1^C. The number of alleles per polymorphic locus ranged from two (*Me*-1, *Mdh*-1, *Mdh*-2) to four (*Gl*-2, *Lgg*-1, *Idh*-1), but none of these loci was consistently polymorphic in every population.

Tests for Hardy-Weinberg equilibrium (using Levene's (1949) correction for small sample size) indicated that of the 72 cases of polymorphism encountered at all loci in all populations, 25 (34.7%) exhibited significant deviations (χ^2, $P<0.05$). When values of genetic variation were averaged over species, *P. compressa* was the most polymorphic taxon for the loci studied (40.6% of loci studied were polymorphic), followed by *P.* cf. *miniata* 37.5%, *P. miniata miniata* 36.5%, *P. miniata sanguinans* 31.3%, with *P. safiana* showing the lowest value of 12.5%.

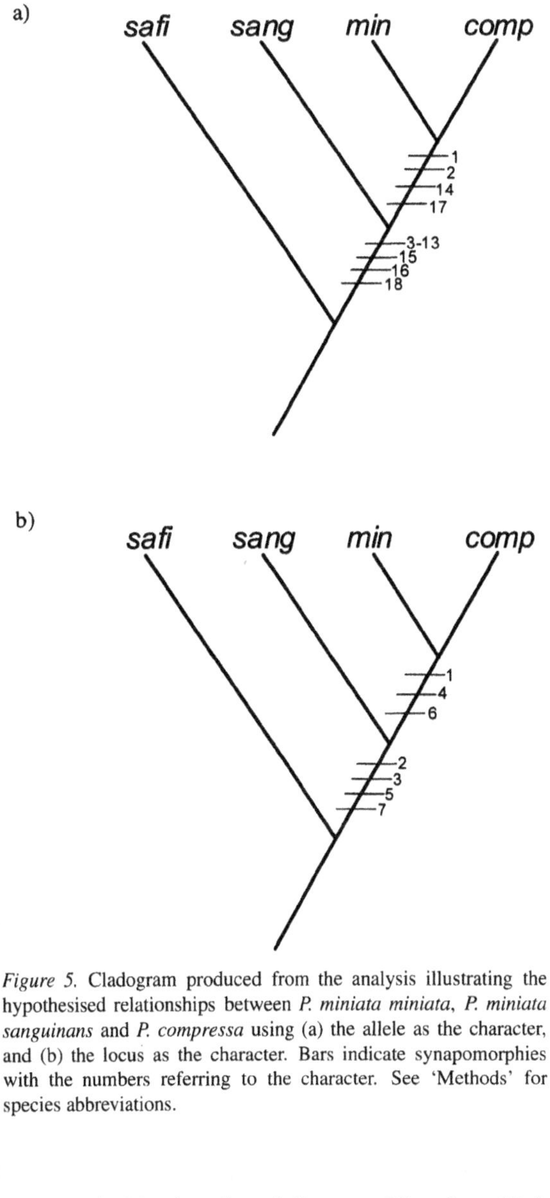

Figure 5. Cladogram produced from the analysis illustrating the hypothesised relationships between *P. miniata miniata*, *P. miniata sanguinans* and *P. compressa* using (a) the allele as the character, and (b) the locus as the character. Bars indicate synapomorphies with the numbers referring to the character. See 'Methods' for species abbreviations.

Genetic identity (*I*) and distance (*D*) values (Nei, 1978) were calculated for each pairwise population comparison (Table 3). The cluster analysis (Figure 4) based on Nei's (1978) *I* revealed four distinctive groupings. The populations of *P.* cf. *miniata* and *P. miniata miniata* clustered together and had a mean intraspecific identity value of 0.979. The *P. compressa* populations yielded an intraspecific *I* value of 0.994, whereas the *P. miniata sanguinans* populations clustered together at a mean intraspecific *I* value of 0.970. *Patella safiana* clustered out by itself to form the fourth grouping. The *P. compressa* group clustered from the *P. miniata miniata* group at an *I* value of 0.851, and the *P. miniata sanguinans* group clustered out at an *I* value=0.797. *Patella safiana* was the most differentiated and separated at an *I* value=0.691.

Using the allele as the character in the phylogenetic analysis, 18 alleles were synapomorphic and were thus phylogenetically informative. Table 4 shows the distribution of these characters among the species. *Patella miniata miniata* and *P.* cf. *miniata* clustered together in Figure 4 at a level indicating conspecificity, therefore *P.* cf. *miniata* is lumped under *P. miniata miniata* in the phylogenetic analysis. The analysis using the 'ie' command in the HENNIG86 programme yielded a single cladogram (Figure 5a), with a CI of 1.00 and RI of 1.00. The characters 3–13, 15, 16, and 18 (Table 4) define *P. miniata miniata*, *P. miniata sanguinans* and *P. compressa* into a monophyletic group. It is evident that *P. compressa* and *P. miniata miniata* are more closely related to each other, than either of them is to *P. miniata sanguinans*. The close relationship between *P. compressa* and *P. miniata miniata* is supported by the characters 1, 2, 14 and 17 (Table 4).

Using the locus as the character, only seven loci yielded phylogenetically informative characters. The data matrix of the allozyme character states are shown in Table 5. The analysis using the 'ie' command yielded a single cladogram (Figure 5b), with a CI of 1.00 and RI of 1.00. The cladogram obtained in Figure 5b is identical to that in Figure 5a. Thus, the cladogram once again shows that *P. miniata miniata*, *P. miniata sanguinans* and *P. compressa* form a monophyletic group, supported by the synapomorphies in characters 2, 3, 5 and 7 (Table 5). The close relationship between *P. compressa* and *P. miniata miniata* are supported by the characters 1, 4 and 6 (Table 5).

Morphometric analysis

Table 6 shows the results obtained from the principal component analysis (PCA). Principal component 1 (PC1) accounted for 81% of the total variation, with the first eigenvector being equally influenced by shell length, shell width, shell height, shell dry weight and shell volume. The equal loading of PC1 suggests that it approximates an isometric size vector. Principal component 2 (PC2) was influenced mainly by shell height, PC3 by shell length, PC4 by shell width and PC5 by shell volume. Principal component 2 accounted for about 14% of the total variation, whereas PC3, PC4 and PC5 collectively accounted for about only 5% of the total variation. The scatterplot of PC2 against PC1 of the populations grouped into species

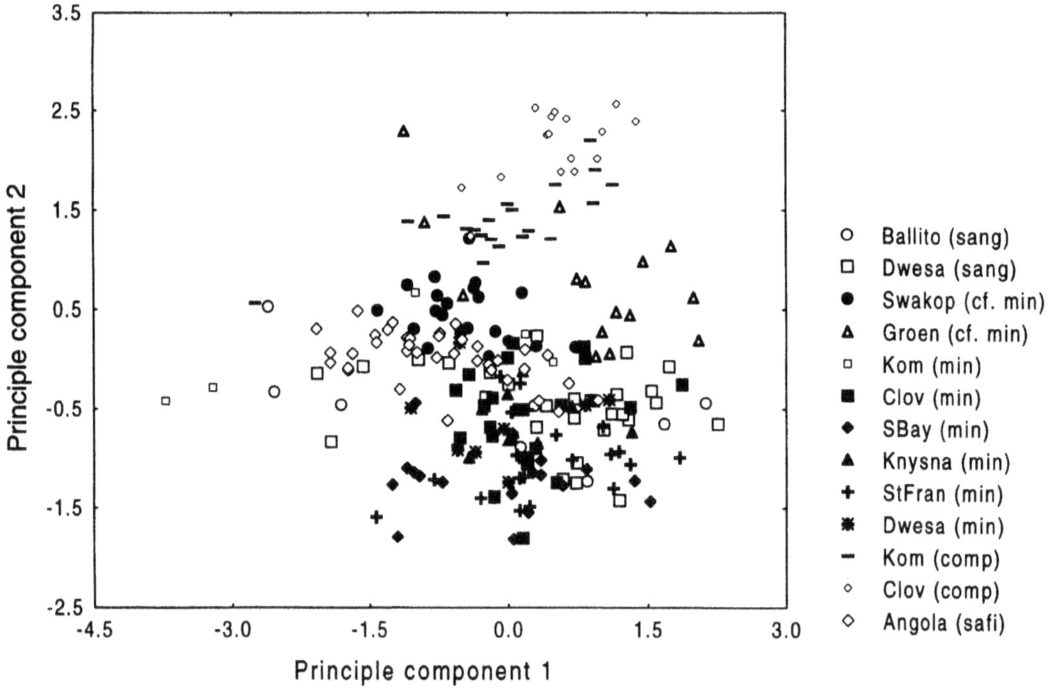

Figure 6. Patella cf. *miniata, P. miniata miniata, P. miniata sanguinans, P. compressa* and *P. safiana*. Scatterplot of the first two principal components based on the five shell morphometric characters. See 'Methods' for site and species abbreviations.

is shown in Figure 6. Overall, groups differ significantly from each other for all five of the principal components (ANOVA, P=0.001). With the exception of the clear separation of the Kommetjie and Clovelly populations of *P. compressa*, which is conspicuously narrower, the scatterplot revealed little differentiation among the other 11 populations. Therefore, the PCA analysis only clearly distinguished the two populations of *P. compressa* from the other populations.

A discriminant analysis using the five morphometric variables (SL, SW, SH, WT, VOL) was performed on all individuals, using the four genetic groupings as the grouping variable. Figure 7 shows a plot of the first two canonical variables for the four genetic groupings. As with the PCA analyses, the *P. compressa* group was distinct. There was a fair amount of overlap between the *P. miniata miniata, P. miniata sanguinans* and *P. safiana* groupings, as indicated by the overlap of the convex hulls, although *P. safiana* could be partially differentiated along the axis for the second canonical variable. Most of the separation between the groups was along the axis of the first and second canonical variables (74% and 20%, respectively). The standardised coefficients of the canonical variables indicated that separation along the first axis was largely due to shell width, whereas along the second axis, the separation was largely due to shell dry weight. The classification function indicated that 81.82% of the individuals could be correctly identified (Table 7). The *P. miniata miniata* and *P. compressa* groupings were correctly reassigned to their respective groupings with >90% accuracy, whereas *P. safiana* was correctly reassigned with 67% accuracy, and the *P. miniata sanguinans* group was correctly reassigned with <50% accuracy.

Morphological analyses

Shell and soft-part morphology

The details of the shell, foot, head and cephalic tentacles are shown in Table 8.

Radular morphology

Figure 8 shows representative samples of the radular structure in *P. miniata miniata, P. miniata sanguinans, P. compressa* and *P. safiana*. All four taxa had radula of similar structure described by the formula 3+1+(2+1+2)+1+3 (Powell, 1973), with three pairs of unmineralised marginal teeth, three pairs of lateral teeth and a single, central rachidian tooth (Figure 8). Despite the arrangement of the teeth being identical,

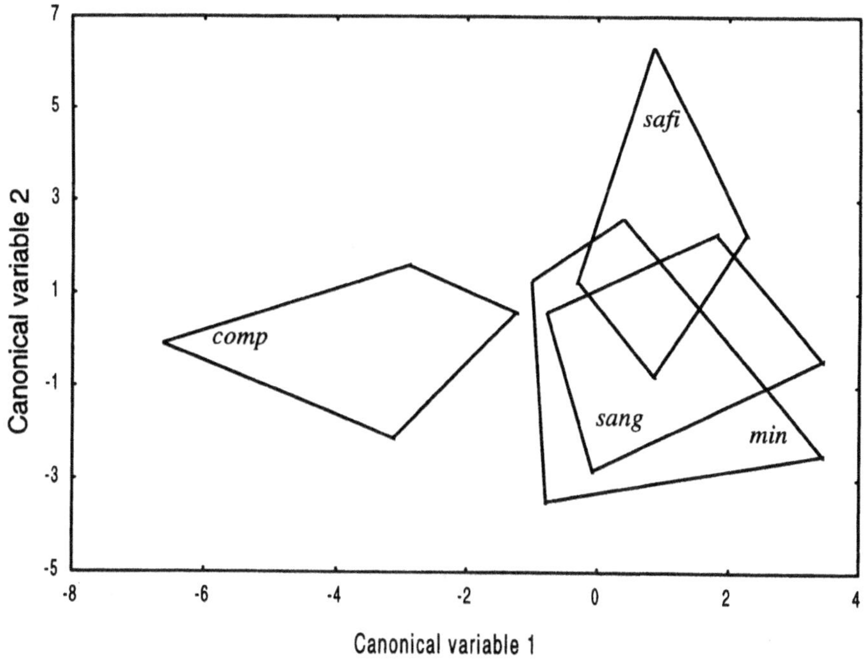

Figure 7. *P. miniata miniata*, *P. miniata sanguinans*, *P. compressa* and *P. safiana*. Plot of the first two canonical variables based on the five shell morphometric characters. See 'Methods' for species abbreviations.

Table 6. *Patella* cf. *miniata*, *P. miniata miniata*, *P. miniata sanguinans*, *P. compressa* and *P. safiana*. Summary of principal component analysis based on five morphometric characters in thirteen populations

	Principal component				
	1	2	3	4	5
Eigenvalue	4.0528	0.7072	0.1071	0.0795	0.0535
(α of variation)	(0.811)	(0.141)	(0.021)	(0.016)	(0.011)
Eigenvectors					
Shell length	0.9637	−0.0169	−0.2418	0.0327	0.1062
Shell width	0.8224	−0.5379	0.0487	−0.1780	−0.0125
Shell height	0.7659	0.6227	0.0855	−0.1288	0.0398
Shell dry weight	0.9569	−0.1319	0.1894	0.1632	0.0652
Shell volume	0.9720	0.1111	−0.0553	0.0590	−0.1903

Table 7. Percent correct a posteriori classification to groups based on morphometric classification functions of shell variables

Group	No. of individuals				% Correctly classified
	A	B	C	D	
A *Patella miniata sanguinans*	17	18	0	1	47.22
B *Patella miniata miniata*	6	113	1	2	92.93
C *Patella compressa*	0	2	34	0	94.44
D *Patella safiana*	0	12	0	25	67.56

Table 8. Shell and soft-part morphology characters. See 'Methods' for species abbreviations

	min	*sang*	*comp*	*safi*
Shell shape	oval	oval	elongate-ovate, narrow with parallel sides	oval
Shell colour: dorsal	reddish brown to pink	speckled brown with thick brown radiations	brown to cream	white with brown radiations
Shell colour: ventral	pale blue-grey to silvery pink-white with fine pink radiations	silvery white with thick brown radiations	pearl grey	silver-grey with black radiations
Foot colour: underside	grey	brownish-grey	grey	brownish-grey
Foot colour: side	purple	purple	purple	purple
Head colour	purple	purple	purple	purple
Cephalic tentacle colour	purple	dark reddish purple	purple	purple

Table 9. Sperm dimensions (mean±SE) of *Patella miniata sanguinans*, *P. miniata miniata* and *P. compressa*

	Total length of nucleus and acrosome (μm)	Nucleus length (μm)	Acrosome length (μm)	Acrosome width (μm)
Patella miniata sanguinans	2.37±0.01	1.64±0.01	1.10±0.01	0.86±0.01
Patella miniata miniata	2.83±0.01	1.94±0.01	1.33±0.01	0.78±0.00
Patella compressa	2.43±0.01	1.70±0.02	1.11±0.01	0.84±0.01

P. compressa differed from the other taxa in that it had straight to concave cutting edges on the cusps, whereas *P. miniata miniata* and *P. miniata sanguinans* had convex cutting edges. All three of these species had lateral cusps that rolled back on themselves, creating the appearance of the teeth having a hollow centre. This feature was absent in *P. safiana* and, indeed, appears in no other *Patella* species.

Sperm microstructure

Sperm from all the taxa examined were of similar shape, all having nuclei which intrude into the subacrosomal space and also possessing complex acrosomes, therefore being classified as Type III sperm (Figure 9; Hodgson et al., 1996). In addition, according to Hodgson et al. (1996), *P. safiana* has sperm of a similar shape to those of the taxa examined. Table 9 shows the means and standard errors of the four measurements taken from the sperm. *Patella miniata miniata* had significantly longer sperm (total length of nucleus and acrosome), longer nuclei, as well as longer and narrower acrosomes (ANOVA, $P<0.05$) than those of *P. miniata sanguinans* and *P. compressa*. The total sperm length and the length of the nucleus of *P. compressa* is significantly longer (ANOVA, $P<0.05$) than those of *P. miniata sanguinans*. Thus, the sperm microstructural dimensions provide useful information to distinguish between *P. miniata miniata*, *P. miniata sanguinans* and *P. compressa*.

Discussion

Fixed allele differences strongly suggest that four taxa should be recognised. *Patella miniata sanguinans* and *P. compressa* were found to be sympatric with *P.*

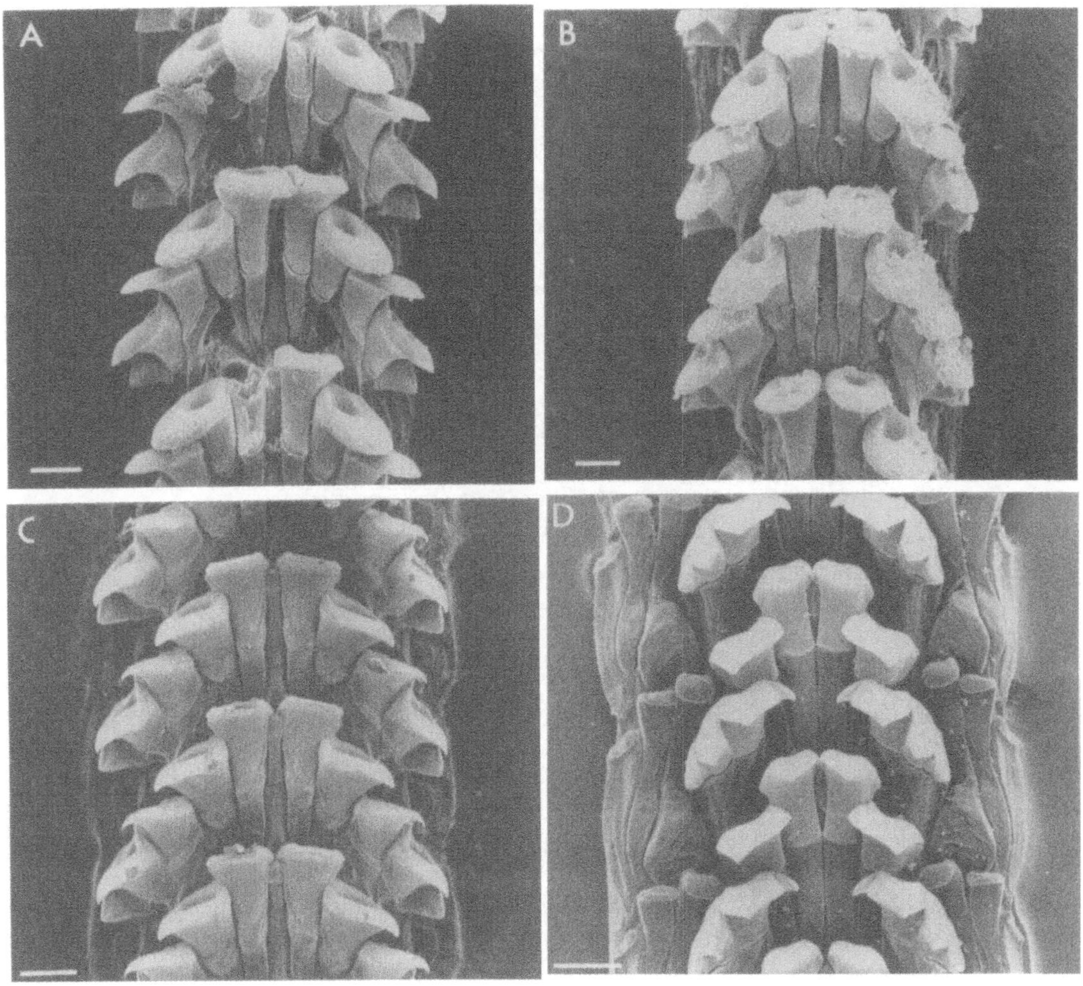

Figure 8. Scanning electron micrograph of the radular morphology of (a) *P. miniata miniata*, (b) *P. miniata sanguinans*, (c) *P. compressa* and (d) *P. safiana*. Scale bar: 100 μm.

miniata miniata at Dwesa, and Kommetjie and Clovelly, respectively. The presence of the fixed allele differences at *Ald*-1 (*P. miniata sanguinans* and *P. miniata miniata*) and *Sod*-1 (*P. compressa* and *P. miniata miniata*), and the fact that no intermediates were found, demonstrates a lack of gene flow and, therefore, indicates that these taxa are reproductively isolated.

Furthermore, populations of *P. miniata sanguinans* cluster out from those of *P. miniata miniata* at a genetic identity value of 0.797, and *P. compressa* populations cluster out at a genetic identity of 0.851 (Figure 4). These genetic identities are similar to those obtained in other studies comparing different species. Thorpe (1982) indicated that most congeneric species have Nei *I* values falling between 0.35 and 0.85, and other studies of patellids have suggested values of *I* between 0.63 and 0.86 (Cretella et al., 1994; Côrte-Real et al., 1996a,b; T. Ridgway et al., 1998). Thus, if these published values and Thorpe's (1982) scale are taken into consideration, *P. miniata sanguinans* and *P. compressa* should be recognised as valid congeneric species, thus elevating *P. miniata sanguinans* to *P. sanguinans*.

Phylogenetic analyses of the genetic data (Figure 5) show that *P. miniata miniata*, *P. miniata sanguinans*, and *P. compressa* form a closely related monophyletic group. More interestingly, however, the results show that *P. miniata miniata* is more closely related to *P. compressa* then it is to *P. miniata sanguinans*. This, therefore, provides additional evidence

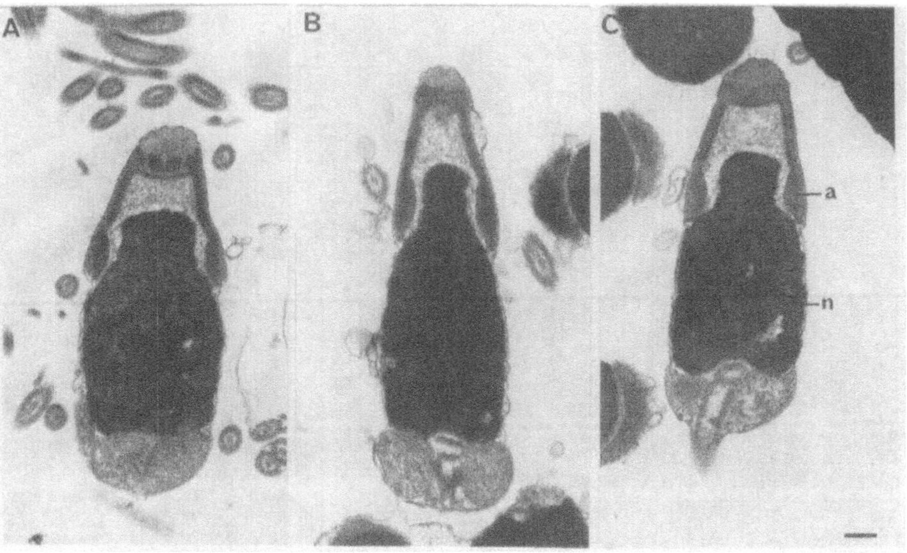

Figure 9. Mid-longitudinal TEM sections through the spermatozoa of (a) *P. miniata miniata*, (b) *P. miniata sanguinans* and (c) *P. compressa*. a, acrosome; n, nucleus. Scale bar: 200 nm.

that *P. miniata sanguinans* should be recognised as a full species (*P. sanguinans*), rather than merely as a subspecies of *P. miniata*.

The genetic data are largely supported by the morphological analyses. In both the PCA (Figure 6) and the discriminant function analysis (Figure 7), the populations of *P. compressa* cluster out from not only the *P. miniata miniata* populations, but also from the populations of *P. miniata sanguinans* and *P. safiana*. The *P. compressa* shells are easily distinguishable in that, due to the habitat of these limpets on the stipes of kelp (*Ecklonia maxima*), the shells are elongate, thin, tall and narrow with parallel sides, whereas those of *P. miniata miniata* are ovate. The radula of *P. compressa*, have straight to concave cutting edges, whereas those of *P. miniata miniata* have convex cutting edges (Figure 8), and the sperm of *P. compressa* is significantly shorter and broader than that of *P. miniata miniata* (Table 9). *Patella miniata sanguinans* shows differences from *P. miniata miniata* in shell and underside foot coloration, and sperm microstructural differences were detected between the two species, with *P. miniata miniata* having significantly longer and narrower sperm than *P. miniata sanguinans* (Table 9).

A further interesting result was observed in the behaviour of *P. miniata sanguinans*, in that at both Dwesa and Ballito, individuals were 'gardening' algal resources (see Branch et al., 1992 for discussion on 'gardening'). No suggestion of 'gardening' was observed in *P. miniata miniata*. Therefore, in this respect *P. miniata sanguinans* and *P. miniata miniata* differ in behaviour. However, no data are available on whether or not the individuals of *P. miniata miniata* are 'gardening' algal resources at Dwesa (where the two species are sympatric), and hence it is difficult to conclude whether this is a fixed difference, or merely a response of the *P. miniata sanguinans* individuals to the lower productivity on the shores of the east coast.

The populations from Swakopmund and Groen River in this study were initially identified by us as *P. adansonii* (following S. Ridgway et al., 1998). However, we were uncertain of their classification, and we therefore referred to them as *P.* cf. *miniata*. In the electrophoretic analyses, the *P.* cf. *miniata* and *P. miniata miniata* populations are almost identical (I=0.979), and it was therefore concluded that the individuals of *P.* cf. *miniata* were in fact *P. miniata miniata*. Furthermore, the sperm of the populations of *P.* cf. *miniata* did not differ significantly from the *P. miniata miniata* populations. In addition, after examination of the original description (Dunker, 1853) and the type material of *P. adansonii* (Museum für Naturkunde der Humboldt-Universität zu Berlin 102150, 102151), we could not find any major differences between the shells of the individuals of *P.* cf. *miniata* and the type of *P. adansonii* (stated to have come from Angola), and thus we propose that *P. adansonii* be synonymised with *P. miniata*.

Thus, based on the evidence from this study, it can be concluded that three species are involved in the '*Patella miniata*' species complex in southern Africa: *P. miniata*, *P. compressa* and the somewhat more distantly related *P. sanguinans*. Following S. Ridgway et al. (1998), we recognise all three as members of the *Cymbula* clade and advocate that they be allocated to the genus *Cymbula* as distinct from the more narrowly defined genus *Patella*.

Acknowledgements

Financial assistance was provided by the University of Cape Town, the Foundation for Research Development (FRD) and the South African Network for Coastal and Oceanographic Research (SANCOR). Thanks are extended to Bronwen Currie (Namibia), Allison and Rico Sakko (Angola) and Theresa Lasiak (Transkei) for the collection of the relevant samples in their areas; to Marvin Randall and Shirley Pinchuck (EM Unit, Rhodes University) for TEM technical support; and to the anonymous reviewers for comments on the manuscript.

References

Bartsch, P., 1915. The report on the Turton collection of South African marine molluscs. Bull. U.S. nat. Mus. 91: 1–305.

Branch, G. M., 1981. The biology of limpets: physical factors, energy flow and ecological considerations. Oceanogr. mar. Biol. annu. Rev. 19: 235–380.

Branch, G. M., J. M. Harris, C. Parkins, R. H. Bustamante & S. Eekhout, 1992. Algal 'gardening' by grazers: a comparison of the ecological effects of territorial fish and limpets. In John, D. M., S. J. Hawkins & J. H. Price (eds), Plant–Animal Interactions in the Marine Benthos. Clarendon Press, Oxford: 405–423.

Branch, G. M., C. L. Griffiths, M. L. Branch & L. E. Beckley, 1994. Two oceans – a guide to the marine life of southern Africa. David Philip, Cape Town.

Buth, D. G., 1984. The application of electrophoretic data in systematic studies. Ann. Rev. Ecol. Syst. 15: 501–522.

Christiaens, J., 1973. Révision du genre *Patella* (Mollusca, Gastropoda). Bull. Mus. natl. Hist. nat. 182: 1306–1392.

Côrte-Real H. B. S. M., S. J. Hawkins & J. P. Thorpe, 1996a. Population differentiation and taxonomic status of the exploited limpet *Patella candei* in the Macaronesian islands (Azores, Madeira, Canaries). Mar. Biol. 125: 141–152.

Côrte-Real H. B. S. M., S. J. Hawkins & J. P. Thorpe, 1996b. An interpretation of the taxonomic relationship between the limpets *Patella rustica* and *P. piperata*. J. mar. biol. Ass., U.K. 76: 717–732.

Cretella, M., G. Scillitani, F. Toscano, P. Turella, D. Picariello & A. Cataudo, 1994. Relationships between *Patella ferruginea* Gmelin, 1971 and other Tyrrhenian species of *Patella* (Gastropoda: Patellidae). J. moll. Stud. 60: 9–17.

Dunker, G., 1853. Index molluscorum, quae in itinere ad guineam inferiorem collegit. Casellis Cattorum.

Farris, J. S., 1988. Hennig86, Version 1.5; manual, software and MSDOS program.

Harris, H. & D. A. Hopkinson, 1976. Handbook of Enzyme Electrophoresis in Human Genetics. Elsevier, Amsterdam.

Hedgecock, D. & N. B. Okazaki, 1984. Genetic diversity within and between populations of American oysters (*Crassostrea*). Malacologia 25: 535–549.

Heller, J. & Y. Dempster, 1991. Detection of two coexisting species of *Oxystele* (Gastropoda: Trochidae) by morphological and electrophoretic analyses. Proc. zool. Soc., Lond. 223: 395–418.

Hodgson, A. N., S. Ridgway, G. M. Branch & S. J. Hawkins, 1996. Spermatozoan morphology of 19 species of prosobranch limpets (Patellogastropoda) with a discussion of patellid relationships. Phil. Trans. r. Soc. Lond. 351: 339–347.

Kilburn, R. & E. Rippey, 1982. Sea shells of Southern Africa. Macmillan, Johannesburg: 1–327.

Koch, H. J., 1949. A review of the South African representatives of the genus *Patella* Linnaeus. Ann. Natal Mus. 11: 487–517.

Levene, H., 1949. On a matching problem arising in genetics. Ann. math. Stat. 20: 91–94.

Markert, C. L. & I. Faulhaber, 1965. Lactate dehydrogenase in isozyme patterns of fish. J. exp. Zool. 159: 319–332.

Mickevich, M. F. & C. Mitter, 1981. Treating polymorphic characters in systematics: a phylogenetic treatment of electrophoretic data. In Funk, V. A. & D. R. Brooks (eds), Advances in Cladistics, Volume 1. Proceedings of the First Meeting of the Willi Hennig Society. New York Botanical Garden, New York: 45–58.

Mickevich, M. F. & C. Mitter, 1983. Evolutionary patterns in allozyme data: a systematic approach. In Platnick, N. I. & V. A. Funk (eds), Advances in Cladistics, Volume 2. Proceedings of the Second Meeting of the Willi Hennig Society. Columbia University Press, New York: 169–176.

Nei, M., 1978. Estimation of average heterozygosity and genetic distance from a small number of individuals. Genetics 89: 583–590.

Pilsbry, H. A., 1891. Acmaeidae, Lepetidae, Patellidae, Titiscaniidae. In Manual of Conchology: Structural and Systematic. Academy of Natural Sciences, Philadelphia.

Powell, A. W. P., 1973. The patellid limpets of the world (Patellidae). Indo-pacif. Mollusca 3: 75–206.

Ridgeway, G. J., S. W. Sherburne & R. D. Lewis, 1970. Polymorphism in the esterases of Atlantic herring. Trans. am. Fish. Soc. 99: 147–151.

Ridgway, S. A., D. G. Reid, J. D. Taylor, G. M. Branch & A. N. Hodgson, 1998. A cladistic phylogeny of the family Patellidae (Mollusca: Gastropoda). Phil. Trans. r. Soc., Lond., Ser. B 353: 1–28.

Ridgway, T. M., B. A. Stewart, G. M. Branch & A. N. Hodgson, 1998. Morphological and genetic differentiation of *Patella granularis* (Gastropoda: Patellidae) along the coast of southern Africa. Proc. zool. Soc. Lond. 245: 317–333.

Shaw, C. R. & R. Prasad, 1970. Starch gel electrophoresis of enzymes – a compilation of recipes. Biochem. Genet. 4: 297–320.

Skibinski, D. O. F., T. F. Cross & M. Ahmed, 1980. Electrophoretic investigation of systematic relationships in the marine mussels *Modiolus modiolus* L., *Mytilus edulis* L. and *Mytilus galloprovincialis* Lmk. (Mytilidae; Mollusca). Biol. J. linn. Soc., Lond. 13: 65–74.

Sokal, R. R. & P. H. A. Sneath, 1973. Numerical taxonomy: the principles and practice of numerical classification. W.H. Freeman, San Francisco.

Swofford, D. L. & G. J. Olsen, 1990. Phylogeny reconstruction. In Hillis, D. M. & C. Moritz (eds), Molecular Systematics. Sinauer Associates, Sunderland, Massachusetts: 411–501.

Swofford, D. L. & R. B. Selander, 1981. BIOSYS-1; a FORTRAN program for the comprehensive analysis of electrophoretic data in population genetics and systematics. J. Hered. 72: 281–283.

Thorpe, J. P., 1982. The molecular clock hypothesis: biochemical evolution, genetic differentiation and systematics. Ann. Rev. Ecol. Syst. 13: 139–168.

Tomlin, J. R. & T. A. Stephenson, 1942. South African Patellidae. Proc. malac. Soc. 25: 4–9.

Ward, R. D. & J. A. Beardmore, 1977. Protein variation in the plaice *Pleuronectes platessa*. Genet. Res. 30: 45–62.

Whitt, G. S., 1970. Developmental genetics of the lactate dehydrogenase isozymes of fish. J. exp. Zool. 175: 1-35.

Hydrobiologia **420:** 119–135, 2000.
A. M. Solé-Cava, C. A. M. Russo & J. P. Thorpe (eds), Marine Genetics.

Techniques and statistical data analysis in molecular population genetics

E. P. Silva[1] & C. A. M. Russo[2]
[1]*Lab. Genética Marinha, Departamento de Biologia Geral, Instituto de Biologia, Universidade Federal Fluminense, Niterói, RJ, Brazil*
[2]*Lab. Biodiversidade Molecular, Departamento de Genética, Instituto de Biologia, Universidade Federal do Rio de Janeiro, Rio de Janeiro, RJ, Brazil.*
E-mail: claudia@biologia.ufrj.br

Key words: molecular methods, review, phylogenetic reconstruction, DNA, allozymes, PCR

Abstract

Following the development of PCR methods, molecular techniques have become widely used for detecting genetic variation in natural populations. Most nucleotide changes can be detected by these techniques. Many of these changes probably reflect silent substitutions that are likely to be selectively neutral, making them particularly suitable to population genetic studies. In this paper, we review the published literature on molecular population genetics, with respect to the genome assayed (nuclear, mitochondrial or chloroplast), the organisms studied, the molecular techniques used, and the biological problems addressed. Several molecular techniques are then compared using experimental results obtained from a population genetic study of the *Mytilus* complex in the North Atlantic and Mediterranean. Finally, the most appropriate theoretical tools to analyse molecular population genetic data are discussed.

Introduction

A major concern in the field of population genetics is to understand the causes of differentiation between populations across ranges of geographic distribution. Thus, for example, one can study evolutionary forces in action (natural selection, genetic drift, etc.) and note phylogenetic and biogeographic patterns among living organisms through geographical space and time. The knowledge produced by the use of theoretical and experimental tools has also proved useful for more applied purposes, such as forensics, fisheries and conservation (reviews by Avise, 1994, 1996; Cipriano & Palumbi, 1999).

Since 1966, when Lewontin & Hubby published their classic paper, allozyme electrophoresis has become very widely used among almost all groups of animals and plants and has been applied to almost every sort of problem in population genetics and its branches (Murphy et al., 1996). Such intense use of the method has revealed its limitations for some groups of living organisms or for specific problems. These limitations, however, do not reflect the exhaustion of the method, but have simply served to narrow the range of possibilities (Murphy et al., 1996).

Currently, DNA techniques seem to be rapidly gaining terrain over allozymes in the field of population genetics. One reason for this is that, thanks to the development of PCR (Saiki et al., 1988), DNA techniques can be applied to tissues with different methods of preservation (e.g. alcohol-fixed, air-dried and even fossils) (Lewin, 1994).

However, DNA techniques are still very expensive and labour-intensive, which is why their use in population genetics remains limited (Skibinski, 1994). These limitations are quite restrictive, because the number of individuals one has to sample is inversely proportional to the genetic difference between Operational Taxonomic Units (OTUs). In the case of population genetic studies, the genetic difference to be detected is often only slightly higher than that between individuals in a single population; thus, for an accurate assessment, the number of individuals sampled should be high (> 30 individuals). Therefore, it is a mistake to think that when molecular techniques become less expensive they will eventually supplant the use of allozymes. Both approaches have their advantages and disadvantages, their limits and their idiosyncrasies. For instance, it has been suggested that allozyme electrophoresis is the best technique for alpha systematics (Hillis et al., 1996a), but DNA sequencing has proved very useful for a wide range of phylogenetic studies (e.g. Holmes et al., 1999; Wang et al., 1999) and microsatellites and SSCP may be the methods of choice for some population genetic studies (Goldstein et al., 1999).

Hopefully, the future may find each of these techniques appropriately used for solving questions about the biology and evolution of living organisms. This paper is intended to review the population genetic literature concerned with molecular methods (i.e. DNA-based). It is also intended to cover briefly the criteria by which we can select from the various molecular techniques and methods of statistical analysis currently available for studying a range of biological problems.

What has been published on population biology?

We have conducted a literature survey of the current use and application of DNA techniques in population biology. BIDS (Bath Information Data Service) was used combining words that are technique-oriented [DNA, RNA, RFLP (Restriction Fragment Length Polymorphism), PCR (Polymerase Chain Reaction), RAPDs (Random Amplified Polymorphic DNA), VNTR (Variable Number of Tandem Repeats), DGGE (Denaturant Gradient Gel Electrophoresis), SSCP (Single-Strand Conformation Polymorphism), minisatellites, microsatellites, fingerprinting, sequencing] with words that are problem-oriented (population genetics, population structure, gene flow, inbreeding, recruitment, dispersal, systematics, taxonomy, speciation, conservation, management). The period covered was from 1979 to 1997; 2490 references were found, of which 987 papers were specifically related to population genetic problems. A further 304 papers applied molecular techniques to problems in population biology, bringing the total to 1291 papers that will be discussed below.

In Figure 1, the number of papers published annually is plotted showing the general trend over the 19-year period. The impact of PCR on research in population biology is clearly shown by the large increase in the annual number of papers in the field, following the publication of the technique (on primer-directed

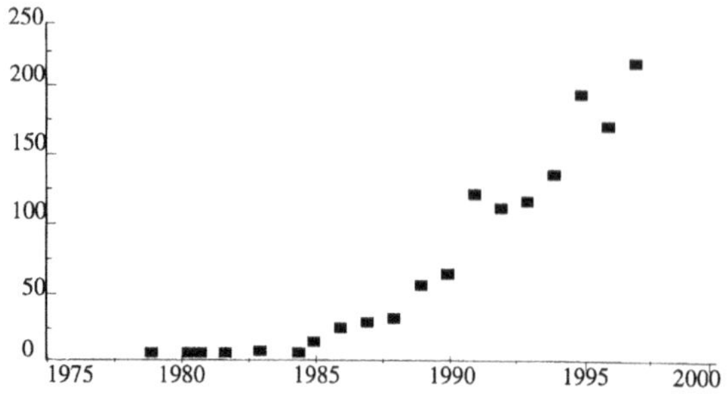

Figure 1. Number of scientific papers using molecular techniques applied to population biology published per year since 1979.

enzymatic amplification of DNA using thermostable DNA polymerase) by Saiki et al., (1988). However, it was only in the 1990s that the impact of DNA technology on population biology began to be seen, and even now it has yet to become fully established (Skibinski, 1994).

Most (54%) of the papers considered in the survey of DNA-based studies (described above) have used mitochondrial DNA, and this seems currently to be the most widely studied molecule for population genetic work. One reason for this preference could be the fact that mitochondrial genes evolve, on average, faster than nuclear structural genes (Gillham, 1994), and this makes them particularly suitable for population level studies. Conversely, 41% of the studies used nuclear DNA and chloroplast DNA was used in the remaining 5% of papers.

In order to have a broader perspective on the literature in this area, we have further analysed our search results with respect to three other aspects: (1) the molecular techniques used; (2) the organisms studied; and (3) the biological problems addressed.

Molecular techniques

Figure 2 shows the proportional use of different techniques in the scientific literature surveyed. Restriction enzymes have been widely used to obtain information on DNA sequence length polymorphism, and a great effort has been put into the sequencing of DNA (both genomic and organellar). Randomly amplified polymorphic DNAs (RAPDs) and multi-locus satellite screening (fingerprinting) techniques have been used mainly for addressing problems related to reproduction and kinship.

The use of techniques involving single-locus minisatellites and microsatellites has recently increased. However, the cost associated with the construction of DNA libraries is still a limiting factor for many labor-

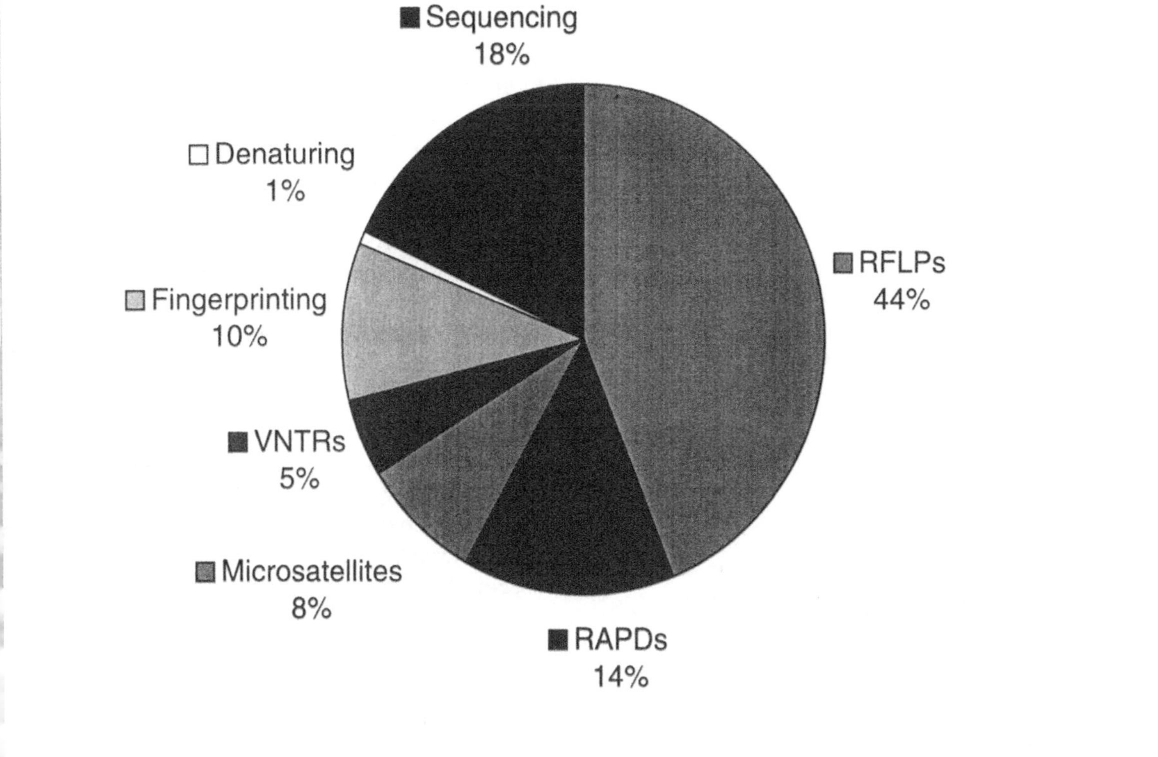

Figure 2. Percentage of scientific papers published in population biology using each of a variety of molecular techniques.

atories wanting to use the techniques. DNA denaturing techniques (e.g. SSCP, DGGE, heteroduplexing) are just coming into use and their application to population genetic studies could provide an interesting new array of molecular markers. Other recently developed methodologies, such as amplified fragment length polymorphisms (AFLP) and sequence-specific oligonucleotides (SSO), among others, were not considered in this survey, since we have concentrated only on techniques widely used in population biology.

Organisms

In order to obtain a better picture of which organisms have been studied, we have divided the publications into groups that reflect the trends in the literature. The results are shown in Table 1 and, in general, demonstrate that scientific efforts are concentrated on groups of economic importance (e.g. fish, molluscs, crustaceans) or others which have been used as a model (e.g. humans, primates, rodents, *Drosophila*). Also apparent is the attention paid to the management and conservation of groups such as aquatic mammals, which have generally shown low levels of variation at the allozyme level (Avise, 1994; Slade et al., 1998).

Biological problems

In general, the application of molecular techniques to population biology involves four main categories of biological problems. The first grouping comprises problems related to the analysis of genetic variation within individuals, tackling areas such as heteroplasmy (Fisher & Skibinski, 1990; Hoech et al., 1991; Zouros et al., 1992, 1994a,b; Magoulas & Zouros, 1993; Skibinski et al., 1994a,b), multi-gene family evolution (Zhang & Nei, 1996), and forensic science (this has provoked a heated debate in scientific journals; Lander & Budowle, 1994; Lewontin & Hartl, 1994; Edwards et al., 1995). We have also included in this group problems related to the identification of scallop meat (Kenchington et al., 1993) and mon-

Table 1. Number and percentage of papers published on DNA technology applied to population biology in different taxonomic groups. (Note that some papers have dealt with more than one major taxonomic group)

Taxonomic group	Number	Percentage
VIRUSES	4	0.30
BACTERIA	35	2.71
PROTISTS	7	0.54
INVERTEBRATES	362	28.04
CNIDARIANS	7	0.54
ROTIFERS	3	0.23
NEMATODS	14	1.08
ARTHROPODS	239	18.51
Crustaceans	48	3.72
Insects	184	14.25
Drosophila	82	6.35
MOLLUSCS	78	6.04
Bivalves	60	4.65
Mussels	38	2.94
Gastropods	16	1.24
ECHINODERMS	18	1.39
CHORDATES	886	68.63
VERTEBRATES	883	68.40
Fishes	217	16.81
Birds	95	7.36
Reptiles	33	2.56
Amphibians	19	1.47
Mammals	302	23.39
Aquatic Mammals	40	3.10
Humans	134	10.38
PLANTS*	175	13.56
FUNGI	42	3.25

*Plants include green plants, algae and phytoplankton.

itoring of commercial whaling practices (Baker & Palumbi, 1994; Cipriano & Palumbi, 1999).

In the second group we have included problems involving variation within populations, such as the effects of inbreeding and bottlenecks on genetic variation (Boulding et al., 1993; Grewe et al., 1993); the neutralism *versus* selectionism debate; and problems involving kinship determination such as nepotism (Pfennig & Reeve, 1993), social structure (Morin et al., 1994; Oldroyd et al., 1994) and reproductive success (Bensche et al., 1994; Lambert et al., 1994; Wetton et al., 1995).

The third group of studies are those directly concerned with genetic variation between populations. These deal mainly with problems of population structuring such as stock definition in fisheries (Silberman & Walsh, 1994; Hall et al., 1995), bioinvasion (Echelle & Dowling, 1992; Geller et al., 1994; Heath et al., 1995) and gene flow (Zink & Dittmann, 1993; Elliot et al., 1994; Hurst & Skibinski, 1995; Palumbi, 1996). This group also encompasses problems related to the taxonomic status of morphotypes (Vuorinen et al., 1993; Rosel et al., 1994) or ecotypes (Hanfstingl et al., 1994; Cluster & Allard, 1995).

Finally, the fourth group of biological problems involves genetic variation above the species level, including the effects of different life cycles on species isolation and differentiation (McMillan et al., 1992), hybridisation and hybrid zones (Edwards & Skibinski, 1987; Moritz et al., 1992; Zink, 1994; Oldroyd et al., 1995), and strain definition in unicellular organisms (Bell & Friedman, 1994). One issue that has received much attention is phylogenetic reconstruction, specifically addressing questions of single *versus* multiple origins of morphological (Collins, 1998), physiological (Block et al., 1993) or behavioural (Sturmbauer et al., 1996) characteristics, more specifically of those used in biological classifications (Avise & Johns, 1999). This latter approach, however, has traditionally been applied to higher levels of divergence than those among populations and so it will be mentioned only briefly in this review.

Which technique is the most appropriate for a specific problem?

There are a number of good reviews of molecular techniques applicable to aspects of population biology (e.g. Karl & Avise, 1993; Lessa & Applebaum, 1993; Queller et al., 1993; Avise, 1994; Skibinski, 1994; Hillis et al., 1996a). Here we will outline the techniques most widely used and give a brief overview of their potential uses and underlining assumptions.

The most widely used technique was RFLP (Karl & Avise, 1993; Skibinski, 1994), in which genomic DNA is cut with restriction enzymes, electrophoresed through agarose gels, blotted on to membranes and probed with cloned radiolabelled DNA that binds to a specific locus. Alleles differing in the presence or absence of nearby restriction sites will produce different fragment sizes. With the advent of PCR (Saiki et al., 1988), RFLP techniques became much simplified, since two small oligonucleotide primers can be used to amplify a region of DNA flanked by primers and this amplified region can then be restricted and analysed on ethidium-bromide or silver-stained gels. Single locus nuclear DNA restriction site polymorphisms behave in a Mendelian fashion. The presence or absence of multiple restriction sequences around one given DNA region constitute haplotypes that can be used to estimate overall levels of gene variation in a similar way to allozymes (Dowling et al., 1996).

Multi-locus minisatellites or DNA fingerprints (Avise, 1994; Skibinski, 1994) are similar to RFLPs, but the probe hybridises to a minisatellite sequence, which consist of tandem repeats of a 10–50 base pair motif and are widespread throughout the genomic DNA. Complex band patterns result from multiple loci with alleles that vary in the number of repeats. Thus, DNA fingerprinting presents some difficulties in interpreting the large number of loci visualised and also the possibility of dominance, resulting from fragments of small size that are usually not visualised.

Very close to DNA fingerprinting is the single-locus minisatellite technique, also called the VNTR technique (Avise, 1994; Skibinski, 1994). Here, the probes used are clones that identify a single minisatellite locus under high stringency conditions. In this case, a pattern is revealed and it is possible to differentiate homozygotes (single bands) from heterozygotes (double bands). These clones are obtained by screening a library with standard minisatellite probes.

Another variety of VNTRs is what are termed microsatellites (Queller et al., 1993). These consist of tandem repeats of very short nucleotide motifs (1–6 bases long). They can be amplified by PCR (once adjacent sequences are known) and alleles can be scored consistently to infer population structure and demographic history of populations (Goldstein & Pollock, 1997; Goldstein et al., 1999; Reich et al., 1999). The main disadvantage of microsatellites is the cost and effort required in developing them.

Randomly amplified of polymorphic DNA (RAPD) (Welsh & McClelland, 1990; Williams et al., 1990) is a technique based in part on the fact that PCR is somewhat tolerant to mismatches between primers and the DNA template. Thus, short (10 bp) arbitrary sequences of nucleotides can be used as primers to amplify random fragments of the DNA molecule. RAPD is similar to DNA fingerprinting in that it is again difficult to differentiate homozygotes and heterozygotes. However, the main problem associated with this technique is the tendency to produce artifacts, due to its high sensitivity to PCR conditions. As a result, reproducibility is a major issue which must be carefully addressed through consistency and precision, as well as through the use of positive and negative controls.

Another group of DNA screening techniques involves distinguishing DNA segments that differ by one or more point mutations (Lessa & Applebaum, 1993). In DGGE (Denaturant Gradient Gel Electrophoresis), polymorphic DNA fragments, amplified via PCR, are run on a gel containing a linear gradient of a denaturing agent (primarily urea). Fragments are separated due to distinct electrophoretic mobilities, caused by differences in the nucleotide sequence of the target fragment and the presence of mispaired bases.

Single-strand conformational polymorphism (SSCP) is related to DGGE but, in this case, heat denaturation associated with a denaturing agent (such as formamide or urea) is used on PCR products, followed by separation via non-denaturing polyacrylamide gels. The two complementary strands of DNA migrate differently and will therefore separate during electrophoresis.

Alternatively, distinct sequences can be identified by heteroduplex analysis, since DNA molecules composed of strands from two different alleles produce mobility shifts during electrophoresis on polyacrylamide gels. Denaturing techniques give complete or nearly complete resolution of the alleles present. The main disadvantage of these techniques lies in the difficulty of developing optimum conditions (Fan et al., 1993).

Nuclear genes which play important metabolic roles are likely to have conserved DNA sequences with low variability, and therefore will provide little information for population and phylogenetic studies (Palumbi, 1996). A solution to this problem is to design primers which lie within an exon (coding region), but amplify non-coding regions (introns), which are more likely to have higher evolutionary rates and are presumably selectively neutral (Lessa, 1992; Palumbi & Baker, 1994). Once the intron is successfully amplified, sequencing, RFLP or SSCP can be applied in order to screen population patterns. Alternatively, intron amplifications can sometimes reveal patterns of length polymorphism, which behave as different alleles at a locus, and can be interpreted as in allozyme studies (Baker & Palumbi, 1994; Côrte-Real et al., 1994; Ohresser et al., 1997). These markers are known as intron length polymorphisms, and are potentially very useful in population genetic studies and for phylogenetic inference.

All molecular techniques discussed here are, in a sense, an indirect means of obtaining sequence information. Direct sequencing techniques represent the

Table 2. Nuclear DNA markers which were used for the comparison

Nuclear DNA Marker	Reference	Polymorphism
ITS (Internal Transcribed Spacer region between 2.8 S &28 S, nrDNA repeat unit)	Dixon et al., 1995	Conformation
Me (Adhesive Protein)	Inoue et al., 1995	Length
PLIIa (coding region of the Protamine-like sperm packing protein)	Heath et al., 1995	Length
CaM-1 (Calmodulin gene, Intron 3)	Côrte-Real et al., 1994	Length
N118 (Anonymous nuclear DNA)	Beynon &Skibinski, 1996	Length/ Conformation
Myo (Intron at the Catch Muscle Myosin)	Silva, 1998	Length

acquisition of genetic information at a maximally detailed level (Avise, 1994; Hillis et al., 1996b). As a result of this, much effort has been expended leading to the rapid development of sequencing techniques (Avise, 1994; Hillis et al., 1996b). The most widely used sequencing procedure is Sanger sequencing, or controlled interruption of enzymatic DNA replication (Sanger et al., 1977). Until recently, DNA sequencing was very labour-intensive; however, with the advent of automated sequencing, it is now possible (albeit still expensive) to quickly produce extensive sequence data for population analysis.

Table 3. Average number of alleles and heterozygosity for each technique over all populations

Technique	Mean number of alleles	Mean heterozygosity
Length Polymorphism	4.03	0.46
RFLPs		
Isolated	2.43	0.19
Combined	10.38	0.57
SSCP	3.45	0.42

What is the best technique to use? A case study using *Mytilus* spp.

One of the main concerns in using molecular techniques is to improve the ability to detect polymorphism. To study this further we have used data obtained in a recent study of the population genetics of mussels (*Mytilus* spp.) (Silva, 1998) using nuclear DNA markers to compare the resolution of some of the techniques available.

Sixteen populations of the genus *Mytilus*, including the nominate species *M. edulis*, *M. galloprovincialis* and *M. trossulus* were sampled in different sites in Britain, Ireland, Spain, Canada and Poland. Genomic DNA was obtained from gill tissue using a phenol-chloroform extraction (for further details see Silva, 1998). Six nuclear DNA markers were employed in a variety of different polymorphism detection techniques (RFLP, SSCP, length polymorphism) (Table 2).

When comparing different techniques, the most critical information concerns the relative efficacy of the different techniques to detect variation, the degree of difficulty in applying these techniques and the relative costs. The resolution of a given marker can be assessed by the number of alleles and heterozygosity shown by that marker (Avise, 1994). Table 3 shows mean values for the number of alleles and heterozygosity presented by each technique over all populations of *Mytilus*.

Length polymorphism and SSCP show only slight differences in numbers of alleles and heterozygosity, and thus overall appear quite similar in terms of resolution. RFLP, using a single restriction enzyme in isolation, showed the lowest detected number of alleles and heterozygosity of the three techniques. This may be of interest since a majority of recent molecular population genetic studies have been based on RFLP analysis (Figure 2).

The use of multiple restriction enzyme patterns to determine haplotypes is a complicated process, since interpretation of multiple heterozygotes can be ambiguous. Solutions to this problem include strategies such as asymmetric PCR (Karl & Avise, 1993) or, in the case of tractable organisms such as *Drosophila*, the use of crosses between isolines (Berry & Kreitman, 1993). Clark (1990) devised an algorithm which can bypass these technical difficulties, but which depends on the presence in the sample of known homozygotes or single-site heterozygotes. The use of this algorithm with isolated RFLP patterns, obtained for the anonymous scnDNA N118, produced single RFLP haplotypes with an increase in the number of alleles and heterozygosity relative to the analysis of isolated RFLP patterns, SSCP, and length polymorphism techniques. Therefore, based on the amount of variation revealed, the best performance of the methods sampled was offered by single RFLP haplotypes produced by combining multiple restriction enzyme patterns.

A second consideration is the effort necessary to get the techniques working. Once DNA fragments are obtained, the simplest technique is to separate length polymorphisms. Only PCR and electrophoresis are necessary to obtain interpretable results, and bands of different mobilities can be separated using gels stained with silver or ethidium bromide (Côrte-Real et al., 1994; Ohresser et al., 1997; Silva, 1998). In the case of intron length polymorphism, the principal limitation is the availability of PCR primers. However, the protocol developed by Lessa (1992) will, apparently, effectively amplify introns with previously unknown sequences.

Additional factors to be considered when developing intron primers relate to the optimisation of PCR conditions (Palumbi, 1996). One strategy is to design primers complementary to the DNA of the species of interest, based on the intron position in other species (Côrte-Real et al., 1994; Silva, 1998). The success of this strategy depends on the availability in genetic databases (GenBank, for example) of DNA sequences from the species of interest.

RFLP analysis is not a demanding technique, once a suitable target DNA fragment is available. The effort required consists of testing a large number of different restriction enzymes in order to establish which ones show repeatable polymorphic patterns (Dowling et al., 1996). Technical complications are few, since the protocols are simple and polymorphism detection is dependent on the presence or absence of the restriction site in the DNA fragment, and not on the technique itself. If the patterns produced by different restriction enzymes are combined, the ability to detect variation can be greatly increased (Table 3; Beynon & Skibinski, 1996; Silva, 1998).

SSCP is the most demanding technique in terms of effort required. A number of PCR and electrophoretic conditions are known to affect SSCP patterns (Hayashi, 1991; Cai & Touitou, 1993; Kasuga et al., 1995), and optimisation of the technique depends on exhaustive tests of several variables. Furthermore, optimal conditions for a given fragment may not apply to a different fragment from the same species, and tests must be repeated. Although some attempts have been made to predict the behaviour of single strands under electrophoresis conditions (Nielsen et al., 1995), this behaviour still remains unpredictable. The main attraction of SSCP analysis is its simplicity once optimal conditions are determined. As with length polymorphisms, the interpretation of SSCP patterns is straightforward and the ability of the technique to detect single point mutations is generally around 90% (Lessa & Applebaum, 1993).

Based on the amount of effort required to get a technique working, length polymorphisms are the simplest of those discussed here. The most expensive technique is undoubtedly RFLP analysis, which involves preliminary tests using a battery of different restriction enzymes, followed by population screening using those enzymes that show polymorphism. On the other hand, length polymorphism and SSCP analysis have no economic requirements beyond the PCR procedure.

At this point the question emerges: what might be the best technique to be used in a particular population genetic study? The results obtained testing six different nDNA markers over 16 populations of *Mytilus* indicate that each approach has the potential to provide informative data. Factors which may determine the choice of one or another technique are: (1) the economic resources of the laboratory; (2) tradition; and (3) the state of advancement reached by the scientific tools. For instance, laboratories which are well supported can use RFLPs routinely. This is not the case for less well funded laboratories for which comparatively inexpensive techniques such as SSCP are a better option, despite the time requirements of the technique. The use of intron polymorphisms, however, depends more on the available genetic information on the organism of interest, i.e. primer availability. Another important factor determining the choice of technique is the tradition, both of laboratories and of the techniques. Techniques which are well established would be a first option, especially for laboratories which are just starting in the field. In the same way, laboratories which pioneered the use of techniques which are now well established may be more conservative concerning the use of less traditional approaches.

The most basic assumptions underlying the use of molecular markers such as RFLPs, length polymorphisms and SSCP variants are that the characters in question are heritable, repeatable and independent (Dowling et al., 1996). Generally these assumptions hold for most studies. However, before drawing general conclusions about biological mechanisms, it is important to bear in mind the assumptions underlying the use of particular molecular markers.

Selection of an appropriate analytical technique

We have described several molecular methods from which raw data can be extracted and have reviewed guidelines for selecting the best molecular method to answer a specific biological problem. However, in order to extract biological information from the raw data produced by these molecular methods, we are often confronted with several different analytic approaches, each with unique advantages and disadvantages. In fact, this stage is by no means trivial since there are several methods of analysis applicable at any given biological level, various of which may even yield conflicting results. Therefore, we shall present in this section a short review of the advantages and the main problems associated with specific methods of analysis, and we will discuss what we feel are the best ways to avoid these problems.

Codominant versus *dominant markers*

There are basically two types of genetic markers: codominant and dominant markers. Codominant markers are those where we can distinguish the heterozygote from the homozygotes (e.g. allozymes, microsatellites, nuclear PCR/RFLP). In this case, allele frequencies can be calculated directly from the raw, genotypic data. Conversely, with dominant markers such as RAPD, the heterozygote has the same pattern as one of the homozygotes, and thus to calculate the allele frequencies one has to assume that the population is in Hardy-Weinberg equilibrium. Obviously, we can test for Hardy-Weinberg equilibrium only if codominant markers are used, but all the analyses described below can be used for any type of molecular data.

Levels of genetic variation

The levels of genetic variation in natural populations are often surprisingly high, but there is still much debate as to how this variation is maintained in natural populations (Orr, 1998; Comeron et al., 1999; Hey, 1999; Quesada et al., 1999). Neutralists argue that random genetic drift is responsible for unstable polymorphisms in populations, maintaining different alleles with very similar selection coefficients (i.e. neutral alleles; Kimura, 1968; Nei, 1988). On the other hand, selectionists see polymorphism as stable, maintained by balancing selection, with the frequency of different alleles varying in different microhabitats.

This debate has been the focus of a substantial amount of attention during the 1970s and 1980s. However, with the advent of molecular techniques resulting in the accumulation of empirical data, both hypotheses (as they were initially proposed) have been disproved. For instance, strictly neutralistic predictions of genetic variation suggest higher levels of genetic variation than those observed in natural populations, indicating that some kind of selection probably operates in most populations (Nei, 1983). This observation led Ohta (1974, 1992) to propose a revised hypothesis to explain how evolution works at the molecular level: the 'nearly neutral' theory. In this case, the vast majority of substitutions at the molecular level are neutral or slightly deleterious, accounting for the lower variation found empirically.

An additional difficulty for neutralism is that several important loci, such as major histocompatibility loci (Hughes & Nei, 1989; Hughes et al., 1994), immunoglobulins (Tanaka & Nei, 1989; Ota & Nei, 1995), and others (Johannesson et al., 1995; Swanson & Vacquier, 1995) have been found to show polymorphisms that are governed by balancing selection. However, a strictly selectionist theory is also difficult to support with empirical data. Following the widespread use of DNA sequencing techniques, it has been concluded that polymorphisms at the vast majority of variable loci are determined by a balance between genetic drift and purifying selection, rather than by balancing selection (Gillespie, 1987; Kimura 1995; but see Comeron et al., 1999). Thus, it seems that any coherent theory of evolution at the molecular level must take into account the ideas of neutralist and selectionist schools of thought.

Initially supporting neutralist predictions, it has become apparent that the level of genetic variation of a given population may be related to effective popu lation size (N_e), where N_e is defined as the number of individuals in an ideal population with the same decrease in heterozygosity due to genetic drift (Hartl & Clark, 1997). In this case, the ideal population has even sex ratio, all individuals are genetically different and potential mates, and population size remains constant through time.

Obviously, not many populations conform to these assumptions and, therefore, N_e is often lower than the actual number of individuals (N) in a population. In some cases it is much lower, as in the case of human populations, where N is about 6 billion individuals, while N_e has been estimated to be only 10 000 individuals (Harpending et al., 1998). In some invertebrates, however, simultaneous hermaphroditism can raise the effective number of individuals to values exceeding the actual number and, indeed, some such species have been found to show some of the highest levels of genetic variation in natural populations (Ritte & Pashtan, 1982; Solé-Cava & Thorpe, 1990, 1991; Thorpe & Solé-Cava, 1994). Heterozygosity is related to N_e and to μ (mutation rate) and it is given by the formula:

$$H = 4N_e\mu/(4N_e\mu + 1).$$

The fact that levels of heterozygosity are expected to be directly related to N_e makes intuitive sense, since the larger the population, the larger will be the number of mutations that appear in this population. More importantly, a large population will have a minimal decrease in heterozygosity due to genetic drift. Note that the number of individuals which should be considered in this case is the effective number of the species, rather than that of the local population (Nei, 1983).

The relationship between N_e and genetic variation has recently led to an increase in consideration of the role of genetic variation in conservation studies (Avise, 1994, 1996). The minimum N_e for an endangered species to retain evolutionary potential has been estimated to be around 500–1000 individuals (Franklin & Frankham, 1998; see also Lynch & Lande, 1998 for larger minimum N_e estimates) depending on the proportion of deleterious substitutions appearing in the population.

More recently, other important issues have been addressed by estimates of genetic variability. One area that has received much attention is the use of DNA fingerprinting to identify endangered species in various fish market surveys (e.g. Cipriano & Palumbi, 1999). Other important issues are those based on the coalescent theory that uses genetic variation to determine several demographic aspects of the population. For instance, patterns of genetic variation have been used to estimate coalescent time, or the time when the common ancestor lived, of a DNA sample (Goldstein et al., 1995; Fu, 1996), to date population expansions and contractions (Harpending et al., 1998; Weiss & Von Haeseler, 1998), and even to reconstruct the ancestral sequence of a given set of DNA samples (Innan & Tajima, 1997). However, because of space limitations we will not address these issues any further in this section.

Estimates of gene flow

Natural populations that are spread across vast areas are seldom homogenous. In most species the population consists of various subpopulations with occasional migration or gene flow between them. The genetic similarity of these populations depends on the number of migrants exchanged per generation. If there is high migration, the entire population approaches panmixia. On the other hand, if populations do not exchange migrants, genetic differentiation occurs due to the forces of mutation and genetic drift (possibly also selection); and, in time, speciation may result.

Particularly in the case of marine invertebrates, most of which are sessile or show low vagility as adults, the main dispersal phase is the larva (Jablonski & Lutz, 1983; Olson & McPherson, 1987). Thus, the amount of differentiation of their populations may depend mainly on the duration of the planktonic period. Marine invertebrates with planktotrophic larvae tend to be more genetically homogeneous over large distances (Russo et al., 1994), whereas taxa with lecithotrophic larvae tend to show higher levels of population structure and thus, over time, will show higher rates of speciation and extinction (Jablonski, 1986; Jackson, 1986).

It has been estimated that if two populations exchange an average of one individual per generation they will not diverge genetically (Allendorf & Phelps, 1981). Interestingly, this number is independent of population size. This occurs because the exchange of a single individual in a large population will be enough to minimise the small effects of genetic drift in large populations. Conversely, small populations are more strongly impacted by genetic drift, but the exchange of a single individual represents the transfer of a larger proportion of the gene pool in every generation. Despite considerable debate as to whether this rule of thumb holds empirically (Varvio et al., 1986), it can prove both interesting and instructive to estimate the number of migrants between populations.

However, gene flow cannot be estimated by counting the number of migrants since it would be an extremely tedious and often impossible process to monitor and count the number of individuals exchanged between subpopulations. Furthermore, the actual number of migrants exchanged may not give an accurate estimate of the number of migrants that contribute to the gene pool of the other subpopulation; that is, the effective number of migrants (N_em). This number can be determined indirectly (Tufto et al., 1998), using other parameters.

The effective number of migrants can be estimated by two different methods: one related to F_{ST} and the other to private alleles. F_{ST} (Wright, 1978) is an index of the amount of variation due to differentiation between subpopulations, and for a two-allele system (low-frequency alleles can be pooled into one class) is given by:

$$F_{ST} = V_p/(p(p-1)),$$

where V_p is the variance of the mean allele frequency p, among all subpopulations. Nei (1977, see also 1987) extended this calculation to a multi-allele system, based on the relationship between the reduction of total expected heterozygosity (H_T) and the average expected heterozygosity for each subpopulation (H_S):

$$F_{ST} = (H_T - H_S)/H_T.$$

Thus, the average number of effective migrants between populations (N_em) can be estimated using the formula:

$$N_e\mathrm{m} = [(1/F_{ST}) - 1]/4$$

or as suggested by Slatkin (1985):

$$N_e\mathrm{m} = \mathrm{e}^{-[\ln p(I)+2.44]/0.505}/(N/25),$$

where $p(I)$ is the average frequency of private alleles (i.e. alleles that are present in a single subpopulation only) and N is the average sample size over all populations.

These calculations can be instructive if we know *a priori* the limits and borders of subpopulations. However, in some cases, a subpopulation can be divided into a few smaller panmitic groups with little gene flow between them. In this case, we would naturally have lower observed heterozygosity than expected if the population behaved as a panmitic unit (i.e. Wahlund effect). (Note that the Wahlund effect is not an evolutionary force, but rather an apparent heterozygote deficiency observed when populations which do not freely exchange individuals are combined.)

Distance measures and divergence time

F_{ST} and N_em are useful indices when describing a population that is subdivided. However, it may also useful to have an estimate of the divergence time of two populations. Although, F_{ST} can be used for two populations (Nei, 1987), the simplest measure of genetic distance between two populations is D_m, or the Minimum Distance (Nei, 1973). This index equals one minus the probability that two randomly chosen alleles from two populations are identical and is given by:

$$D_m = 1 - \Sigma x_i y_i,$$

where x_i and y_i are the frequencies of the ith allele for both populations.

However, if polymorphic populations are considered, heterozygosities should be subtracted for the populations, and the distance becomes defined by the minimum distance minus the average heterozygosity for the populations:

$$D = 1 - \Sigma x_i y_i - [(1 - \Sigma x_i^2) + (1 - \Sigma y_i^2)]/2.$$

One advantage of using molecular data is that one can assume the number of substitutions to be time-dependent, clearly not a justified assumption when working with morphological data, and thus the amount of genetic divergence can be used to estimate the actual divergence time between populations. Of course, there are several assumptions that might not hold for specific cases (Brohman et al., 1996), but estimates are still likely to be more reliable than those derived from morphological characters or from the fossil record. Thus, as previously mentioned, the first step is to find a neutral molecular marker which has a level of genetic variability suitable for the problem at hand. The second step is data analysis. In order to estimate divergence times one has to use a distance measure which is appropriate for estimating the divergence time of the populations. Nei (1972) developed the distance measure given below, which, given certain assumptions, is proportional to the divergence time of two populations. The standard genetic distance is:

$$D_s = -\ln I,$$

where I is the Genetic Identity Index of Nei (1972) (i.e. the normalised probability that randomly drawn alleles from each population are identical by descent).

This measure is given by the formula:

$$I = Jxy/JxJy^{1/2},$$

where

$$Jxy = \Sigma_{i=1}^{n} x_i y_i$$

and,

$$J_x = \Sigma x_i^2,$$
$$J_y = \Sigma y_i^2,$$

where x_i and y_i are the frequencies for the ith allele in populations X and Y, respectively. This identity index has been used in population genetics. In fact, it has been so widely used that Thorpe (1983) plotted I values for conspecific populations, congeneric species and genera within a family. While this is an interesting approach and provides some objective basis as to how much genetic differentiation can be present within single species, it has certain problems with circularity. That is, Thorpe's distribution is not a theoretical distribution that provides actual probabilities, but rather it provides empirical estimates for what systematists define as different populations, species or genera. Of course, defining criteria vary from one taxonomic group to another, making it difficult to draw conclusions based on all comparisons. Interestingly, however, we now have so many estimates for these comparisons that it is possible to generate specific, accurate, and thus useful estimates for selected taxonomic groups.

There are so many different distance measures available for current use that one is often confronted with the problem of having to choose one. Computer simulations have shown that certain distance measures are more reliable than others, depending on the data. For instance, Takezaki & Nei (1996), using the infinite allele model (appropriate for allozyme data), have suggested that Nei et al.'s (1983) D_{A} distance measure is the most appropriate:

$$D_{\text{A}} = 1 - \frac{1}{r}\sum_{j}^{r}\sum_{i}^{m_j}\sqrt{x_{ij}y_{ij}},$$

where r is the number of loci assayed and m_j is the number of alleles for the jth locus.

More recently, Goldstein et al. (1995) developed a distance measure, based on the stepwise mutation model (appropriate for microsatellite data), which can be used to estimate divergence times when the mutation rate is known. The greatest advantages of this distance measure are that it requires no external calibration for the rate of molecular evolution and it is independent of population size. Computer simulations have also shown that this distance measure $(\delta\mu^2)$ performs quite well under the stepwise mutation model (Takezaki & Nei, 1996) when the number of loci assayed is high. The measure of Goldstein et al. (1995) is given by:

$$(\delta\mu^2) = (\mu_{\text{A}} - \mu_{\text{B}})^2,$$

where μ_{A} and μ_{B} are the means of allele sizes for populations A and B, respectively.

Phylogenetic reconstruction methods

At present there are numerous phylogenetic reconstruction methods (Nei, 1991, 1996) but, due to space limitations, we will focus on the two algorithms most often used in population genetics, unweighted pair group mean analysis (UPGMA; Sneath & Sokal, 1973) and neighbour-joining (NJ; Saitou & Nei, 1987) algorithms.

UPGMA

UPGMA, or unweighted pair-group method with arithmetic means, is one of the most widely used and simplest methods of topological reconstruction. Computer simulations suggest that this method performs poorly when the evolutionary rates are not constant, i.e. if we cannot assume a constant molecular clock (Saitou & Imanishi, 1989; Nei, 1991).

The UPGMA method was described in 1973 for constructing taxonomic phenograms, or trees that group OTUs by phenotypic similarity, and has been shown to reconstruct phylogenies fairly well when the molecular clock assumption holds. In fact, when the molecular clock functions perfectly, this is the method which performs best (Nei, 1991). Other advantages of this method are (1) that it is extremely fast (a tree with hundreds of species can be constructed by a program in a few seconds) and (2) that, since it averages out the branch lengths, it diminishes associated errors. This latter point is important when the data set is subject to large errors (e.g. number of loci or sites is small), and UPGMA seems to be particularly useful in these cases.

Given a distance matrix, the UPGMA algorithm groups the pair of taxa (for instance, species A and B) with the smallest distance (branch lengths are always half the distance between OTUs), then recalculates the distance matrix treating A and B as a single OTU and the distance between AB and C as the average distance between A and C, and B and C.

Neighbour-joining
This method is one of the minimum evolution methods (Cavalli-Sforza & Edwards, 1967; Saitou & Imanishi, 1989), where the tree with the smallest sum of branch lengths is found. The minimum evolution algorithm originally described is computationally intensive, since it searches the minimum branch length tree among all possible trees. Obviously, the number of possible trees increases with the number of OTUs, and rapidly reaches a number impossible to compute.

For this reason, Saitou and Nei (1987) proposed an heuristic algorithm for this minimum evolution tree and called it neighbour-joining, where the minimum evolution principle is implicit on each step of the algorithm, producing a single bifurcating tree. The neighbour-joining algorithm starts with a star-tree topology and sequentially chooses the pair of OTUs (or neighbours) that minimise the total length of the tree. This process is continued until the tree is completely resolved. It has been shown to perform quite well (that is, it recovers the correct tree most of the time), when unbiased distances are used (Rzhetsky & Nei, 1993).

Tests of tree reliability

Bootstrap test
Any given data set, regardless of its phylogenetic signal content, will yield a tree. Thus, one has to be extremely careful when drawing conclusions simply based on a tree. For this reason, various statistical tests have been developed in order to provide estimates of the reliability of phylogenetic trees. The most commonly used statistical test in phylogenies is the bootstrap (Efron, 1982). The bootstrap technique uses resampling with replacement and was developed for phylogenetic studies by Felsenstein (1985). In this case the total number of residues is drawn with replacement and all sites have the same probability of being sampled. Since it is a replacement technique, some sites will be sampled more than once, whereas others will not be sampled. On each replicate a tree is built. After all replications the trees are compared and the proportion of trees that produced a particular grouping is assigned to the corresponding interior branch of the tree.

Hedges (1992) has estimated that, if we perform 2000 bootstrap replicates, there will be a 1% error in the precision (that is, the error of the bootstrap proportion when compared with an infinite number of replicates) of a 95% bootstrap value. Therefore, he suggests that a minimum of 2000 bootstrap replicates should be performed before drawing conclusions based on bootstrap proportions.

The logic behind the bootstrap technique is that if we resample the data, leaving out some part of them, and the result is a completely different topology, the data should not be trusted in the first place. The bootstrap provides a measure of the internal consistency of the data set. In a more theoretical sense, the bootstrap test does not yield the probability that a certain group is present on the true tree (accuracy) but, rather, they give the probability that the same group will be present if we use a completely different data set (repeatability). The assumption behind the bootstrap approach is that the data evolve independently and are identically distributed (*iid*). For this reason resampling techniques should always resample sites and not OTUs, since, because of their common phylogenetic history, the latter do not satisfy the assumptions.

The bootstrap technique has been widely used on DNA and amino acid sequence data, specifically when the branch lengths are quite large, that is, when dealing with highly divergent species. For non-sequence types of molecular data, the bootstrap technique may not be appropriate because the *iid* assumptions do not apply (Van Dongen, 1995).

Most of the time bootstrap values are interpreted as the probability that a certain group belongs to the true phylogeny. However, the bootstrap technique has been shown to provide conservative estimates of this probability (Hillis & Bull, 1993; Sitnikova et al., 1995). For example, when the true tree is a star phylogeny (i.e. no phylogenetic resolution), 5% of the tests are expected to show values that are higher than 95%. However, in this case, the actual proportion of bootstrap values higher than 95% is much smaller, and it decreases dramatically with an increase in the number of OTUs on the tree (Sitnikova et al., 1995). Thus, more recently, other tests have been developed in order to estimate this probability more accurately. These are discussed below.

Confidence probability tests

As mentioned above, the bootstrap technique is a statistical test that estimates the internal consistency of the data, and the repeatability of the data set. However it is also useful to compute probabilities that certain groups are present in the true tree, since we are often concerned with unfolding the phylogenetic relationships of our data set. The first approach to determine these probabilities, proposed by Nei et al. (1985)

provides a method to compute the standard errors of branch points of an evolutionary tree.

Another approach is to estimate the probability of positiveness of interior branches (Rzhetsky & Nei, 1992). This test estimates the type I error on the assumption that the interior branch has a length different from zero. Thus, the null hypothesis in this case is that the branch has zero length (i.e. has no phylogenetic resolution) and the alternative hypothesis is that the branch length is different from zero (i.e. has phylogenetic resolution). The formula for this confidence probability (*CP*) is given by:

$$CP = (1 - a) \times 100,$$

where a is the type I error associated with the hypothesis that the branch length is positive. Note that this test can also be used to compare alternative phylogenies, since the various internal nodes that separate different trees can be tested to establish whether their lengths are different from zero.

Computer simulations have shown that this test, unlike bootstrap tests, does not underestimate the probability that a particular group is present in the true tree (Sitnikova et al., 1995). However, using a known vertebrate phylogeny, Russo (1997) performed empirical tests of the CP and bootstrap tests, demonstrating that both tests give robust estimates of the reliability of phylogenies. In that study, which used all protein coding genes of the mitochondrial DNA, the percentage of wrong partitions significantly supported (type I error) by these tests was below that expected by chance (i.e. $< 5\%$; Russo, 1997). This provides a strong indication that these tests are good estimators of the reliability of the tree and hence their use is to be encouraged.

Conclusions

In this review we have indicated that special attention should be paid to underlying assumptions when choosing molecular techniques and procedures for statistical analysis. In the same way, the appropriate technique for a specific problem should be that which can produce information compatible with the required method of analysis. The application of molecular techniques to studies of population biology is expanding rapidly. Consequently, the amount of scientific literature in this field is enormous and encompasses a broad range of biological problems. The diversity of molecular techniques used is considerable, as are the analytical methods. Our review is intended to provide a brief overview of the state of the field, adding some guidelines to aid the choice of techniques and of analytical methods to tackle problems likely to be encountered in studies of molecular population genetics.

Acknowledgements

The authors wish to thank Antonio B. Carvalho, Brenden S. Holland, Antonio M. Solé-Cava, Naoko Takezaki and John P. Thorpe for helpful discussions and comments on a previous version of the manuscript. This work was funded by CNPq from Brazilian Education Ministry and Fapers.

References

Allendorf, F. W. & S. R. Phelps, 1981. Use of allelic frequencies to describe population structure. Can. J. Fish. aquat. Sci. 38: 1507–1514.

Avise, J. C., 1994. Molecular Markers, Natural History and Evolution. Chapman & Hall, London.

Avise, J. C., 1996. Introduction: the scope of conservation genetics. In Avise, J. C. & J. L. Hamrick (eds), Conservation Genetics. Chapman & Hall, New York: 1–9.

Avise, J. C. & G. C. Johns, 1999. Proposal for a standardised temporal scheme of biological classification for extant species. Proc. natn. Acad. Sci. U.S.A. 96: 7358–7363.

Baker, C. S. & S. R. Palumbi, 1994. Which whales are hunted? A molecular genetic approach to monitoring whaling. Science 265: 1538–1539.

Bell, J. & S. B. Friedman, 1994. Genetic structure and diversity within local populations of *Bacillus mycoides*. Evolution 48: 1698–1714.

Bensche, S., D. Hasselquist & T. V. Schants, 1994. Genetic similarity between parents predicts hatching failure: Nonincestuous inbreeding in the great reed warbler? Evolution 48: 317–326.

Berry, A. & M. Kreitman, 1993. Molecular analysis of an allozyme cline: Alcohol dehydrogenase in *Drosophila melanogaster* on the east coast of North America. Genetics 134: 869–893.

Beynon, C. M. & D. O. F. Skibinski, 1996. The evolutionary relationships between three species of mussel (*Mytilus*) based on anonymous DNA polymorphisms. J. exp. mar. Biol. Ecol. 203: 1–10.

Block, B. B., A. F. R. Finnerty, A. F. R. Stewart & J. Kidd, 1993. Evolution of endothermy in fish: mapping physiological traits on a molecular phylogeny. Science 260: 210–214.

Boulding, E. G., J. D. G. Boom & A. T. Beckenbach, 1993. Genetic variation in one bottlenecked and two wild populations of the Japanese scallop (*Patinopecten yessoensis*): empirical parameter estimates from coding regions of mitochondrial DNA. Can. J. Fish. aquat. Sci. 50: 1147–1157.

Brohman, L., A. Rambaut & P. H. Harvey, 1996. Determinants of rate variation in mammalian DNA sequence evolution. J. mol. Evol. 43: 610–621.

Cai, Q.-Q. & I. Touitou, 1993. Excess PCR primers may dramatically affect SSCP efficiency. Nucl. Acids Res. 21: 3909–3910.

Cavalli-Sforza, L. L. & A. W. F. Edwards, 1967. Phylogenetic analysis: models and estimation procedures. Am. J. hum. Genet. 19: 233–257.

Cipriano, F. P. & S. R. Palumbi, 1999. Genetic tracking of a protected whale. Nature 397: 307–308.

Clark, A. G., 1990. Inference of haplotypes from PCR-amplified samples of diploid populations. Mol. Biol. Evol. 7: 111–122.

Cluster, P. D. & R. W. Allard, 1995. Evolution of ribosomal DNA (rDNA) genetic structure in colonial Californian populations of *Avena barbata*. Genetics 139: 941–954.

Collins, A. G., 1998. Evaluating multiple alternative hypotheses for the origin of Bilateria: An analysis of 18S rRNA molecular evidence. Proc. natn. Acad. Sci. U.S.A. 95: 15458–15463.

Comeron, J., M. Kreitman & M. Aguadé, 1999. Natural selection on synonymous sites is correlated with gene length and recombination in *Drosophila*. Genetics 151: 239–249.

Côrte-Real, H. B. S. M., D. R. Dixon & P. W. H. Holland, 1994. Intron-targeted PCR: a new approach to survey neutral DNA polymorphism in bivalve populations. Mar. Biol. 120: 407–413.

Dowling, T. E., C. Moritz, J. D. Palmer & L. H. Rieseberg, 1996. Nucleic acids III: Analysis of fragments and restriction sites. In Hillis, D. M., C. Moritz & B. K. Mable (eds), Molecular Systematics. Sinauer Associates, Sunderland, Massachusetts: 249–320.

Echelle, A. A. & T. E. Dowling, 1992. Mitochondrial DNA variation and evolution of the death valley pupfishes (Cyprinidon, Cyprinodontidae). Evolution 46: 193–206.

Edwards, C. A. & D. O. F. Skibinski, 1987. Genetic variation of mitochondrial-DNA in mussel (*Mytilus edulis* and *Mytilus galloprovincialis*) populations from south-west England and south Wales. Mar. Biol. 94: 547–556.

Edwards, J. H., D. G. Boyd, C. M. Strom, D. Goldman & J. Long, 1995. More on DNA typing dispute. Nature 373: 98–99.

Efron, B., 1982. The jackknife, the bootstrap and other resampling plans. Society for Industrial and Applied Mathematics, Philadelphia, Pennsylvania.

Elliott, N. G., A. J. Smolenski & R. D. Ward, 1994. Allozyme and mitochondrial DNA variation in orange roughy, *Haplostethus atlanticus* (Teleostei: Trachichthyidae): little differentiation between Australian and North Atlantic populations. Mar. Biol. 119: 621–627.

Fan, E., D. B. Levin, B. W. Glickman & D. M. Logan, 1993. Limitations in the use of SSCP analysis. Mutat. Res. 288: 85–92.

Felsenstein, J., 1985. Confidence limits on phylogenies: an approach using the bootstrap. Evolution 39: 783–791.

Fisher, C. & D. O. F. Skibinski, 1990. Sex-biased mitochondrial DNA heteroplasmy in the marine mussel *Mytilus*. Proc. r. Soc., Lond. B 242: 149–156.

Franklin, I. R. & R. Frankham, 1998. How large must populations be to retain evolutionary potential? Anim. Conserv. 1: 69–70.

Fu, Y. X., 1996. Estimating the age of the common ancestor of a DNA sample using the number of segregating sites. Genetics 144: 829–838.

Geller, J. B., J. T. Carlton & D. A. Powers, 1994. PCR-based detection of mtDNA haplotypes of native and invading mussels on the northeastern Pacific coast: latitudinal pattern of invasion. Mar. Biol. 119: 243–249.

Gillespie, J. H., 1987. Molecular evolution and the neutral allele theory. Oxford Surv. evol. Biol. 4: 10–37.

Gillham, N. W., 1994. Organelle Genes and Genomes. Oxford University Press, New York.

Goldstein, D. B. & D. D. Pollock, 1997. Launching microsatellites: a review of mutation processes and methods of phylogenetic inference. J. Hered. 88: 335–342.

Goldstein, D. B., A. R. Linares, L. L. Cavalli-Sforza & M. W. Feldman, 1995. Genetic absolute dating based on microsatellites and the origin of modern humans. Proc. natn. Acad. Sci. U.S.A. 92: 6723–6727.

Goldstein, D. B., G. W. Roemer, D. A. Smith, D. E. Reich, A. Bergman & R. K. Wayne, 1999. The use of microsatellite variation to infer population structure and demographic history in a natural model system. Genetics 151: 797–801.

Grewe, P. M., C. C. Krueger, C. F. Aquadro, E. Bermingham, H. L. Kincaid & B. May, 1993. Mitochondrial DNA variation among lake trout (*Salvelinus namaycush*) strains stocked into Lake Ontario. Can. J. Fish. aquat. Sci. 50: 2397–2403.

Hall, H., J. Beardmore & D. O. F. Skibinski, 1995. A study of mitochondrial DNA variation in brown trout in Wales. In Sakagawa, G. (ed.), Assessment Methodologies and Management. Proceedings of the World Fisheries Congress, theme 5. Oxford and IBH Publishing, New Delhi: 193–195.

Hanfstingl, U., A. Berry, E. A. Kellogg, J. T. Costa III, W. Rudiger & F. M. Ausubel, 1994. Haplotypic divergence coupled with lack of diversity at the *Arabidopsis thaliana* alcohol dehydrogenase locus: roles for both balancing and directional selection? Genetics 138: 811–828.

Harpending, H. C., M. A. Batzer, M. Guiven, M., L. B. Jorde, A. R. Rogers & S. T. Sherry, 1998. Genetic traces of ancient demography. Proc. natn. Acad. Sci. U.S.A. 95: 1961–1967.

Hartl, D. L. & A. G. Clark, 1997. Principles of population genetics. Sinauer Associates, Sunderland, Massachusetts.

Hayashi, K., 1991. PCR-SSCP: a simple and sensitive method for detection of mutations in the genomic DNA. PCR Meth. Applicat. 1: 34–38.

Heath, D. D., P. D. Rawson & T. J. Hilbish, 1995. PCR-based nuclear markers identify alien blue mussel (*Mytilus* spp.) genotypes on the west coast of Canada. Can. J. Fish. aquat. Sci. 52: 2621–2627.

Hedges, S. B., 1992. The number of replications needed for accurate estimation of the bootstrap P value in phylogenetic studies. Mol. Biol. Evol. 9: 366–369.

Hey, J., 1999. The neutralist, the fly and the selectionist. Trends Ecol. Evol. 14: 35–38.

Hillis, D. M. & J. Bull, 1993. An empirical test of bootstrapping as a method for assessing confidence limits in phylogenetic analysis. Syst. Biol. 42: 182–192.

Hillis, D. M., C. Moritz & B. K. Mable, 1996a. Molecular Systematics. 2nd edition. Sinauer Associates, Sunderland, Massachusetts.

Hillis, D. M., B. K. Mable, A. Larson, S. K. Davis & E. A. Zimmer, 1996b. Nucleic acids IV: Sequencing and cloning. In Hillis, D. M., C. Moritz & B. K. Mable (eds), Molecular Systematics. 2nd edn. Sinauer Associates, Sunderland, Massachusetts: 321–381.

Hoech, W. R., K. H. Blakley & W. M. Brown, 1991. Heteroplasmy suggests limited biparental inheritance of *Mytilus* mitochondrial DNA. Science 251: 1488–1490.

Holmes, E. C., M. Worobey & A. Rambaut, 1999. Phylogenetic evidence for recombination in Dengue virus. Mol. Biol. Evol. 16: 405–409.

Hughes, A. L. &. M. Nei, 1989. Nucleotide substitution at major histocompatibility complex class II loci: evidence for overdominant selection. Proc. natn. Acad. Sci. U.S.A. 86: 958–962.

Hughes, A. L., M. K. Hughes, C. Y. Howell & M. Nei, 1994. Natural selection at the class II major histocompatibility complex loci of mammals. Phil. Trans. r. Soc. Lond. B 345: 359–367.

Hurst, C. D. & D. O. F. Skibinski, 1995. Comparison of allozyme and mitochondrial DNA spatial differentiation in the limpet *Patella vulgata*. Mar. Biol. 122: 257–263.

Innan, H. T. & F. Tajima, 1997. The amounts of nucleotide variation within and between allelic classes and the reconstruction of the common ancestral sequence in a population. Genetics. 147: 1431–1444.

Jablonski, D., 1986. Larval ecology and macroevolution in marine invertebrates. Bull. mar. Sci. 39: 565–587.

Jablonski, D. & R. A. Lutz, 1983. Larval ecology of marine benthic invertebrates: paleobiological implications. Biol. Rev. 58: 21–89.

Jackson, J. B. C., 1986. Modes of dispersal of clonal benthic invertebrates: consequences for species' distributions and genetic structure of local populations. Bull. mar. Sci. 39: 58–606.

Johannesson, K., B. Johanesson & U. Lundgren, 1995. Strong natural selection causes microscale allozyme variation in a marine snail. Proc. natn. Acad. Sci. U.S.A. 92: 2602–2606.

Karl, S. A. & J. C. Avise, 1993. PCR-based assays of Mendelian polymorphisms from anonymous single-copy nuclear DNA: Techniques and applications for population genetics. Mol. Biol. Evol. 10: 342–361.

Kasuga, T., J. Cheng & K. R. Mitchelson, 1995. Metastable single-strand DNA conformational polymorphism analysis results in enhanced polymorphism detection. PCR Meth. Applicat. 4: 227–233.

Kenchington, E., K. S. Naidu, D. L. Roddick, D. I. Cook & E. Zouros, 1993. Use of biochemical genetic markers to discriminate between adductor muscles of the sea scallop (*Placopecten magellanicus*) and the Iceland scallop (*Chlamys islandica*). Can. J. Fish. aquat. Sci. 50: 1222–1228.

Kimura, M., 1968. Evolutionary rate at the molecular level. Nature 217: 624–626.

Kimura, M., 1995. Limitations of Darwinian selection in a finite population. Proc. natn. Acad. Sci. U.S.A. 92: 2343–2344.

Lambert, D. M., C. D. Millar, K. Jack, S. Anderson & J. L. Craig, 1994. Single- and multilocus DNA fingerprinting of communally breeding pukeko: do copulations or dominance ensure reproductive success? Proc. natn. Acad. Sci. U.S.A. 91: 9641–9645.

Lander, E. S. & B. Budowle, 1994. DNA fingerprinting dispute laid to rest. Nature 371: 735–738.

Lessa, E. P., 1992. Rapid surveying of DNA sequence variation in natural populations. Mol. Biol. Evol. 9: 323–330.

Lessa, E. P. & G. Applebaum, 1993. Screening techniques for detecting allelic variation in DNA sequences. Mol. Ecol. 2: 119–129.

Lewin, R., 1994. Fact, fiction and fossil DNA. New Scientist 141: 38–41.

Lewontin, R. C. & D. L. Hartl, 1994. Forensic DNA typing dispute. Nature 372: 398–399.

Lewontin, R. C. & J. L. Hubby, 1966. A molecular approach to the study of genetic heterozygosity in natural populations. II: Amount of variation and degree of heterozygosity in natural populations of *Drosophila pseudoobscura*. Genetics 54: 595–609.

Lynch, M. & R. Lande, 1998. The critical effective size for a genetically secure population. Anim. Conserv. 1: 70–72.

Magoulas, A. & E. Zouros, 1993. Restriction-site heteroplasmy in anchovy (*Engraulis encrasicolus*) indicates incidental biparental inheritance of mitochondrial DNA. Mol. Biol. Evol. 10: 319–325.

McMillan, W. O., R. A. Raff & S. R. Palumbi, 1992. Population genetic consequences of developmental evolution in sea urchins (genus *Heliocidaris*). Evolution 46: 1299–1312.

Morin, P. A., J. J. Moore, R. Chakraborty, L. Jin, J. Goodall & D. S. Woodruff, 1994. Kin selection, social structure, gene flow, and the evolution of chimpanzees. Science 265: 1193–1201.

Moritz, C., J. W. Wright & W. M. Brown, 1992. Mitochondrial DNA analyses and the origin and relative age of parthenogenetic *Cnemidophorus*: phylogenetic constraints on hybrid origins. Evolution 46: 184–192.

Murphy, R. W., J. W. Sites Jr., D. G. Buth & C. H. Haufler, 1996. Proteins: isozyme electrophoresis. In Hillis, D. M., C. Moritz & B. K. Mable (eds), Molecular Systematics. Sinauer Associates, Sunderland, Massachusetts: 51–120.

Nei, M., 1972. Genetic distance between populations. Am. Nat. 106: 283–292.

Nei, M., 1973. The theory and estimation of genetic distance. In Norton, N. E. (ed.), Genetic Structure of Populations. University Press of Hawaii, Honolulu: 45–54.

Nei, M., 1977. F-statistics and analysis of gene diversity in subdivided populations. Ann. hum. Genet. 41: 225–233.

Nei, M., 1983. Genetic polymorphism and the role of mutation in evolution. In Nei, M. & R. Koehn (eds), Evolution of Genes and Proteins. Sinauer Press, Sunderland, Massachusetts: 165–190.

Nei, M., 1987. Molecular Evolutionary Genetics. Columbia University Press, New York.

Nei, M., 1988. Relative roles of mutation and selection in the maintenance of genetic variability. Phil. Trans. r. Soc., Lond. B 319: 615–629.

Nei, M., 1991. Relative efficiencies of different tree-making methods for molecular data. In Miyamoto, M. M. & J. Cracraft (eds), Phylogenetic Analysis of DNA Sequences. Oxford University Press, New York: 90–128.

Nei, M., 1996. Phylogenetic analysis in molecular evolutionary genetics. Ann. Rev. Genet. 30: 371–403.

Nei, M., J. C. Stephens & N. Saitou., 1985. Methods for computing the standard errors of branching points in an evolutionary tree and their application. Mol. Biol. Evol. 2: 66–85.

Nei, M., F. Tajima & Y. Tateno, 1983. Accuracy of estimated phylogenetic trees from molecular data. II. Gene frequency data. J. mol. Evol. 19: 153–170.

Nielsen, D. A., A. Novoradovsky & D. Goldman, 1995. SSCP primer design based on single-strand DNA structure predicted by a DNA folding program. Nucl. Acids Res. 23: 2287–2291.

Ohresser, M., P. Borsa & C. Delsert, 1997. Intron-length polymorphism at the actin gene locus mac-1: A genetic marker for population studies in the marine mussels *Mytilus galloprovincialis* Lmk. and *M. edulis* L. Mol. mar. Biol. Biotechnol. 6: 123–130.

Ohta, T., 1974. Mutational pressure as the main cause of molecular evolution and polymorphism. Nature 252: 351–354.

Ohta, T., 1992. The nearly neutral theory of molecular evolution. Ann. Rev. Ecol. Syst. 23: 263–286.

Oldroyd, B. P., A. J. Smolenski, J.-M. Cornuet & R. H. Crozier, 1994. Anarchy in the beehive. Nature 371: 749.

Oldroyd, B. P., J.-M. Cornuet, D. Rowe, T. E. Rinderer & R. H. Crozier, 1995. Racial admixture of *Apis mellifera* in Tasmania, Australia: similarities and differences with natural hybrid zones in Europe. Heredity 74: 315–325.

Olson, R. R. & R. McPherson, 1987. Potential *vs.* realised larval dispersal: fish predation on larvae of the ascidian *Lissoclinum patella* (Gottschaldt). J. exp. mar. Biol. Ecol. 110: 245–256.

Orr, H. A., 1998. Testing natural selection *vs.* genetic drift in phenotypic evolution using quantitative trait locus data. Genetics 149: 2099–2104.

Ota, T. & M. Nei, 1995. Evolution of immunoglobulin V_H pseudogenes in chickens. Mol. Biol. Evol. 12: 94–102.

Palumbi, S. R., 1996. Nucleic acids II: the polymerase chain reaction. In Hillis, D. M., C. Moritz & B. K. Mable (eds), Molecular Systematics. Sinauer Associates, Sunderland, Massachusetts: 205–247.

Palumbi, S. R. & C. S. Baker, 1994. Contrasting population structure from nuclear intron sequences and mtDNA of humpback whales. Mol. Biol. Evol. 11: 426–435.

Pfennig, D. W. & H. K. Reeve, 1993. Nepotism in a solitary wasp as revealed by DNA fingerprinting. Evolution 47: 700–704.

Queller, D. C., J. E. Strassmann & C. R. Hughes, 1993. Microsatellites and kinship. Trends Ecol. Evol. 8: 285–288.

Quesada, H., M. Warren & D. O. F. Skibinski, 1999. Nonneutral evolution and differential mutation rate of gender-associated mitochondrial DNA lineages in the marine mussel *Mytilus*. Genetics 149: 1511–1526.

Reich, D. E., M. Feldman & D. B. Goldstein, 1999. Statistical properties of two tests that use multilocus data sets to detect population expansions. Mol. Biol. Evol. 16: 453–466.

Ritte, U. & A. Pashtan, 1982. Extreme levels of genetic variability in two Red Sea *Cerithium* species (Gastropoda: Cerithidae). Evolution 36: 403–407.

Rosel, P. E., A. E. Dizon & J. E. Heyning, 1994. Genetic analysis of sympatric morphotypes of commom dolphins (genus *Delphinus*). Mar. Biol. 119: 159–167.

Russo, C. A. M., 1997. Efficiencies of different statistical tests in supporting a known vertebrate phylogeny. Mol. Biol. Evol. 14: 1078–1080.

Russo, C. A. M., A. M. Solé-Cava & J. P. Thorpe, 1994. Population structure and genetic variation in two tropical sea anemones (Cnidaria, Actiniidae) with different reproductive strategies. Mar. Biol. 119: 267–276.

Rzhetsky, A. & M. Nei, 1992. A simple method for estimating and testing minimum-evolution trees. Mol. Biol. Evol. 9: 945–967.

Rzhetsky, A. & M. Nei, 1993. Theoretical foundation of the minimum-evolution method of phylogenetic inference. Mol. Biol. Evol. 10: 1073–1095.

Saiki, R., D. H. Gelfand, S. Stoffel, S. Scharf, R. Higuchi, G. T. Horn, K. B. Mullis & H. A. Erlich, 1988. Primer directed enzymatic amplification of DNA with a thermostable DNA polymerase. Science 239: 487–491.

Saitou, N. & T. Imanishi, 1989. Relative efficiencies of the Fitch-Margoliash, maximum parsimony, maximum-likelihood, minimum-evolution, and neighbor-joining methods of phylogenetic tree construction in obtaining the correct tree. Mol. Biol. Evol. 6: 514–525.

Saitou, N. & M. Nei, 1987. The neighbor-joining method: a new method for reconstructing phylogenetic trees. Mol. Biol. Evol. 4: 406–425.

Sanger, S., S. Nicklen & A. R. Coulson, 1977. DNA sequencing with chain-terminating inhibitors. Proc. natn. Acad. Sci. U.S.A. 74: 5463–5467.

Silberman, J. D. & P. J. Walsh, 1994. Population genetics of the spiny lobster, *Panulirus argus*. Bull. mar. Sci. 54: 1084.

Silva, E. P., 1998. Population genetic studies of the mussel *Mytilus* using nuclear DNA. Unpublished Ph.D. thesis. University of Wales, Swansea.

Sitnikova, T., A. Rzhetsky & M. Nei, 1995. Interior branch and bootstrap tests of phylogenetic trees. J. mol. Evol. 12: 319–333.

Skibinski, D. O. F., 1994. The potential of DNA techniques in the population and evolutionary genetics of aquatic invertebrates. In Beaumont, A. R. (ed.), Genetics and Evolution of Aquatic Organisms. Chapman & Hall, London: 177–199.

Skibinski, D. O. F., C. Gallagher & C. M. Beynon, 1994a. Mitochondrial-DNA inheritance. Nature 368: 817–818.

Skibinski, D. O. F., C. Gallagher & C. M. Beynon, 1994b. Sex-limited mitochondrial DNA transmission in the marine mussel *Mytilus edulis*. Genetics 138: 801–809.

Slade, R. W., C. Moritz, A. R. Hoelzel & H. R. Burton, 1998. Molecular population genetics of the southern elephant seal *Mirounga leonina*. Genetics 149: 1945–1957.

Slatkin, M., 1985. Rare alleles as indicators of gene flow. Evolution 39: 53–65.

Sneath, P. H. A. & R. R. Sokal, 1973. Numerical taxonomy. W. F. Freeman, San Francisco.

Solé-Cava, A. M. & J. P. Thorpe, 1990. High levels of genetic variation in marine sponges. In Rutzler, K. (ed.), New Perspectives in Sponge Biology. Smithsonian Institution Press, Washington: 332–337.

Solé-Cava, A. M. & J. P. Thorpe, 1991. High levels of genetic variation in natural populations of marine lower invertebrates. Biol. J. linn. Soc. 44: 65–80.

Sturmbauer, C., J. C. Levinton & J. Christy, 1996. Molecular phylogeny analysis of fiddler crabs: test of the hypothesis of increasing behavioral complexity in evolution. Proc. natn. Acad. Sci. U.S.A. 93: 10855–10857.

Swanson, W. J. & V. D. Vacquier, 1995. Extraordinary divergence and positive Darwinian selection in a fusagenic protein coating the acrosomal process of abalone spermatozoa. Proc. natn. Acad. Sci. U.S.A. 92: 4957–4961.

Takezaki, N. & M. Nei, 1996. Genetic distances and reconstruction of phylogenetic trees from microsatellite DNA. Genetics 144: 389–399.

Tanaka, T. & M. Nei, 1989. Positive Darwinian selection observed at the variable-region genes of immunoglobulins. Mol. Biol. Evol. 6: 447–459.

Thorpe, J. P., 1983. Enzyme variation, genetic distance and evolutionary divergence in relation to levels of taxonomic separation. In Oxford, R. S. & D. Rollison (eds), Protein Polymorphism: Adaptive and Taxonomic Significance. Academic Press, London: 131–152.

Thorpe, J. P. & A. M. Solé-Cava, 1994. The use of allozyme electrophoresis in invertebrate systematics. Zool. Scr. 23: 3–18.

Tufto, J. R., A. F. Raybould, K. Hindar & S. Engen, 1998. Analysis of genetic structure and dispersal patterns in a population of sea beet. Genetics 149: 1975–1985.

Van Dongen, S., 1994. How should we bootstrap allozyme data? Heredity 74: 445–447.

Varvio, S. L., R. Chakraborty & M. Nei, 1986. Genetic variation in subdivided populations and conservation genetics. Heredity 57: 189–192.

Vuorinen, J. A., R. A. Bodaly, J. D. Reist, L. Bernatchez & J. J. Dodson, 1993. Genetic and morphological differentiation between dwarf and normal size forms of lake whitefish (*Coregonus clupeaformis*) in Como Lake, Ontario. Can. J. Fish. aquat. Sci. 50: 210–216.

Wang, D. Y.-C., S. Kumar & S. B. Hedges, 1999. Divergence time estimates for the early history of animal phyla and the origin of plants, animals and fungi. Proc. r. Soc., Lond. B 266: 163–171.

Weiss, G. H. & A. Von Haesler, 1998. Inference of population history using a likelihood approach. Genetics 149: 1539–1546.

Welsh, J. & M. McClelland, 1990. Fingerprinting genomes using PCR with arbitrary primers. Nucl. Acids Res. 18: 7213–7218.

Wetton, J. H., T. Burke, D. T. Parkin & E. Cairns, 1995. Single-locus DNA fingerprinting reveals that male reproductive sucess increases with age through extra-pair paternity in the house sparrow (*Passer domesticus*). Proc. r. Soc., Lond. B 260: 91–98.

Williams, J. G. K., A. R. Kubelik, K. J. Livak, J. A. Rafalski & S. V. Tingey, 1990. DNA polymorphisms amplified by arbitrary primers are useful as genetics markers. Nucl. Acids Res. 18: 6531–6535.

Wright, S., 1978. Evolution and the genetics of populations. Vol. 4. Variability within and among natural populations. University of Chicago Press, London.

Zhang, J. & M. Nei, 1996. Evolution of antennapedia-class homeobox genes. Genetics 142: 295–303.

Zink, R. M., 1994. The geography of mitochondrial DNA variation, population structure, hybridization, and species limits in the fox sparrow (*Passerella iliaca*). Evolution 48: 96–111.

Zink, R. M. & D. L. Dittmann, 1993. Gene flow, refugia and evolution of geographic variation in the song sparrow (*Melospiza melodia*). Evolution 47: 717–729.

Zouros, E., A. O. Ball, C. Saavedra & K. R. Freeman, 1994a. Mitochondrial DNA inheritance. Nature 368: 817–818.

Zouros, E., A. O. Ball, C. Saavedra & K. R. Freeman, 1994b. An unusual type of mitochondrial DNA inheritance in the blue mussel *Mytilus*. Proc. natn. Acad. Sci. U.S.A. 91: 7463–7467.

Zouros, E., K. R. Freeman, A. O. Ball & G. H. Pogson, 1992. Direct evidence for extensive paternal mitochondrial-DNA inheritance in the marine mussel *Mytilus*. Nature 359: 412–414.

Hydrobiologia **420:** 137–152, 2000.
A. M. Solé-Cava, C. A. M. Russo & J. P. Thorpe (eds), Marine Genetics.

DNA tests of neutral theory: applications in marine genetics

D. O. F. Skibinski
School of Biological Sciences, University of Wales, Swansea SA2 8PP, U.K.

Key words: DNA sequences, molecular markers, tests of neutrality, positive selection, purifying selection

Abstract

The principal methods of using DNA sequence information to test the neutral theory of evolution and polymorphism are described. These include the use of synonymous and nonsynonymous substitutions for detecting purifying and positive selection, the analysis of nucleotide diversity, mismatch analysis and the HKA, McDonald-Kreitman, Tajima and Ewens-Watterson tests. Analysis of the covariation of different kinds of molecular markers and the relationship between genetic variation and fitness is also considered. Examples of the use of these approaches in a wide variety of marine organisms are described. It is emphasised that tests of neutral theory, in addition to providing important fundamental knowledge about the action of evolutionary forces, provide valuable information about the influence of environmental and demographic factors.

Introduction – DNA sequences and evolutionary forces

The neutral theory of molecular evolution maintains that most evolutionary change, and polymorphism, at the molecular level is due to the random genetic drift of neutral mutations (Kimura, 1983). Neutral theory is a simple theory in that it contains few parameters – population size (N) or effective population size (N_e), mutation rate (μ), migration (m) and divergence time of species (t). For this reason, it is easy to formulate predictions that can be tested using empirical data such as DNA sequences. In practice, neutral theory is quite often rejected. It then proves its utility by providing information about those factors, principally natural selection, than can cause deviations from neutrality. The alternative to neutral theory is selection theory, which holds that patterns of genetic variation within and between species are adaptive and modulated by natural selection. It is more difficult to test selection theory directly because selection can work in diverse ways and values for the additional selection parameters to be included can rarely in practice be determined easily and accurately. Nevertheless, some progress has been made towards formulating selection theories with relatively few parameters (e.g. Gillespie, 1989). There are some complications in distinguishing between the two theories. For example, neutral predictions are most often derived from equilibrium neutral theory where a constant population size is assumed. If population size fluctuates, patterns of variation similar to those expected from selection can be produced. Another consideration is that environments change, and thus selection coefficients are likely to fluctuate over time. Gillespie (1991) has shown that many models with fluctuating selection give predictions that are indistinguishable from those of neutral theory.

It is recognised in neutral theory that many new mutations will impair function and be removed quickly from populations and thus contribute neither to genetic polymorphism nor to evolutionary change in any important way. The selective removal of these disadvantageous mutations is called purifying selection. The bulk of the remaining new mutations will be neutral, although a very small minority will be selectively advantageous. A more recent version of the neutral theory, that is now widely recognised as an improvement, is called the nearly neutral theory (see Ohta, 1992). This argues that most new mutations have a slight effect either positive or negative, on fitness. Because most mutations that have a positive effect on fitness are fixed quickly, the bulk of these nearly neutral mutations are likely to be slightly disadvantageous. In large populations, purifying selection will tend to reduce their frequency although they might occasionally go to fixation: in small populations they will be effectively neutral and their dynamics determined entirely by drift. An important point to note is that in the original neutral theory, the mutations that are subject to purifying selection contribute little to polymorphisms or evolutionary change, whereas in the nearly neutral theory they can contribute in a substantial way. Although nearly neutral theory has moved neutral theory nearer to selection theory, there remains a fundamental distinction. Selection theory focuses on the concept that most genetic variation is adaptive, whereas both the original neutral theory and the nearly neutral theory hold that most genetic variation is either neutral or impairs optimal function to some extent.

The testing of neutral theory through the analysis of DNA sequences leads to valuable fundamental knowledge relating to evolutionary forces. However, the approach can also provide useful information of practical importance to environmental biologists and conservationists. For example, evidence of selection on DNA sequences might be indicative of environmental stresses on populations. Deductions about dispersal using genetic markers are dependent to a large extent on the assumption of selective neutrality of the markers. The justification for conservation policies aimed at preserving genetic diversity within populations might be weakened if the variation is neutral rather than adaptive. Rand (1996), focusing on the use of the Tajima and HKA tests, makes a case for the application of tests of neutrality in conservation genetics. He argues that tests of neutral theory can provide powerful indicators of demographic parameters beyond traditional measures of population structure.

It is intended that this review should explain the most commonly used tests of neutrality based on DNA data and show how these tests can be and have been used in marine organisms. Some explanation of the underlying theory of each of the tests or approaches is explained first. Then there follows a sample of diverse case studies in which combinations of these tests have been applied to particular species. DNA tests of neutral theory have been used quite widely in terrestrial organisms, particularly *Drosophila* and primates, over the last 10–15 years, but have been applied much less frequently in marine species. This might reflect the lag in time required for the transfer of DNA methods to marine species, for which unexpected technical problems can occur. It is hoped that this review will stimulate biologists unfamiliar with the methods described to gain further information and deeper insight from molecular data, gathered perhaps for other purposes.

Tests and approaches

Synonymous and nonsynonymous substitution rates

By aligning two homologous coding DNA sequences obtained either from the same or different species, it is possible to estimate the number of nonsynonymous substitutions (D_n) and synonymous substitutions (D_s) per nucleotide site that have occurred over the evolutionary time that separates the sequences. It is assumed generally that synonymous substitutions are close to being neutral, thus a difference in D_s between two genes should reflect a difference in neutral mutation rate. Valuable evolutionary information can be obtained by comparing D_n and D_s. If the sequences have diverged solely as the result of genetic drift then $D_n/D_s = 1$, assuming that the neutral mutation rates for nonsynonymous and synonymous sites are the same. If purifying selection has acted on the sequences, D_n will be reduced because deleterious nonsynonymous mutations will increase in frequency less readily within populations and are thus less likely to become fixed between lineages. There will be less or no effect on D_s, thus with purifying selection, $D_n/D_s < 1$. If positive selection has acted to increase the frequency of some beneficial new amino acid changing, nonsynonymous, mutations within populations, D_n but not D_s, will be elevated. Thus, if $D_n/D_s > 1$, this can taken as good evidence of positive selection.

Many of the genes where $D_n/D_s > 1$ code for the antigenic surface proteins of parasites and viruses (Endo et al., 1996), presumably reflecting the race to escape the immune system of their hosts. Endo et al. (1996) examined 3595 groups of homologous sequences from the DNA databases and found only 17 (0.5%) candidate loci for positive selection which had $D_n/D_s > 1$ over the entire gene sequence. Because positive selection at some sites might be masked by purifying selection at others, the observation that $D_n/D_s < 1$ cannot be taken as evidence that positive selection is absent from particular regions within a gene. When the analysis was focused at a finer level on regions within genes, 5% of genes gave some evidence of positive selection. It should be noted that the ratio D_n/D_s could change over time if population size changes. For example, when purifying selection, as in the nearly neutral theory, is operating in a large population, D_n/D_s will be less than one. If population size then decreases, the nonsynonymous variation becomes more effectively neutral, the evolutionary rate becomes equal to the mutation rate (μ)

and both D_n and D_n/D_s will increase. With positive direction selection, the rate of evolution is directly related to population size (Kimura, 1983), thus a reduction in population size will result in a slowing of nonsynonymous evolution and D_n/D_s will decrease.

Nucleotide diversity

The amount of genetic diversity within populations can be measured in numerous ways. With data in the form of allele and genotype frequencies, as with allozyme data for example, observed heterozygosity, or that expected assuming Hardy-Weinberg equilibrium, is frequently used. In theory, heterozygosity (h) is a simple function of $N_e\mu$. For example, for the infinite allele neutral model, $h = M/(1+M)$, where $M = 4N_e\mu$ (Kimura & Crow, 1964). For DNA, nucleotide diversity (k) can be calculated as the number of nucleotide differences between a pair of sequences, averaged across all possible pairs of sequences in the sample. Tajima (1983) showed that $M = k$. Frequently nucleotide diversity is expressed on a per site basis to control for differences in sequence length between genes being compared. If neutrality is assumed, a difference in k between two gene regions within a population or between the same gene in different populations might be interpreted as a difference in neutral mutation rate. However, the estimate of k will be subject, not only to sampling error of sequences, but also to the drift variance due to the fluctuation of k over time as the result of genetic drift. The drift variance will usually be substantially greater than the sampling error. Repeated sampling either over evolutionary time or over replicate loci evolving independently could reduce the drift variance, but this cannot often be done in practice. Increasing the length of the DNA sequence under study will help. However, increasing the sample size of sequences studied is of little use, as this would mainly affect only the sampling error. More recently, estimators of M, with lower errors than k, have been developed that make use of phylogenetic information in a sample of sequences (e.g. Fu, 1994). A significant difference in contemporary genetic diversity or allele frequencies between two populations, taking account of sampling error alone, might be useful if the main focus were the immediate consequences of genetic conservation or stock assessment policy, but would provide little information on underlying evolutionary forces.

The Tajima test

Increased useful information about evolutionary forces can be obtained by comparing the results of different estimators of M. This is the basis of the test of Tajima (1989). Watterson (1975) showed that in neutral theory $M = S/a$, where S is the number of sites along a DNA sequence that are segregating, that is not fixed, within a sample of sequences, and a is a parameter related to sample size. The Tajima test statistic, D, has the quantity $(k - S/a)$ as numerator. In neutral theory, it is expected that $D = 0$. Purifying selection, which results in many rare variants and a single common variant within a population, would elevate the number of segregating sites (S), but depress the average number of differences between sequences (k), giving $D < 0$. By contrast, balancing selection operating over a long period of time and resulting in the maintenance of just two or a few diverged sequences within a population would elevate the value of k, but depress the value of S, giving $D > 0$. Deviation from neutrality in the Tajima test could be caused by other factors. For example, a low level of immigration from a genetically distinct population would introduce rare sequences and make D negative. Population expansion would make previously rare sequences more common and lead to a deficit of rare variants, making D more positive in value. Population expansion after a bottleneck would be followed by the accumulation of many new rare variants by mutation reducing the value of D. A selective sweep, fixing a single advantageous sequence, would have a similar effect as new mutant sequences begin to accumulate.

The Ewens–Watterson test

One of the earliest tests of neutral theory compares the expected homozygosity in a sample of alleles with that expected under the infinite allele equilibrium neutral model. This is the Ewens–Watterson test of selective neutrality (Watterson, 1978). Given a sample size and the number of alleles observed in the sample, the test determines the probability of observing an expected homozygosity higher (or lower) than that actually observed, under equilibrium neutral theory. Improbably high homozygosity is associated with having a single allele at high frequency, many rare alleles and a deficit of alleles of intermediate frequency, a distribution expected with purifying selection. Improbably low homozygosity is associated with several alleles at high frequency and few rare or intermediate frequency alleles, an expected consequence of balancing selection. The Ewens–Watterson test, like the Tajima test, is also sensitive to migration and to non-equilibrium conditions associated with population expansion and contraction. It regards individual DNA sequences as alleles and thus, unlike the Tajima test, does not make use of sequence information. Moreover, as the test is normally applied it assumes the infinite allele model, loci conforming to a stepwise model, for example microsatellites, might give a deficit of homozygosity even if the variation is neutral.

The HKA test

The HKA test of neutral evolution (Hudson et al., 1987) considers the relationship between polymorphism and divergence at two loci in two related species (or populations). In the infinite allele neutral model, the amount of polymorphism is determined by the parameter $N_e\mu$, whereas the amount of divergence between populations is determined by the parameter μt. In the test, the average N_e in the descendent species is assumed to equal that of the ancestor, and of course t is fixed for two species. Thus both N_e and t can be regarded as fixed parameters. In this circumstance, both polymorphism and divergence increase with increasing μ, and the ratio polymorphism/divergence should have similar values for two genes with different values of μ. The test involves the comparison of the number of polymorphic and fixed sites in each of two DNA sequences studied in the two species. An excess of polymorphism at one locus might then indicate balancing selection. This excess could, however, also be interpreted as an excess of fixed differences at the second locus, indicative of positive directional selection. These two possibilities can be distinguished if neutrality can be assumed at one of the loci, for example this might be a non-coding region or pseudogene.

The test has been applied widely in *Drosophila* species. For example, Moriyama & Powell (1996) carried out 700 tests using 21 gene loci in *D. melanogaster*. Over half of the tests conform to neutrality, but there are also some highly significant results, and the authors conclude that no simple all encompassing explanation accounts for the data satisfactorily. When positive directional selection drives an advantageous mutation to fixation, alleles linked in disequilibrium will also tend to fix through the 'hitchhiking effect'. Such a selective sweep will tend to associate high divergence with low polymorphism leading to rejection of neutrality by the HKA test. The effect is likely to be most marked in regions of low recombination, and empirical evidence does exist for a correlation between recombination rate and nucleotide diversity (e.g. Begun & Aquadro, 1992). As pointed out above, selective sweeps should lead to negative values of Tajima's D, but this has not been found in empirical studies of reduced diversity in regions of low recombination (Aguade & Langley, 1994). Gillespie (1994) argued that in most realistic models, selection is rarely likely to be sufficiently strong or last long enough to produce selective sweeps. Charlesworth et al. (1995) have suggested that selection against deleterious mutations can also contribute to the reduction of variation at linked neutral sites, a process known as background selection. They pointed out that background selection need not lead to negative values of Tajima's D and is thus consistent with empirical observations.

In common with other tests, deviation from neutrality in the HKA test does not necessarily prove selection. The test assumes no recombination, both within and between loci, but failure of data to meet these criteria actually leads to a lower chance of a false positive result, in other words the test is conservative when trying to identify selection. Systematic deviations from equilibrium neutral expectations are unlikely to be caused by changes in population size, as these should have similar effects on variation at both loci. Rejection of neutrality can be caused by changes in mutation rate in the descendent species. This is most likely to occur in the nearly neutral model when a reduction in population size increases the effective neutral mutation rate, as deleterious mutations become effectively neutral. In principle, a variation of the HKA test might be applied using allele frequencies at two allozyme loci in descendent species. However, the large drift variance in this circumstance should render detection of a significant result impossible.

The McDonald–Kreitman test

The McDonald–Kreitman test (McDonald & Kreitman, 1991), like the HKA test, also considers the relationship between polymorphism and divergence, but focuses on nonsynonymous and synonymous variation within a single gene region. This gives a distinct advantage over the HKA test. Because nonsynonymous and synonymous sites are intermingled within the same gene sequence, drift variance should not be a major contributor to differences in diversity between these types of site. In addition, violations of equilibrium conditions, due to changes in populations size that affect both the Tajima and HKA tests, do not affect the McDonald–Kreitman test (Brookfield & Sharp, 1994). In essence, the test is applied to a phylogenetic tree with two major clades or branches separated by a single internal branch linking through a common ancestor. The numbers of nonsynonymous and synonymous changes are counted within the clades (polymorphisms) and on the internal branch (fixations). There has been controversy about how nonsynonymous and synonymous changes should be counted (Graur & Li, 1991; Whittam & Nei, 1991). However, Quesada et al. (1998) showed that different proposed methods lead to almost identical results and conclusions in applications of the test. In neutral theory, evolutionary forces should influence nonsynonymous and synonymous changes similarly. Thus, the ratio of nonsynonymous to synonymous changes should be the same for polymorphisms and fixations. The ratios can be compared in a 2×2 contingency test. An excess of nonsynonymous polymorphisms can be explained as the result of accumulation of mutations that are slightly deleterious within populations but, because they are deleterious, unlikely ever to become fixed between populations. An excess of nonsynonymous fixations can be explained by positive divergent selection. Rand & Kann (1998) applied the McDonald–Kreitman test to mtDNA using ten species pairs from diverse animal groups. All tests showed an excess of nonsynonymous polymorphism and in five tests this was significant. These results provide general support for the nearly neutral theory and for the widespread occurrence of slightly deleterious mutations in mtDNA.

Mismatch analysis

As explained above, nucleotide diversity (k) can be calculated as the number of nucleotide differences between a pair of sequences averaged across all possible pairs of sequences in the sample. The distribution of the individual nucleotide difference values between all pairs of a non-recombining sequence, such as mtDNA, that are averaged to give k, is called a mismatch distribution. The shape of this distribution provides information about the action of evolutionary and historical forces acting on a DNA sequence (Slatkin & Hudson, 1991; Rogers & Harpending, 1992; Rogers, 1995). Analysis of mismatch distributions has been applied extensively to human mtDNA data. In the equilibrium neutral model, mismatch distributions have a bimodal or multimodal distribution. The observed marked unimodal distribution for human data fits the equilibrium neutral model poorly but a model of sudden expansion well (Rogers & Harpending, 1992). However, unimodality may also result from selection or a selective sweep (Rogers, 1995), or in some cases from pooling data from different parts of a subdivided population (Bertorelle & Slatkin, 1995). Although the interpretation of mismatch distributions is controversial, they clearly provide information beyond that contained in a single nucleotide diversity value.

Covariation of molecular markers

Geographic variation in allele frequencies provides valuable information about evolutionary forces, particularly natural selection. The forces of neutral theory, drift, migration and mutation are expected to affect all loci equally, whereas natural selection might act differently at different loci. Thus, greater than expected geographic variation under neutral theory might provide evidence for directional selection, whereas greater than expected geographic homogeneity might suggest that balancing selection is maintaining similar allele frequencies in different populations. An example of this approach applied initially to allozyme loci is the test of Lewontin & Krakauer (1975). The standardised variance in allele frequencies among populations (F_{ST}) is computed for each locus and then the variance of these estimates is compared with a theoretical prediction derived from neutral theory. A difficulty with the approach is that the theoretical variance of F_{ST} is dependent on breeding structure and historical factors which are unknown generally (Nei & Maruyama, 1975; Robertson, 1975).

An alternative approach, now that distinctively different kinds of molecular markers are available, is to compare F_{ST} values for one set of nuclear markers with those for another. The theoretical variance of F_{ST} is not then an issue. Thus, within a single species, a lower than average value of F_{ST} for allozyme loci compared, for example, with nuclear RFLP markers in non-coding regions or with microsatellite loci, might be interpreted either as balancing selection at the allozyme loci or diversifying directional selection at the DNA loci. It is difficult to distinguish between these hypotheses unless it can be assumed that the variation in one set of markers is neutral. Moreover, mtDNA markers cannot usefully be employed because of the confounding effect of the assumed lower effective population size. Nevertheless, a difference between different types of nuclear markers usually indicates selection of some kind. However, some consideration must also be given to the possibility of statistical artifacts. McDonald (1994) argued that F_{ST} values should be used rather than genetic distance estimates and that sample sizes per population should be the same for the different marker sets. Also, to avoid errors associated with global differences in the number of alleles between marker sets, that might, for example, exist between microsatellite and RFLP data, pooling should be used to standardise all loci to two-allele polymorphisms.

With allozyme data, F_{ST} values are frequently converted to estimates of migration (N_m) using an equation derived from equilibrium neutral theory. This assumes an infinite allele model, and also that the mutation rate is negligible compared to the migration rate. These assumptions are not appropriate for microsatellites, which have a much higher mutation rate, and evolve according to a stepwise mutation model. Thus differences in F_{ST} between microsatellites and allozymes might relate to mutational properties rather than selection. Models that take a stepwise model into account have been developed for estimating migration, using microsatellites. For example, the R_{ST} statistic of Slatkin (1995) for microsatellites is analogous to the conventional F_{ST} statistic applied to allozyme data.

One terrestrial example of this approach is interesting as it demonstrates a test of a specific hypothesis derived from selection theory rather than neutral theory. Pyrethroid insecticides act on voltage-gated sodium channels in nerve membranes. Resistance to the pyrethroid permethrin in the tobacco budworm (*Heliothis virescens*), a cotton pest, is linked to a DNA marker for a sodium channel locus *hscp*. It is known that levels of the insecticide in the environment show marked geographic differentiation in North America. Taylor et al. (1995) predicted that tobacco budworm populations would be highly differentiated at the *hscp* locus as a result of geographic variation in the intensity of selection due to pyrethroids. The value of F_{ST} averages 0.002 for 13 allozyme loci and equals 0.001 for a DNA marker at an esterase locus. By contrast, the value is 0.041 at the *hscp* locus, more than an order of magnitude greater. Taylor et al. (1995) interpreted this as confirming an a priori hypothesis about the influence of selection, contrasting this with the usual approach where evidence for selection is obtained by refuting the null hypothesis of neutrality.

Correlation of heterozygosity and fitness

There is a substantial literature on the correlation of allozyme heterozygosity with growth or other measures of fitness (e.g. Zouros & Pogson, 1994; Britten, 1996). Two main hypotheses have been used to explain the correlation. The selection hypothesis holds that the marker loci themselves are functionally important and that heterozygotes exhibit superior fitness. By contrast, the associative overdominance hypothesis holds that the markers are neutral, but in linkage disequilibrium with recessive deleterious alleles at other loci. The negative fitness effects of these recessive alleles will tend to be masked in individuals heterozygous at the marker loci. Pogson & Zouros (1994) proposed that comparing the correlation between fitness and heterozygosity at allozyme and noncoding DNA loci could test the two hypotheses. The selection hypothesis predicts a correlation for the allozyme loci, but not for the DNA loci, as these can be presumed to be neutral. By contrast, the associative overdominance hypothesis predicts correlations for both the DNA and the allozyme loci.

Case studies

Positive selection on sperm proteins in abalones and sea urchins

Some of the best examples of positive selection for nonsynonymous substitutions are found in sperm proteins in abalones (*Haliotis*). Higher concentrations of sperm are needed in egg sperm mixtures to achieve cross-species hybridisations than intra-species fertilisations, suggesting species-specificity of egg-sperm interactions. Such specificity, which might be of importance in general in marine invertebrates using external fertilisation, could be achieved by coevolution of gamete recognition proteins in eggs and sperm. The abalone egg is contained within a glycoproteinaceous vitelline envelope (VE). On attachment of the sperm to the egg, the protein lysin, contained within the sperm acrosome, creates a hole in the VE by a non-enzymatic process allowing the sperm to pass through and fertilise the egg.

Lee & Vacquier (1992) analysed lysin from sever abalone species from California. The protein consists of a conserved signal peptide of 18 amino aci residues followed by a mature sequence of 136–13 residues. It was found that the majority of amino acic replacements are nonconservative, that is involving functionally different side groups, and that the number of nonsynonymous substitutions exceeds the number of synonymous substitutions in 20 out of 21 pairwise species comparisons. The ratio D_n/D_s ranges from 0.95 to 3.57, but is higher for the more closely related abalone species, suggesting a high level of nonsynonymous substitution accompanying initial species divergence. Lee et al. (1995) extended the study to 20 species worldwide. D_n/D_s is always greater than 1 in closely related species (whenever D_s is less than 20%) and the highest value observed is 4.10. All but two of 29 pairwise species comparisons where $D_n/D_s>1$ are within the two major clades on a phylogenetic tree of lysins separating North Pacific species from those in the rest of the world. Many comparisons between sympatric species give high values of the ratios. Abalone embryos tend to settle near their parents and genetic structure can occur over distances of a few km giving the opportunity for relatively localised differentiation and speciation. The reinforcement of postzygotic isolation barriers through selection against hybrid embryos could favour rapid fixation of nonsynonymous mutations that isolate the gamete recognition systems of closely related species.

Various other hypotheses have been proposed to explain the rapid nonsynonymous evolution of lysin. Vacquier et al. (1997) showed that a second protein of 18-kDa occurs in the abalone sperm acrosome in addition to lysin. This also has a high D_n/D_s ratio, suggesting positive selection. Comparison of DNA sequences and of secondary and tertiary structure suggests that the two proteins arose by a gene duplication. Thus, it is possible that strong positive selection for some diversification of function occurred immediately after duplication. This is supported by the observation that lysin appears to be more effective in dissolving the VE, whereas the 18-kDa protein is more effective in causing fusion of the two gametes. Another possibility is sexual antagonism where, to avoid polyspermy, natural selection in females favours the slowing down of sperm fusion by altering egg surface ligands. This would result in rapid selection for alteration in the sperm recognition proteins. Sexual selection, which is thought to be able to drive rapid evolution for reproductive traits, could also be involved. Finally, rapid sperm protein evolution might be associated with self-nonself recognition and immunological defence. Intracellular microbes would evolve rapidly to escape recognition by egg cells. Ligands on the egg surface might evolve rapidly to keep ahead of this microbe evolution and simultaneously drive rapid evolution of recognition proteins in sperm.

Confirmation of the rapid evolution of abalone sperm proteins was obtained by Metz et al. (1998a). Between species nonsynonymous differences in lysin and the 18-kDa protein exceed nucleotide differences in introns from these genes by 3.5–24-fold. In some species comparisons, nonsynonymous substitutions in the sperm proteins also exceed synonymous substitutions in mtDNA. To survey within species variation, Metz et al. (1998a) sequenced exon/intron segments for between 6 and 19 individuals of red abalone (*H. rufescens*). Only two nucleotide differences are found out of over 29 000 nucleotides surveyed. By contrast, in a G protein intron and in mtDNA COI, divergence between individuals average 1.4% and 0.4%, respectively. It was suggested that the low level of variation for sperm proteins might be the result of selective sweeps.

Recently, Swanson & Vacquier (1998) have proposed a somewhat unexpected explanation for the rapid evolution of abalone lysin. The cDNA sequences encoding the tandemly repeated motif of the lysin egg receptor (VERL) were studied in seven abalone species. VERL is an unbranched glycoprotein in the VE and consists of about 28 repeats of 459 bp with two lysin molecules binding for each repeat. Some nonsynonymous substitutions in the VERL repeats are species-specific. Within and between species D_s and D_n values were determined for the repeat sequence and it was observed that in general $D_n< D_s$ suggesting purifying selection. In addition, the repeats are relatively homogeneous within but divergent between species, suggesting concerted evolution. This was supported by further evidence, for example a *Hind*III site occurs in all red abalone VERL repeats but is completely absent from pink abalone repeats. Concerted evolution of the VERL repeats could cause rapid evolution of lysins by the following mechanism. A mutation in a single VERL repeat might cause suboptimal binding of lysin. If the new VERL repeat spreads by concerted evolution, natural selection would modify lysin to improve the binding. Because concerted evolution, caused by unequal crossing over, is a rapid process, it could drive parallel rapid nonsynonymous evolutionary changes in lysin.

Evidence for positive selection was also found in bindin, the acrosomal sperm protein of the sea urchin *Echinometra* (Metz & Palumbi, 1996). In comparisons of 13 bindin sequences from three Indo-Pacific species, values of D_n = 0 occur in one part of the coding region suggesting purifying selection. In an adjacent region of the sequence, D_n/D_s has a value of about 3 indicating positive selection. Bindin is highly polymorphic within species, as well as showing fixed differences between species. By contrast, for mtDNA, polymorphisms are shared between species, despite the smaller effective population size. This difference is consistent with the action of positive selection on bindin. Despite the evidence of positive selection on bindin alleles, the McDonald–Kreitman test gives a non-significant result consistent with neutrality. This could be the result of the high level of nonsynonymous polymorphism elevating the nonsynonymous:synonymous ratio within species to a value similar to that between species.

Echinometra bindin differs from abalone lysin where polymorphism is at a low level. It is not clear why positive selection should favour polymorphism of bindin. Balancing selection seems an unlikely explanation, because alleles are clustered within species on a bindin phylogenetic tree with no evidence of long lived balanced polymorphism predating speciation events (Metz & Palumbi, 1996). It is possible that the egg receptor, while showing highest affinity for one form of bindin, might nevertheless accept a range of slightly different bindin structures. Variation could thus accumulate through some relaxation of purifying selection on bindin within species. Sexual isolation is favoured in models where there are multiple alleles at male and female recognition loci in partially isolated subpopulations. However, *Echinometra* does not show the evidence of bindin allele frequency differences between populations that such models would predict.

Contrasting results pointing away from positive selection were found in four allopatric species of sea urchin of the genus *Arbacia* (Metz et al., 1998b). Fertilisation experiments between Pacific *A. incisa* and Atlantic *A. punctuata*, species which are visibly quite distinct, revealed no evidence of differentiation of the gamete recognition systems. Sequence differences in bindin coding regions between species range between 1.6% and 4.1%, several times lower than for mtDNA coding regions (9%–13%) or bindin introns (7%–18%). D_n/D_s ratios between species range between 0.02 and 0.24 suggesting purifying selection. Counting total nucleotide changes over all four species, ratios of nonsynonymous:synonymous are 9:17 for fixations between species and 6:27 for polymorphisms within species. Although there is a tendency towards an excess of nonsynonymous fixations in the McDonald-Kreitman test, this is not significant. Thus, despite 4–9 Myr of independent evolution, there is no evidence that positive selection has acted within *Arbacia*. There are several possible explanations. Unlike abalones, where congeneric species occur in sympatry, high gene flow over widely distributed allopatric populations might have inhibited any local differentiation by positive selection. In *Arbacia*, bindin might have less freedom to vary since protein structure analyses reveal regions similar to transmembrane domains in which selective constraint could be expected. Finally, there might simply be reduced selective pressure for divergence of gamete recognition systems in widely distributed allopatric species lacking fertilisation barriers.

Purifying selection on mtDNA in marine mussels

The three species in the mussel genus *Mytilus* (*M. edulis*, *M. trossulus* and *M. galloprovincialis*) are widely distributed in cold and temperate regions of Europe and America. *Mytilus* is unusual in having two distinct mtDNA genomes. The F genome is inherited maternally, the M genome is inherited paternally. The three species share haplotypes, although American *M. trossulus* has distinct lineages. In European *M. trossulus*, the M genome is replaced by an F genome variant. *Mytilus* provides a good model for applying tests of neutrality because for many tests one mtDNA genome can be used as a control against the other. Following this approach, Stewart et al. (1996) studied 813 bp of the COIII gene in two American *M. trossulus* and two American *M. edulis* individuals. Similarly, Quesada et al. (1998, 1999) studied 59 sequences of the mtDNA COIII gene from different European populations of all three taxa and compared these data with seven 308 bp sequences from American *M. trossulus* M and F haplotypes studied by Stewart et al. (1995).

In comparisons both within and between continents, the values of D_s are higher for the M than F genome consistent with a higher mutation rate for the M genome. For both genomes, values of D_n/D_s are much higher within European populations (0.092–0.157) than for comparisons between continents (0.010–0.038) or between highly diverged American *M. edulis* and *M. trossulus* (0.011–0.027). In comparisons between genomes, values of the ratio are consistently lower for the F genome. These observations suggest a relaxation of purifying selection within Europe and more intense purifying selection on the F genome. In European *Mytilus* populations, nucleotide diversity calculated using the coalescent method of Fu (1994) tends to be higher for synonymous than nonsynonymous substitutions. Nucleotide diversity is also higher for the M than for the F genome, and this difference is particularly marked at synonymous sites. This suggests a higher mutation rate for the M than the F genome. For all three taxa in Europe, the values of Tajima's D are negative for both genomes. The values are consistently higher for the M genome than for the F genome. Although the test statistic is rarely significantly different from zero, the trends in the data are consistent with the action of purifying selection that is less intense on the M genome. In general, male mussels are heteroplasmic for both F and M genomes, whereas females are homoplasmic for the F genome. Deleterious mtDNA mutations in the M genome might thus be masked leading to a lower intensity of purifying selection on this genome.

When comparing F and M lineages for American taxa, Stewart et al. (1996) observed nonsynonymous:synonymous ratios of 5:6 for fixations between species and 15:108 for polymorphisms within species, giving a significant excess of nonsynonymous substitutions in the McDonald-Kreitman test. However, in European taxa, excesses of nonsynonymous polymorphisms occur when comparing F and M lineages, although the excess is small and non-significant in *M. edulis*. Excesses of nonsynonymous polymorphisms also occur in inter-continental within genome tests. For example, for the F genome, the nonsynonymous:synonymous ratios are 0:27 for fixations and 21:48 for polymorphisms, a highly significant excess of nonsynonymous polymorphisms. Moreover, all the nonsynonymous F genome polymorphisms occur in the European populations. An excess of nonsynonymous polymorphism also occurs for the M genome, but is non-significant. This complex pattern is consistent with the nearly neutral model of molecular evolution (Ohta, 1992). This would involve somewhat more intense purifying selection on the F than the M genome, but less intense purifying selection in European taxa particularly *M. edulis* compared with American taxa. This picture is consistent with the values observed for the D_n/D_s ratios. Recent relaxation in selective constraint and changes in effective population size might have played a role in causing the geographic and species-specific differences.

Selection and neutrality in Atlantic cod

Arnason & Palsson (1996) studied sequence variation in nine populations of Atlantic cod, *Gadus morhua*, from different localities around Norway, using sequence analysis of the mtDNA cytochrome b gene. Nine base changes occur among 100 sequences, all except one at third position synonymous sites. Results for both the Ewens-Watterson test and Tajima test are perfectly in accord with neutrality, as expected for variation occurring largely at synonymous sites. The observed ratio of nonsynonymous:synonymous polymorphism is 1:9, whereas a 1:2 ratio is expected considering possible mutational events within the sequence. Moreover, the single nonsynonymous change is for a functionally similar amino acid. These data suggest that purifying selection has removed many amino acid replacements. The operation of purifying selection is confirmed by a nonsynonymous:synonymous ratio of 0:15 for fixations between Atlantic cod and a related species *Gadus ogac*. The evidence for purifying selection is not necessarily inconsistent with the non-significant results in the neutrality tests. Such a situation can arise when strong purifying selection removes deleterious mutations rapidly, the remaining variation being neutral.

Pogson et al. (1995) carried out a study that suggests neutrality of nuclear DNA markers, but selection acting on allozymes in *Gadus morhua*. RFLP variation at 11 anonymous single copy nuclear DNA loci was examined in six populations of cod from the north Atlantic Ocean and compared with previously published data on patterns of variation at ten allozyme loci from geographically similar locations. All the DNA polymorphisms are located within the flanking regions of single copy genes and are thus good candidates for neutral markers. F_{ST} values at the allozyme loci range from 0.002 to 0.031 with a mean of 0.014. Values at the DNA loci range from 0.008 to 0.309 with a mean of 0.069. Two of the allozyme values and ten of the DNA values are significantly different from zero. As a group, the allozyme loci have significantly lower F_{ST} values than the DNA loci. The DNA and allozyme data are not contemporaneous, but previous studies had shown little temporal variation in allozyme frequencies.

Expected heterozygosity is greater at the DNA (0.373) than allozyme (0.134) loci. However, it was considered unlikely that the higher F_{ST} value at the DNA loci was the result of a higher mutation rate. Differences in mutation rates were thought to be small and in any case dominated by migration at both types of loci. Moreover, there is no evidence of a correlation between heterozygosity and F_{ST} across individual loci. Balancing selection was invoked to explain geographic uniformity at the allozyme loci and it was suggested that the DNA loci are largely neutral. This explanation for the DNA markers was supported by evidence of a highly significant positive correlation between F_{ST} and geographic distance, consistent with the action of drift and limited gene flow. Both types of markers show an excess of rare alleles compared with expectations of the infinite allele model. This is most marked for the DNA loci where 129 of 172 alleles detected have frequencies lower than 0.01. This appears difficult to explain by the nearly neutral model, which would predict closer fit to neutrality for the DNA than the protein loci. Another possibility is that Atlantic cod have recently experienced an expansion in population size. Such a non-equilibrium situation would result in the accumulation of many young and rare alleles.

Pogson & Fevolden (1998) tested the associative overdominance hypothesis by examining the relationship between growth rate and heterozygosity, at eight of the noncoding DNA markers used by Pogson et al. (1995), in two populations of *Gadus morhua*. A large and highly significant correlation occurs in a population from Balsfjord (0.420) but the correlation is small and non-significant in the Barents Sea (0.003). The existence of the significant correlation supports the associative overdominance hypothesis. However, Pogson & Fevolden (1998) pointed out that an unrealistically high frequency of deleterious alleles throughout the genome is needed to give such a high correlation through random associations of functional and neutral alleles. The higher correlation in the Balsfjord population was attributed to the more stressful environmental conditions compared with the Barents Sea, fish of the same age tend to be reduced in size by 30–60% in the fjord. Thus, mutations that are deleterious or slightly deleterious in Balsfjord might be effectively neutral in the Barents Sea. The study draws attention to the possibility of conditional neutrality of mutations depending on environmental conditions.

Genetic differentiation in bivalves

Although oyster larvae are planktonic for several weeks giving the possibility of extensive gene flow, mtDNA restriction fragment polymorphisms show a phylogeographic breakpoint between the Gulf of Mexico and the Atlantic Ocean in the American oyster, *Crassostrea virginica* (Reeb & Avise, 1990). Over the same region, allele frequencies at 14 allozyme loci show little variation (Buroker, 1983). The greater mtDNA differentiation might in this circumstance, might simply be attributed to the smaller effective population size for mtDNA. However, Karl & Avise (1992) provided evidence that RFLP alleles at four anonymous nuclear DNA loci also show significant differentiation between the Gulf of Mexico and the Atlantic Ocean. It was suggested that the nuclear DNA variants are neutral and their differentiation caused by genetic drift, whereas the allozyme alleles are kept at more uniform frequencies by balancing selection. McDonald et al. (1996) pointed out that this hypothesis requires that similar environmental forces must act on all the 14 allozyme loci in these two regions, which differ in many environmental variables. Thus, they set out to repeat the study using six additional anonymous nuclear DNA loci. The F_{ST} values range from −0.005 to 0.163 for the allozyme loci, and from −0.006 to 0.116 for the six new DNA loci. The two sets of loci do not differ significantly, but do both differ significantly from the original set of four DNA loci for which F_{ST} varies between 0.224 and 0.583. McDonald et al. (1996) discussed several possible explanations for the different results obtained for the two sets of DNA loci. Hare et al. (1996) reported failure with some of the PCR primers used by Karl & Avise (1992) and outlined potential artifacts that might impinge on the interpretation of heterozygote deficit phenomena in marine bivalves. However, McDonald et al. (1996) considered that scoring artifacts cannot easily explain the difference. They also rejected inadvertent sampling of transplanted animals, a difference in cloning methods used, a difference in strategy in initial screening for RFLPs, and even subconscious or conscious bias in the selection of loci. The differences in results between the studies thus remain somewhat of a mystery.

Hare & Avise (1998) examined three nuclear DNA phylogenies, including in their study one of the loci used by Karl & Avise (1992). Unlike mtDNA, the haplotypes on these nuclear trees do not align phylogenetically into distinct Atlantic and Gulf lineages. This highlights the danger of drawing inferences about the phylogeny of species or populations from studies of single loci. One possible explanation is that the smaller effective population size of mtDNA compared with nuclear DNA has facilitated more rapid sorting out to reciprocal monophyly in the two geographic regions. Of relevance to phylogenetic studies, however, is that selection can radically affect the structure of evolutionary trees. Hare & Avise (1998) emphasised that balancing selection can delay allelic extinction and reduce the time to reciprocal monophyly, whereas directional selection can have the opposite effect.

Another example of covariation of molecular markers suggesting relative homogeneity of allozyme allele frequencies is provided by the study of Karl et al. (1996) on the hydrothermal vent clam, *Calyptogena magnifica*. Patterns of variation at four allozyme loci and three single copy nuclear DNA loci were examined in four widely separated populations from the eastern Pacific. F_{ST} values are in general quite low, consistent with high dispersal. The four allozyme loci have F_{ST} values between 0.002 and 0.007 and the DNA loci have F_{ST} values between 0.008 and 0.113. Thus, as in the American oyster, the tendency towards greater geographic homogeneity of allozyme compared with DNA loci might be indicative of balancing selection on the allozyme loci.

Boom et al. (1994) studied mtDNA variation in five Canadian populations of the introduced Pacific oyster, *Crassostrea gigas*. A total of 44 haplotypes occurred in 141 individuals. Two of these haplotypes were at high frequency in all populations and the remainder rare and usually unique to individual populations. The frequency distribution of haplotypes is similar to that observed by Reeb & Avise (1990) in *C. virginica*, for which there are two high frequency and 81 rare haplotypes in a sample of 212 individuals. The two common haplotypes are highly diverged, separating Gulf and Atlantic populations in *C. virginica*, but differ by only a single cut site in *C. gigas*. The Ewens-Watterson test gave a highly significant result for *C. gigas* indicating an excess of homozygosity and more rare alleles than expected by neutral theory given the sample size. This observation is consistent with a population expansion following introduction, and shows how the Ewens-Watterson test might be used to provide confirmatory evidence of such historical changes.

Studies on bivalves have figured prominently in the analysis of the relationship between heterozygosity and fitness (Zouros & Pogson, 1994). A critical test of the selection and associative overdominance hypotheses was carried out by Pogson & Zouros (1994) who examined the relationship between heterozygosity and growth rate in a sample of 222 scallops, *Placopecten magellanicus*, using nine electrophoretic protein loci and eight DNA loci derived from a cDNA library. Seven of the electrophoretic loci are for enzymes the other two for non-enzymatic proteins. Five of the DNA loci were detected through variable tandem repeats (VNTR), the other four by conventional restriction site differences. Observed correlations with growth are 0.148 (P = 0.028) for the enzyme loci, 0.119 (P = 0.077) for all the protein loci, −0.019 (P = 0.778) for the DNA loci, and −0.048 (P = 0.480) for the VNTR loci alone. Thus, there is a clear indication of a positive correlation for the protein loci, but not the DNA loci. Since the DNA loci were derived from a cDNA library they are no less likely to be situated close to, and be in linkage disequilibrium, with deleterious alleles than the protein loci. Thus, the data better support the selection hypothesis, that is that the effects on fitness derive directly from the protein loci themselves. This study thus contrasts with the study in cod (Pogson & Fevolden, 1998) which gave support to the associative overdominance hypothesis.

Tests of neutrality in Antarctic krill, coconut crabs and starfish

Antarctic krill (*Euphausia superba*) are an important component in the food web of the Southern Ocean. Krill are highly abundant and sustain the largest single species crustacean fishery in the world. Although krill have a discontinuous distribution influenced by oceanic gyres, homogeneity of allozyme frequencies suggests a single interbreeding unit, rather than the existence of discrete stocks (Fevolden & Schneppenheim, 1989). Zane et al. (1998) sequenced 154 bp from the mtDNA ND1 gene in individuals from four Antarctic populations. Sixty-three different haplotypes occur in a sample of 249 krill. Three neutral theory tests were applied to the sequence data. The Tajima test was carried out separately for each population and for all populations pooled. The test statistic varies between −1.740 and −1.085 with two values statistically significant. This indicates an excess of segregating sites compared with the average number of differences between sequences, as expected for purifying rather than balancing selection. However, all of the sequence differences except one are at synonymous sites, suggesting that population expansion rather than selection might be the cause. Excess homozygosity in the Ewens-Watterson test occurs in all individual samples of krill except one. For example, a sample from around South Georgia has the greatest proportional excess, with observed and expected homozygosity being 0.189 and 0.071, respectively. These results are consistent with those of the Tajima test in suggesting an excess of the most common and the rarest variants.

This study on krill provides one of only a few examples of mismatch analysis in marine organisms. The observed mismatch distribution in krill is unimodal with a peak at 1–2 nucleotide differences per sequence. The calculated expected distribution for equilibrium neutral theory for these data falls away from a peak at 0 differences per sequence. The difference between observed and expected can be explained if it is assumed that krill have undergone a demographic expansion. It was estimated from the data that the expansion began around 0.25 millions years ago, much earlier than the last maximal glaciation. Thus, the results of all three DNA tests of neutral theory are consistent with population expansion and show how such tests can provide information that could conceivably be of importance for management of krill fisheries.

Two other examples of mismatch analysis lead to diverse conclusions about population size changes. Lavery et al. (1996) studied mtDNA variation in the coconut crab, *Birgus latro*, using RFLP analysis. A mismatch distribution was obtained for Pacific populations indicative of a rapidly expanding population. However, independent evidence suggested that population sizes have actually decreased over the last millennium. It was concluded that the mismatch distribution reflects more ancient expansion events and is robust to the influence of more recent changes in population size. In contrast to the results obtained for Pacific populations, coconut crab populations in the Indian Ocean show the multimodal mismatch distributions characteristic of stable populations and equilibrium neutral theory. Williams & Benzie (1997) studied differentiation in four populations of the starfish *Linckia laevigata* from the Indo-West Pacific using RFLP analysis of the mtDNA control region. Bimodal mismatch distributions occur showing no indication of the unimodal distribution expected from a population that has recently undergone exponential expansion.

Covariation of molecular markers in sea beet and turtles

Two final examples of covariation of molecular markers are of interest as they involve the comparison of microsatellite variation with other markers and suggest some potential difficulties caused by the mutational properties of microsatellites. Raybould et al. (1996) studied genetic structure of sea beet (*Beta vulgaris maritima*) in ten coastal localities in Dorset, England using seven allozyme and six single copy nuclear DNA RFLP loci. Pairwise F_{ST} estimates were transformed and plotted against log distance between localities. For the RFLP loci a significant regression ($P = 0.001$) occurs indicating increasing F_{ST} with increasing distance as expected for neutral markers in the isolation by distance model. However, no such relationship is found for the allozyme loci ($P = 0.491$). It was proposed either that balancing selection was maintaining relatively uniform allele frequencies, or that the RFLP loci were subject to disruptive selection. Raybould et al. (1998) extended this work in a subset of these populations by including three microsatellite loci and comparing analyses using F_{ST} and the analogous statistic R_{ST} (Slatkin, 1995), that takes account of stepwise mutation and the high mutation rate relative to migration, considered appropriate for microsatellite variation. Using F_{ST}, allozyme and RFLP loci both show isolation by distance, but this is not now more marked for the RFLP loci. For the microsatellite loci, significant isolation by distance is observed when using R_{ST}, but not F_{ST}. This confirms the appropriateness of using R_{ST} for microsatellite loci. In addition, it is shown that with F_{ST}, genetic similarity of plants within close neighbourhood patches can be detected with microsatellites, but not with the other loci. It appears that F_{ST} can detect the local differentiation caused by the high mutation rate of microsatellites, and for which R_{ST} corrects.

FitzSimmons et al. (1997) compared population genetic structure in Australian populations of the green turtle (*Chelonia mydas*) in anonymous single copy DNA loci and microsatellites. The proportion of variation between geographic regions is greater for the anonymous DNA loci ($F_{ST} = 0.027$) than for microsatellite loci, both under the infinite allele model ($F_{ST} = 0.014$) and stepwise mutation model ($R_{ST} = 0.007$), respectively. The variation in F_{ST} among individual anonymous DNA loci is also higher than among microsatellite loci. It was suggested that this higher variation might be the result of linkage disequilibrium between alleles at these loci and those at other selected loci. Such linkage effects would be eroded more rapidly at microsatellite loci because of their higher mutation rate.

DNA variation in Tigriopus *and* Fundulus

The final examples in this section on case studies concerns two paradigms of the search for selection acting at single allozyme loci. These are provided by studies on the marine copepod *Tigriopus californicus* and the coastal fish *Fundulus heteroclitus*. The application of DNA approaches to these species has thrown further light on the selective forces that might be acting on their allozymes.

Burton & Lee (1994) studied the genetic structure of populations of *Tigriopus californicus* along the California coast using eight allozyme loci, a 500 bp region of the mtDNA COI gene, and approximately 500 bp of the nuclear histone gene H1 including coding and flanking regions. This species inhabits ephemeral intertidal rockpools and has a natural history dominated by dispersal, local extinction and recolonisation. Despite this, marked allozyme frequency differences and hybrid breakdown are observed between populations, suggesting that local populations persist long enough for genetic differentiation to occur. The two sets of DNA data give phylogenetic trees that are concordant in showing a phylogeographic split between central and southern Californian populations in a region that includes Point Conception, a recognised zoogeographic boundary. A UPGMA tree of populations constructed from the allozyme data also has a major break but differs from the gene trees by including the two most southern of the central Californian populations in the central branch. This copepod thus resembles the American oyster, *Crassostrea virginica*, in possessing a mtDNA breakpoint (Reeb & Avise, 1990). However, it has concordant mtDNA and nuclear DNA phylogenies unlike the American oyster (Hare & Avise, 1998). In addition, the marked allozyme differentiation contrasts strikingly with the uniform allele frequencies observed in the American oyster (Karl & Avise, 1992). Burton & Lee (1994) suggest that the discordant allozyme and histone gene phylogenies in *T. californicus* might be explained by natural selection in favour of typically central Californian allozyme alleles in the more northerly of the southern populations, or by introgression of these alleles into the southern populations.

The second example concerns the study by D.A. Powers and colleagues (reviewed by Powers & Schulte, 1998) of the lactate dehydrogenase-B locus (*Ldh-B*) in *Fundulus heteroclitus*. Two alleles, a and b, at this locus show significant clinal variation along the Atlantic coast of North America. Multidisclipinary approaches, including enzyme kinetics, physiological studies and whole organism performance analysis, have established functional differences between alleles in relation to temperature which are consistent with the latitudinal clinal variation. A mtDNA phylogenetic tree based on RFLP data shows a deep phylogeographic break between northern and southern populations. Recent work has also shown that variation in the *Ldh-B* flanking regulatory sequence is associated with differences in transcription rate between fish from northern and southern populations. A phylogenetic tree of flanking sequences also shows a break between sequences from northern and southern populations. Evidence for selection acting on the flanking regions was obtained from the HKA test comparing polymorphism within and divergence between northern and southern populations for the *Ldh-B* coding region, the *Ldh-B* flanking region and the mtDNA cyt-*b* gene. In the three pairwise comparisons between these three regions, only the two involving the flanking region give a significant deviation from neutrality, due to excess divergence between northern and southern populations in the flanking region. Further support for non-neutrality of *Ldh-B* was provided by the McDonald–Kreitman test, which shows a significant excess of nonsynonymous differences between cDNA sequences from northern and southern populations. The neutrality tests thus provide evidence that selection might be acting on the flanking regulatory region as well as on the coding region responsible for the allozyme polymorphism. Clearly, this complicates any simple interpretation that focuses on selection acting on the coding region alone.

Conclusions

Making a strong contrast between neutral and selection theory is probably helpful for heuristic purposes. However, the experiments carried out to date on marine organisms using DNA sequences suggest a complex interplay of evolutionary forces that vary both within and between species. These include examples of conformity to equilibrium neutral theory, of population expansion, of strong positive directional selection and of purifying selection. If any particular force shows signs of dominating in natural populations it might well be purifying selection viewed within the ambit of the nearly neutral theory. Also several studies, for example those on the coconut crab, krill and Pacific oyster, provide evidence of changes in population size. This is of general relevance to the testing of even the original version of neutral theory as such changes in population size can result in variations in polymorphism and evolutionary rate not easily distinguishable from those brought about by selection.

It may be easy to accept that selection or neutrality can show varying levels of dominance between species. However, an important result that appears to be emerging from DNA studies is that even within a single species some populations show evidence of selection on the same DNA markers and others do not. This is seen, for example, in the studies on mussels and cod. Although questions have frequently been raised in the past about the appropriateness of assuming neutrality of allozyme markers in gene flow studies, difficulties are also becoming apparent with DNA markers. Purifying selection on mtDNA appears to be a common phenomenon and, as for example, in the study of sea beet, the peculiarities and high mutation rate at microsatellite loci complicate their use for measuring gene flow. Whitlock & McCauley (1999) reviewed the difficulties that selection and mutation bring to the application of simple models for estimating gene flow and entitled this section of their paper "The Fantasy Island Model"!

The individual tests of neutrality discussed here are similar to each other in that they generally use the same basic data – DNA sequences. Although the tests differ to a greater or lesser extent in the way information is extracted and manipulated statistically, they are clearly not totally independent in a statistical sense. This raises the question of how results or conclusions drawn from different tests might be integrated or synthesised into an overall picture. One approach is to build an evolutionary and population model that mirrors the experimental system under study. Parameters that then give the best fit of the model to the empirical data can be identified. Further simulation using these parameters and approaches such as maximum likelihood or bootstrapping can then be used to assess statistical significance or the goodness of fit of the model. Recent examples of such an approach are associated with the coalescent method (Hudson, 1991) in which population parameters are estimated from the phylogenetic tree of sequences. For example, computer modelling of coalescent structure have been developed for estimating population sizes and migration rates for empirical DNA or allozyme data assuming neutrality (Beerli, 1998). Ford (1998) has also applied this modelling approach to data on six nuclear DNA genes from nine populations of chinook salmon (*Oncorhynchus tschawytscha*). Models with constant population size and with either no migration or equal rates of migration could be rejected. However, a variety of models with unequal migration, varying population size and selection could all account for the data. Conclusions about historical changes in population size could be inferred if selective neutrality was assumed, but it was not possible simultaneously to make deductions about population structure and natural selection. No doubt, continued development in the efficiency of molecular techniques in the 21st century will provide increasing volumes of sequence data for analysis by population biologists. Advances in understanding of evolutionary forces might then derive from parallel developments of more sophisticated modelling techniques, facilitated by the evolution of increasingly powerful computers.

Acknowledgements

I thank my students and colleagues, particularly Humberto Quesada, for helping me to improve my understanding of molecular evolution. I thank Antonio Solé-Cava for inviting me to a beautiful workshop on marine genetics in Brazil, and later for his patience.

References

Aguade, M. & C. H. Langley, 1994. Polymorphism and divergence in regions of low recombination in *Drosophila*. In Golding, B. (ed.), Non-Neutral Evolution: Theories and Data. Chapman & Hall, New York: 67–76.

Arnason, E. & S. Palsson, 1996. Mitochondrial cytochrome b DNA sequence variation of Atlantic cod *Gadus morhua*, from Norway. Mol. Ecol. 5: 715–724.

Beerli, P., 1998. Estimation of migration rates and populations sizes in geographically structured populations. In Carvalho, G. (ed.), Advances in Molecular Ecology. NATO Science, Series A: Life Sciences. IOS Press, Amsterdam: 39–53.

Begun, C. J. & C. F. Aquadro, 1992. Levels of naturally occurring DNA polymorphism correlate with recombination rates of *Drosophila melanogaster*. Nature 356: 519–520.

Bertorelle, G. & M. Slatkin, 1995. The number of segregating sites in expanding human populations, with implications for estimates of demographic parameters. Mol. Biol. Evol. 12: 887–892.

Boom, J. D. G., E. G. Boulding & A. T. Beckenback, 1994. Mitochondrial DNA variation in introduced populations of pacific oyster, *Crassostrea gigas*, in British Columbia. Can. J. Fish. aquat. Sci. 51: 1608–1614.

Britten, H. B., 1996. Meta-analysis of the association between multilocus heterozygosity and fitness. Evolution 50: 2158–2164.

Brookfield, J. F. Y & P. M. Sharp, 1994. Neutralism and selectionism face up to DNA data. Trends Genet. 10: 109–111.

Buroker, N. E., 1983. Population genetics of the American oyster *Crassostrea virginica* along the Atlantic coast and the Gulf of Mexico. Mar. Biol. 75: 99–112.

Burton, R. S. & B. -N. Lee, 1994. Nuclear and mitochondrial gene genealogies and allozyme polymorphisms across a major phylogeographic break in the copepod *Tigriopus californicus*. Proc natn. Acad. Sci. U.S.A. 91: 5197–5201.

Charlesworth, D., B. Charlesworth & M. T. Morgan, 1995. The pattern of neutral molecular variation under the background selection model. Genetics 141: 1605–1617.

Endo, T., K. Ikeo & T. Gojobori, 1996. Large-scale search for genes on which positive selection may operate. Mol. Biol. Evol. 13: 685–690.

Fevolden, S. E. & R. Schneppenheim, 1989. Genetic homogeneity of krill (*Euphausia superba* Dana) in the Southern Ocean. Polar. Biol. 9: 533–539.

FitzSimmons, N. N., C. Moritz, C. J. Limpus, L. Pope & R. Prince, 1997. Geographic structure of mitochondrial and nuclear gene polymorphisms in Australian green turtle populations and male-biased gene flow. Genetics 147: 184–1854.

Ford, M. J., 1998. Testing models of migration and isolation among populations of chinook salmon (*Oncorhynchus tschawytscha*). Evolution 52: 539–557.

Fu, Y. X., 1994. A phylogenetic estimator of effective population size or mutation rate. Genetics 136: 685–692.

Gillespie, J. H., 1989. Could natural selection account for molecular evolution and polymorphism? Genome 31: 311–315.

Gillespie, J. H., 1991. The causes of molecular evolution. Oxford University Press, New York.

Gillespie, J. H., 1994. Alternatives to the neutral theory. In Golding, B. (ed.), Non-Neutral Evolution: Theories and Data. Chapman & Hall, New York: 1–17.

Graur, D. & W. H. Li, 1991. Neutral mutation hypothesis test. Nature 354: 114–115.

Hare, M. P. & J. C. Avise, 1998. Population structure in the American oyster as inferred by nuclear gene genealogies. Mol. Biol. Evol. 15: 119–128.

Hare, M. P., S. A. Karl & J. C. Avise, 1996. Anonymous nuclear DNA markers in the American oyster and their implications for the heterozygote deficiency phenomenon in marine bivalves. Mol. Biol. Evol. 13: 334–345.

Hudson, R. R., 1991. Gene genealogies and the coalescent approach. Oxford Surv. evol. Biol. 7: 1–44.

Hudson. R. R., M. Kreitman & M. Aguade, 1987. A test of neutral molecular evolution based on nucleotide data. Genetics 116: 153–159.

Karl, S. A. & J. C. Avise, 1992. Balancing selection at allozyme loci in oysters: implications from nuclear RFLP's. Science 256: 100–102.

Karl, S. A., S. Schultz, D. Desbruyeres, R. Lutz & R. C. Vrijenhoek, 1996. Molecular analysis of gene flow in the hydrothermal vent clam (*Calyptogena magnifica*). Mol. mar. Biol. Biotech. 5: 193–202.

Kimura, M., 1983. The neutral theory of molecular evolution. Cambridge University Press, London.

Kimura, M. & J. F. Crow, 1964. The number of alleles that can be maintained in a finite population. Genetics 49: 725–738.

Lavery, S., C. Moritz & D. R. Fielder, 1996. Genetic patterns suggest exponential population growth in a declining species. Mol. Biol. Evol. 13: 1106–1113.

Lee, Y. -H. & V. D. Vacquier, 1992. The divergence of species-specific abalone sperm lysins is promoted by positive Darwinian selection. Biol. Bull. 182: 97–104.

Lee, Y. -H., T. Ota & V. D. Vacquier, 1995. Positive selection is a general phenomenon in the evolution of abalone sperm lysin. Mol. Biol. Evol. 12: 231–238.

Lewontin, R. C. & J. Krakauer, 1975. Testing the heterogeneity of F values. Genetics 80: 397–398.

McDonald, J. H., 1994. Detecting natural selection by comparing geographic variation in protein and DNA polymorphisms. In Golding, B. (ed.), Non-Neutral Evolution: Theories and Data. Chapman & Hall, New York: 88–100.

McDonald, J. H. & M. Kreitman, 1991. Adaptive protein evolution at the *Adh* locus in *Drosophila*. Nature 351: 652–654.

McDonald, J. H., B. C. Verrelli & L. B. Geyer, 1996. Lack of geographic variation in anonymous nuclear polymorphisms in the American oyster, *Crassostrea virginica*. Mol. Biol. Evol. 13: 1114–1118.

Metz, C. E., G. Gomez-Gutierrez & V. D. Vacquier, 1998b. Mitochondrial DNA and bindin gene sequence evolution among allopatric species of the sea urchin genus *Arbacia*. Mol. Biol. Evol. 15: 185–195.

Metz, E. C. & S. R. Palumbi, 1996. Positive selection and sequence rearrangements generate extensive polymorphism in the gamete recognition protein bindin. Mol. Biol. Evol. 13: 397–406.

Metz, C. E., R. Robles-Sikisaka & V. D. Vacquier, 1998a. Nonsynonymous substitution in abalone sperm fertilization genes exceeds substitution in introns and mitochondrial DNA. Proc. natn. Acad. Sci. U.S.A. 95: 10676–10681.

Moriyama, E. N. & J. R. Powell, 1996. Intraspecific nuclear DNA variation in *Drosophila*. Mol. Biol. Evol. 13: 261–277.

Nei, M. & T. Maruyama, 1975. Lewontin-Krakauer test for neutral genes. Genetics 80: 395.

Ohta, T., 1992. The nearly neutral theory of molecular evolution. Ann. Rev. Ecol. Syst. 23: 263–286.

Pogson, G. H. & S. E. Fevolden, 1998. DNA heterozygosity and growth rate in the Atlantic cod *Gadus morhua* (L). Evolution 52: 915–920.

Pogson, G. H., K. A. Mesa & R. G. Boutilier, 1995. Genetic population structure and gene flow in the Atlantic cod *Gadus morhua*: a comparison of allozyme and nuclear RFLP loci. Genetics 139: 375–385.

Pogson, G. H. & E. Zouros, 1994. Allozyme and RFLP heterozygosities as correlates of growth rate in the scallop *Placopecten magellanicus*: a test of the associative overdominance hypothesis. Genetics 137: 221–231.

Powers, D. A. & P. M. Schulte, 1998. Evolutionary adaptations of gene structure and expression in natural populations in relation to a changing environment: a multidisciplinary approach to address the million-year saga of small fish. J. exp. Zool. 282: 71–94.

Quesada, H., M. Warren & D. O. F. Skibinski, 1998. Nonneutral evolution and differential mutation rate of gender-associated mitochondrial DNA lineages in the marine mussel *Mytilus*. Genetics 149: 1511–1526.

Quesada, H., R. Wenne & D. O. F. Skibinski, 1999. Interspecies transfer of female mitochondrial DNA is coupled with role-reversals and departure from neutrality in the mussel *Mytilus trossulus*. Mol. Biol. Evol. 16: 655–665.

Rand, D. M., 1996. Neutrality tests of molecular markers and the connection between DNA polymorphism, demography and conservation biology. Conserv. Biol. 10: 665–671.

Rand, D. M. & L. M. Kann, 1998. Mutation and selection at silent and replacement sites in the evolution of animal mitochondrial DNA. Genetica 102/103: 393–407.

Raybould, A. F., R. J. Mogg & R. T. Clarke, 1996. The genetic structure of *Beta vulgaris* ssp. *maritima* (sea beet) populations: RFLPs and isozymes show different patterns of gene flow. Heredity 77: 245–250.

Raybould, A. F, R. J. Mogg, C. Aldam, C. J. Gliddon, R. S. Thorpe & R. T. Clarke, 1998. The genetic structure of sea beet (*Beta vulgaris* ssp. *maritima*) populations. III. Detection of isolation by distance at microsatellite loci. Heredity 80: 127–132.

Reeb, C. A. & J. C. Avise, 1990. A genetic discontinuity in a continuously distributed species: mitochondrial DNA in the American oyster, *Crassostrea virginica*. Genetics 124: 397–406.

Robertson, A., 1975. Remarks on the Lewontin-Krakauer test. Genetics 80: 396.

Rogers, A. R., 1995. Genetic evidence for a Pleistocene population explosion. Evolution 49: 608–615.

Rogers, A. R. & H. Harpending, 1992. Population growth makes waves in the distribution of pairwise genetic differences. Mol. Biol. Evol. 9: 552–569.

Slatkin, M., 1995. A measure of population subdivision based on microsatellite allele frequencies. Genetics 139: 457–462.

Slatkin, M. & R. R. Hudson, 1991. Pairwise comparisons of mitochondrial DNA sequences in stable and exponentially growing populations. Genetics 129: 555–562.

Stewart, D. T., E. R. Kenchington, R. K. Singh & E. Zouros, 1996. Degree of selective constraint as an explanation of the different rates of evolution of gender-specific mitochondrial DNA lineages in the mussel *Mytilus*. Genetics 143: 1349–1357.

Stewart, D. T., C. Saavedra, R. R. Stanwood, A. O. Ball & E. Zouros, 1995. Male and female mitochondrial DNA lineages in the blue mussel (*Mytilus edulis*) species group. Mol. Biol. Evol. 12: 735–747.

Swanson, W. J. & V. D. Vacquier, 1998. Concerted evolution in an egg receptor for a rapidly evolving abalone sperm protein. Science 281: 710–712.

Tajima, F., 1983. Evolutionary relationship of DNA sequences in finite populations. Genetics 105: 437–460.

Tajima, F., 1989. Statistical method for testing the neutral mutation hypothesis by DNA polymorphism. Genetics 123: 585–595.

Taylor, M. F. J., Y. Shen & M. E. Kreitman, 1995. A population genetic test of selection at the molecular level. Science 270: 1497–1499.

Vacquier, V. D., W. J. Swanson & Y. -H. Lee, 1997. Positive Darwinian selection on two homologous fertilization proteins: what is the selective pressure driving their divergence? J. mol. Evol. 44: S15–S22.

Watterson, G. A., 1975. On the number of segregation sites. Theor. Pop. Biol. 7: 256–276.

Watterson, G. A., 1978. The homozygosity test of neutrality. Genetics 88: 405–417.

Whitlock, M. C. & D. E. McCauley, 1999. Indirect measures of gene flow and migration. $F_{ST} \neq 1/(4N_m + 1)$. Heredity 82: 117–125.

Whittam, T. S. & M. Nei, 1991. Neutral mutation hypothesis test. Nature 354: 115–116.

Williams, S. T. & J. A. H. Benzie, 1997. Indo-West Pacific patterns of genetic differentiation in the high-dispersal starfish *Linckia laevigata*. Mol. Ecol. 6: 559–573.

Zane, I., L. Ostellari, L. Maccatrozzo, L. Bargelloni, B. Battaglia & T. Patarnello, 1998. Molecular evidence for genetic subdivision of Antarctic krill (*Euphausia superba* Dana) populations. Proc. r. Soc., Lond. B 265: 2387–2391.

Zouros, E. & G. H. Pogson, 1994. The present status of the relationship between heterozygosity and heterosis. In Beaumont, A. R. (ed.), Genetics and the Evolution of Aquatic Animals. Chapman and Hall, London: 135–146.

Hydrobiologia **420:** 153–164, 2000.
A. M. Solé-Cava, C. A. M. Russo & J. P. Thorpe (eds), Marine Genetics.

Molecular genetics and the management and conservation of marine organisms

N. A. Sweijd[1], R. C. K. Bowie[1], B. S. Evans[2,3] & A. L. Lopata[4]
[1]*Department of Zoology, University of Cape Town, Private Bag Rondebosch, 7701, Cape Town, South Africa*
E-mail: nsweijd@botzoo.uct.ac.za
[2]*School of Zoology, University of Tasmania, GPO Box 252C, Hobart, Tasmania, 7001 Australia*
[3]*Present address: CRC for aquaculture, CSIRO Marine Research, GPO Box 1538, Hobart, Tasmania, 7001 Australia*
[4]*Department of Immunology, Groote Schuur Hospital, Anzio Rd, Observatory, 7925, Cape Town, South Africa*

Key words: biochemical genetics, molecular genetics, marine species, identification

Abstract

Biochemical and molecular species identification techniques have a broad range of applications in the management and conservation of marine organisms. While species boundaries are not always clearly defined, phylogeneticists utilise autapomorphic characters to distinguish phylogenetic species. Genetic markers discriminate between marine taxa when traditional morphological distinctions are unclear. The applications of these techniques can be divided into four general categories. Firstly, compliance enforcement, which often depends on genetic identification techniques to enable officials to identify the species to which regulations pertain. Secondly, quality control applications, to allow for the testing of marine products to guard against fraudulent substitution with less valuable species, which is particularly pertinent since processing often obliterates identifiable features. Thirdly, a variety of applications to ecological and life-history studies and conservation management are reported. Here, the genetic identification techniques of species from cryptic life-cycle stages or of morphologically indistinct species are an indispensable tool for marine scientists, conservators and managers. Lastly, the application of genetic techniques for sourcing population origin is briefly discussed. The biochemical and molecular techniques applied to species identification all exploit phenotypic or genotypic polymorphisms that are sampled using either tertiary level protein based methods or primary level DNA based methods. In this review, examples of the applications along with the total protein, allozyme, serological, PCR and other DNA based methodologies are briefly described and some generalities with regard to their use are presented.

Introduction

The primary endeavour of biochemical and molecular phylogenetics has been to discover mutable, heritable, taxon-specific genetic characters (usually primary level nucleotide sequence or tertiary level protein polymorphisms) that would delineate, and contribute to, an understanding of the phylogenetic relationships between taxa (De Queiroz & Gauthier, 1992; Haig, 1998). Whilst the distribution of synapomorphic characters is used to reconstruct phylogenetic relationships, the autapomorphic characters are diagnostic of phylogenetic species (Baum, 1992). Where these discrete, fixed and exclusive characters are genetically detectable, they fortuitously provide a utility where they might be applied to the problem of species identification.

Biochemical and molecular methods of species identification have been applied in many diverse fields such as medical (Van Belkum et al., 1998) and veterinary diagnostics (e.g. Taylor et al., 1996), food science and safety (Meyer & Candrian, 1996; Armstrong et al., 1997; Hunt et al., 1997), forensic science (in species other than humans, Bartlett & Davidson, 1992; Murray et al., 1995; O'Ryan & Harley, 1998) and conservation (Haig, 1998). A substantial body of literature is accumulating that pertains to the identity of marine species (see Knowlton, 1993; Ward & Grewe,

1994 and Burton, 1996 for previous reviews). This review discusses some of the studies that use species identification techniques in four areas where they have been applied to marine species and sub-populations of species. These include compliance enforcement and commercial quality control, both of which are forensic applications. A broad category of applications that include investigations of the life-history, ecology and conservation of marine species follows. Lastly, the application of biochemical and molecular techniques to the determination of source origin (sub-population) of sub-specific samples is outlined. In this review, we describe these four categories and briefly cover the commonly used methodologies as well as some likely to be used in the future. Finally, we will attempt to draw out some generalities and provide guidelines for application of these techniques in the future.

Operationally distinct taxa

The species concept is the subject of much debate, particularly in the marine environment, which is replete with examples of unclear species boundaries and species complexes (Knowlton, 1993; Avise, 1994; Gosling, 1994). For the purposes of management of marine organisms, or their ecological co-habitors, one is usually not directly concerned with their species status, but rather their status as operationally distinct taxa – thus obviating the need to enter into the species debate in most cases (see Nixon & Wheeler, 1990; Baum, 1992; Rojas, 1992). Nevertheless, the phylogenetic species concept, which considers a phylogenetic species as "an irreducible cluster of organisms possessing at least one diagnostic character" (Baum, 1992), provides parameters within which to operate and is applicable at several levels in the phylogenetic hierarchy, including the sub-specific level. Here, particularly for forensic applications, we concern ourselves with single, species-specific genetic characters that discriminate between taxa that are legally 'generally accepted' as diagnosable taxa (Harvey, 1990; Baum, 1992), often relying on previously peer reviewed and published data. This is a prudent and practical approach to many of the problems described below, although not ubiquitously so.

Compliance enforcement

Most commercially and recreationally exploited marine species are subject to fishery regulations or international agreements (e.g. Palumbi & Cipriano, 1998). Several marine species fetch very high retail prices internationally and this exerts substantial pressure on them. Some examples of these highly prized marine animals are marine mammals, or their processed organs, (whale meat – up to $44 per kg, Mills et al., 1997; dolphin meat – up to $3000 per dolphin, Earthtrust, in litt.), fish (for example: tuna – up to $60 000 per fish, Ward et al., 1995; sturgeon eggs – $2400 per kg, De Salle & Birstein, 1996) and invertebrates (abalone – up to $500 per kg dried abalone, pers. obs). Many of these high values are in inverse proportion to the availability of the products and as a result some of the highest priced marine products are those of threatened species (Baker & Palumbi, 1994; De Salle & Birstein, 1996; Rehbein, 1997a; Birstein et al., 1998; Palumbi & Cipriano, 1998). Due to the rewards attainable, these species are often the subject of illegal exploitation in contravention of regional and international agreements (e.g. Barlett & Davidson, 1991; Utter, 1991; Baker et al., 1996; Palumbi & Cipriano, 1998).

Much of the problem lies in the fact that in all international agreements and statutory regulations pertaining to fisheries, reference is made to the Latin or scientific name of the species or genus concerned in order to remove the confusion caused by common names of the species (e.g. Annex.: Nomenclature of whales, The International Whaling Commission (IWC) conference in Washington, 1946). Consequently, in a legal dispute, the burden is placed on compliance officials to prove the specific identity of a species or product, since the inability to do this would render the law useless (e.g. Bartlett & Davidson, 1992; Baker et al., 1996; Palumbi & Cipriano, 1998; Sweijd et al., 1998). Where this occurs, a legal loophole is created and can result in the acquittal of alleged poachers due to lack of evidence (Sweijd et al., 1998). For example, in a case where South African abalone (*Haliotis midae*) were allegedly marketed in cans falsely labelled as 'Product of Australia' (ostensibly containing another *Haliotis* species), the defence contended that South African regulations did not have jurisdiction over the case (Sweijd et al., 1998). The contents were proved to be of South African origin using a DNA-based species identity test. As a result of this case, the South African police now have a molecular tool to help curb the precipitous decline of this abalone species, due to over-exploitation.

Thus, the requirement for methods that unequivocally identify samples of processed marine products to species level has, in many instances, become a conservation priority. Recognising this, the IWC Scientific

Committee has discussed the possibility of adopting a programme of sampling whale meat from the controversial Norwegian and Japanese commercial and scientific whaling operations (Palumbi & Cipriano, 1998). This action was intended to create a genetic profile register to allow better monitoring and tracking of whale meat sales in Japan and Norway. However, these two member nations have opposed the motion. Norway and Japan do not recognise the competence of the IWC on matters of trade, and are voluntarily implementing their own official whale DNA database projects independently (IWC, 1998; Walloe pers. comm.; Nobuyuki, pers. comm.).

Quality regulation and enforcement

Certain fish products and species have a higher value than closely related congeners and typically, many of the identifying morphological features of these species are lost during processing and packaging (e.g. Bartlett & Davidson, 1991; Sotelo et al., 1993; Ram et al., 1996; Quintero et al., 1998; Sweijd et al., 1998). This provides an opportunity for fraudulent substitution of lower value products for high value products and/or the mislabelling of produce. For example, cheaper smoked trout (*Oncorhynchus mykiss*) is difficult to distinguish by taste and appearance alone from more expensive smoked salmon (*Salmo salar*), for which it has been substituted (Carrera et al., 1996, 1998). Similar practices occur with flatfish, where European plaice (*Pleuronectes platessa*) and flounder (*Platichthys flesus*) are substituted for the more valuable Dover sole (*Solea solea*; Céspedes et al., 1998). Huang et al. (1995b) describe how sought-after fillets of the red snapper (*Lutjanus campechanus*) were substituted with congeneric species in 70% of samples from across Florida between 1988 and 1992. Canned tuna is subject to fraudulent mislabelling since prices of several species vary on different markets (Quintero et al., 1988).

These problems are not confined to fish species alone. For example, An et al. (1990) developed a technique for identifying and quantifying the degree of adulteration of Rock Shrimp (*Sicyonia brevirostris*) in seafood products. Using their 'portable laboratory', Baker et al. (1996) were able to show that what was ostensibly whale meat on a Japanese market was actually dolphin meat. When disputes arise an inability to identify products morphologically to species level would destroy any case, and hence the need for forensic genetic species identification methods that can be employed on processed marine products.

In addition to the forensic application of phylogenetic data, genetic species identification has further utility in fisheries management. The literature bears many examples of cryptic species, which are distinguished solely on the basis of genetic data since they are morphologically inseparable from their congeners in both economically important species and others (see Knowlton, 1993; Gosling, 1994; Ward & Grewe, 1994; Shaklee & Bentzen, 1998 for a variety of examples). Fisheries that target a particular species may inadvertently catch a second species that is either cryptic or morphologically so similar that they are mistaken for the original species. Northern and southern bluefin tuna (*Thunnus* spp.) are morphologically similar and can be mistaken for each other, which poses problems for the management of the depleted southern bluefin. This problem was resolved by developing species-specific allozyme and mtDNA markers (Ward et al., 1995). Similarly, Van der Bank & Kirchner (1997) were able to show that fixed allozyme allele differences occurred between morphologically similar species of a popular game fish (*Argyrosomus* spp.), which are regularly mistaken for each other. These examples illustrate the potentially negative economic (undervalued or overvalued catch) or ecological (detrimental effects of inadvertent over-exploitation) consequences, which could be obviated if the species identity and their respective life histories were better understood. The potential for misidentification has prompted Daley et al. (1997) to produce an identification guide for marine quota species in Australia with photographs of cellulose acetate allozyme gels that demonstrate species-specific electrophoretic allozyme markers.

Life-history, ecological and conservation applications

For the many microbiological taxa that constitute the marine picoplankton, biochemical and molecular methods are a fundamental tool for species identification (Schmidt et al., 1991; Wimpee et al., 1991). In macrofauna, certain life history stages, such as invertebrate larvae, are morphologically cryptic (in relation to larvae from other species and even other classes of invertebrate) and yet are found co-existing in the water column (Hu et al., 1992; Ward & Grewe, 1994; Burton, 1996; Cary, 1996; Bilodeau et al., 1999).

This has presented fisheries managers and marine conservation managers with several problems, rendering the life history and ecology of such species very difficult to study during the larval stages (Coffroth & Mulawka, 1995; Evans et al., 1998). Thus, alternatives to morphological methods of species identification are required. For instance, juvenile stages or non-reproductive males of crayfish species are morphologically indistinguishable and, even though identification keys exist for adults, isoelectric focusing (IEF) of soluble proteins has proved the most effective way to separate them (Powell et al., 1995). Similarly, Silberman & Walsh (1992) used the polymerase chain reaction (PCR) and restriction fragment length polymorphisms (RFLP) data from 28S RNA genes to identify phyllosoma larvae of three lobster species (*Panulirus argus*, *P. guttatus* and *P. laevicauda*).

Olson et al. (1991) used PCR and sequencing of the 16S RNA gene to identify sea cucumber larvae, revealing the spatial and temporal patterns of settlement of the sea cucumbers (Medeiros-Bergen et al., 1995). Similarly, Bell & Grassle (1998) have used a combination of molecular probes and RFLPs of amplified fragments of 18S RNA to distinguish surfclam (*Spisula solidissima*) veliger larvae from other pelagic bivalve larvae. This work facilitated the study of settlement and dispersal of this commercially exploited species on the continental shelf off New Jersey. Similar techniques have been employed to study the dispersal and colonisation of organisms inhabiting hydrothermal vents, which are notoriously difficult to culture and study (Cary, 1996).

Toxigenic dinoflagellate blooms are renowned for their potentially disastrous effects in both natural habitat and aquaculture (Shumway & Cambella, 1993; Mendoza et al., 1995; Walsh et al., 1998). In order to study these organisms and develop an early warning system, species or strain identity is critical. Morphological characters have been used to distinguish toxic marine dinoflagellate taxa from the less virulent congeners, but during culture these plastic morphological features transform (Mendoza et al.,1995; Adachi et al., 1996). As a result, there is considerable controversy and confusion with regard to species identity and it is likely that a molecular or biochemical approach will provide the solution (Scholin & Anderson, 1996; Walsh et al., 1998). To this end, Mendoza et al. (1995) developed species- and genus-specific polyclonal antibodies for the identification of species of *Gymnodinium* and *Alexandrium* which provided a means of identifying natural, cultured and even formalin-preserved samples. Walsh et al. (1998) used rRNA gene sequences to address the same problem. Recently, new DNA based techniques such as immunofluorescent molecular beacon technology have been developed to study harmful algal bloom dynamics (Tyrrell et al., 1997).

Problems also occur at the other end of the size scale in the marine environment, with misidentification of beached beaked whales being a common occurrence (Dalebout et al., 1998). Using a database of reference sequences of beaked whales, Dalebout et al. (1998) were able to show that some beachings comprised new distribution records for some species, some extremely rare species and even one suspected new species of beaked whale. This example illustrates the importance of species identity in conservation and raises several issues. The U.S. Endangered Species Act, among other similar statutes, while protecting 'distinct' species, leaves the problem of making the distinctions to biologists and legal experts (Bowen et al., 1991; Avise, 1994; Haig, 1998). Where a species' status is in question, it can have a significant consequence for the conservation of the taxon concerned (Rojas, 1992; Haig, 1998). A good example of this is the case of Kemp's ridley sea turtle (*Lepidochelys kempi*); its status as a distinct endangered species, as opposed to a sub-species or hybrid, was finally established using mtDNA markers, thus affording it protection under the above-mentioned act (Bowen et al., 1991 and see Avise, 1998).

An important application of species-identification in the marine environment is to recognise the occurrence of natural or artificially produced hybrids (Knowlton, 1993; Ward & Grewe, 1994). Several examples of invertebrate hybrids have been discovered, but their true status can only be confirmed if they possess species-specific genetic markers from both contributing species. Mitochondrial DNA data was used to confirm the natural hybridisation of *Mytilus edulis* and *M. galloprovincialis* (Skibinski, 1985). Similarly, Brown (1995) found hybrid abalone of *Haliotis laevigata* and *H. rubra* using allozyme markers. The desire to increase production and combine desirable traits of two oyster (*Crassostrea*) species by hybridisation has had mixed success (Allen & Gaffney, 1993). These authors underline the need for genetic methods to assess success due to the infiltration of broodstock contaminants that could lead to spurious results.

Often hybridisation and introgression of species is the result of the introduction of exotic species into a new habitat (Geller, 1994; Heath et al., 1995; Geller et al., 1997; Evans et al., 1998). This occurs commonly in the marine environment due to the exchange of the ballast water in ports; this often contains a wide variety of viable planktonic organisms (Carlton & Geller, 1993; Geller et al., 1994). Geller and his colleagues (Geller & Powers, 1994; Geller et al., 1994) developed a PCR based technique to study this problem in the native North American mussel *Mytilus trossulus* and the invading *M. galloprovincialis*. In this example, using species-specific markers, they were able to distinguish between the morphologically similar *Mytilus* species, and reveal a complex pattern of invasion and introgression between the two. Geller et al. (1997) used partial 16S RNA mitochondrial gene sequences to show that European crabs of the genus *Carcinus* had invaded both the South African and Japanese coasts, where this had previously gone undetected. Evans et al. (1998) used mtDNA data to show that approximately 80% of planktonic sea-star larvae sampled from the Derwent River estuary in Tasmania were from the recently introduced northern Pacific *Asterias amurensis*.

Another intriguing application of molecular and biochemical species identification is that of identifying prey species from their remains in the gut of predators and thus resolving interactions between species that are otherwise impossible to discern due to the degradation of visually identifiable morphological features (e.g. Pickavance, 1970; Feller et al., 1979; Theilacker et al., 1993; Arnold et al., 1996; Scribner & Bowman, 1998). Using immunological techniques that recognise species-specific proteins in the partially digested gut contents of prey species, Feller et al. (1979) were able to determine some trophic relationships in benthic communities that would otherwise have gone undetected. Using a similar technique, Theilacker et al. (1993) were able to quantify the degree of predation of anchovy eggs and larvae (*Engraulis mordax*) by detecting their remains in the gut of euphausiacean predators and mathematically modelling the results. A study evaluating the success of stock enhancement of red drum (*Sciaenops ocellatus*), by release of hatchery-reared juveniles, was attempted by examining predation of this fish by predatory needlefish (*Strongylura notata*) and barracuda (*Sphyraena barracuda*, Arnold et al., 1996). These researchers showed that the needlefish was a far greater threat to success of red drum stock enhancement than the barracuda. A recent contribution from South Africa reported on an immunoassay used to identify zooplanktonic prey in the gut of commercially exploited squid (*Loligo vulgaris reynaudii*) paralarvae, providing a tool with which to study the high mortality from starvation that occurs at this life-stage (Venter et al., 1999). This application is limited by the degree of digestion of the prey species which will vary according to circumstances and sampling techniques (Scribner & Bowman, 1998). To end this, Scribner & Bowman (1998) used species-specific microsatellite DNA allele markers of less than 210 bp to identify the species of waterfowl prey (sub-family Anserinae) from the gut contents of predatory glaucous gulls (*Larus hyberboreus*) in western Alaska, where some of the prey species have experienced precipitous declines in number.

Sub-specific identification

When dealing with sub-specific identity problems, there is an overlap with the field of population genetics which falls outside the scope of this review. Nevertheless, where enough measurable genetic differentiation is detected between sub-populations or stocks, sourcing of specimens (as originating from sub-specific populations) is possible (Haig, 1998). Indeed, sourcing of individuals has become a necessity in many fisheries. Closed areas or reserves provide rich hunting grounds for poachers and in order to prove the origin of disputed samples (in these cases morphological distinctions are even less likely) a genetic method is required. Seeb et al. (1990) used protein markers to show that a consignment of King crab (*Paralithodes camtschatica*) was illegally fished from a closed area within their distribution. Harvey (1990) reports on a case where protein polymorphisms were used to show that a large mouth bass specimen (*Micropterus salmoides*) was fraudulently brought into a lake from a separate source to ensure winning an angling competition. In the abovementioned examples, the authorities were fortunate that the amount of genetic variability detectable using protein electrophoresis was sufficient to prove guilt.

With the advent of highly polymorphic DNA markers, such as microsatellite DNA, enough intraspecific variability is detectable to generate composite genotypes that can be used to prove population identity statistically (Haig, 1998; Smouse & Chevillon, 1998). These and other techniques such as mixed stock analysis (Carvalho & Hauser, 1994; Wirgin et al., 1997) could help to enforce compliance to a sub-specific level. It is likely that in the future a need for individual-specific characters will arise. For example, in South Africa, proof that abandoned sacks of abalone were illegally fished has been required for a conviction in some cases. Here, the desire is to match abalone tissue collected from the suspect's person or equipment to the individual abalone in question. In fish, the potential for individual specific genotyping (fingerprinting) for assessing mating success has already been tested (see Colbourne et al., 1996) using microsatellite markers. The use of sub-specific markers, like genetic tags, of sub-specific cohorts has an important utility in aquaculture and fisheries (Seeb et al., 1990; Ferguson, 1994; Ferguson & Danzmann, 1998).

The above section deals with a very wide variety of species identification techniques to the many issues and problems that fishery managers, marine conservators and scientists are required to deal with. There is almost as wide a scope of biochemical and molecular genetic methodologies that have been applied to these issues and problems, some of which is reviewed in the following sections.

Biochemical and molecular genetic methodologies for species identification

Phylogenetic studies on many commercially exploited marine taxa have been published and thus data from which species-specific markers can be obtained are often already available in the literature (e.g. abalone, Youn-Ho & Vacquier, 1995; whales, Arnason et al., 1993), obviating the need to develop phylogenies as a basis for a species identification technique.

However, where these data are not available, the first step in resolving identity problems is the selection of an appropriate marker that is most likely to provide the degree of resolution required. Which of the many commonly used polymorphic genetic markers is appropriate will depend on two main factors: the cost of development of the marker and the degree of relatedness of the species, or individuals within a species, concerned (see Meyer, 1994; Skibinski, 1994; Parker et al., 1998; Shaklee & Bentzen, 1998 for reviews).

The costs of developing biochemical and molecular methods are enormous, but can be reduced by integrating the work into existing laboratories where equipment and reagents are available. It is also in the interest of the state and fishing companies to protect the resources of commercially exploited marine species, and thus partnerships or contracts may be easier to engage than in many other cases. Some states have developed dedicated laboratories to deal with non-human forensic problems, like the U.S. Fish and Wildlife Service Laboratory (Avise, 1994), the Texas Parks and Wildlife Department's Forensic Program (Harvey, 1990) and the South African Police Services Forensic Laboratory (Hennop, pers. comm.). Despite these options, the more traditional methods (protein based) are quickly becoming as expensive as the more modern DNA based methods, and thus cost is largely an administrative problem that should be weighed up against the benefits that such techniques will confer on the problem at hand.

Of primary importance in selecting a marker is a consideration of the relationship among the taxa concerned. Closely related species require a marker that provides a higher degree of resolution than for less closely related species. In some cases, discrimination between sub-species is required (for example the abalone *Haliotis discus discus* and *Haliotis discus hannai*, Naganuma et al., 1998), whereas in others, species from different genera require separation (for example, salmon, *Salmo salar*, and trout, *Oncorhynchus mykiss*, Carrera et al., 1998).

Selection criteria might be based on available literature (data from studies on related taxa) or otherwise should be based on the degree of resolution expected from a particular marker. Generally these have been of two main types: protein and DNA based methods.

Protein based methods

While protein based methods are generally regarded as less robust than DNA based methods, they nevertheless have several benefits and are still widely applied (Sotelo et al., 1993; Huang et al., 1995a; Burton, 1996; Hsieh et al., 1997). Since the value of seafood is linked to its freshness, many of the cytosolic proteins are usually intact when they reach the market and in some cases the specimens are still alive. However, in highly processed products, such as smoked, canned and dried seafood, most of the useful proteins are denatured and rendered useless for analysis, although this is not always the case (see Hsieh et al., 1997). At the species level, a relatively high degree of interspecific polymorphism is apparent and thus, where the tissue is in a suitable condition, useful distinguishing characters are readily available from proteins (Powell et al., 1995). With proteins, the principle is that different species will possess unique proteins or be differentiated on the basis of structural polymorphisms of a protein. Three types of protein based techniques have been used to assay these polymorphisms in marine species, namely total protein analyses, serological methods and locus specific allozyme markers (Cunniff, 1998).

The earliest use of genetic techniques for species identification (Lundstrom, 1977) used total protein analysis in thin-layer polyacrylamide gels and/or isoelectric-focusing (IEF) to resolve total protein extracts from fish muscle. Whitmore (1986) was able to separate nine species of sunfish using putative parvalbumin proteins that were separated by IEF. Total protein extracts of the appropriate tissues are made in a suitable buffer, the crude extract is electrophoresed on thin polyacrylamide gels and visualised by applying an appropriate protein stain. These are either denaturing gels (SDS–PAGE) which separate the proteins on the basis of size or they are equilibrated with an empirically determined pH range of ampholytes which separates them on the basis of charge (IEF, Whitmore, 1990; Sotelo et al., 1993; Hsieh et al., 1997). The migration of these proteins to their respective positions on the gel separates them into a species-specific banding pattern, which can be regarded as phenotypic expressions of genetic polymorphisms and which consistently differentiate the species concerned (see Whitmore, 1990; Sotelo et al., 1993; Yman, 1993, for more detailed methods).

Serological methods have been employed for species identification purposes in several of the examples cited above. These methods enhance the basic protein-based techniques by detecting and visualising species-specific proteins in samples, using antibodies raised specifically against them (e.g. An et al., 1990; Sotelo et al., 1993; Carrera et al., 1996; Venter et al., 1999). In general, a species-specific protein or group of proteins are isolated and used as antigens to immunise mammalian hosts in order to raise mono or polyclonal antibodies. A number of analytical techniques can be employed to detect the reaction between the antigen and antibody in order to test for the presence of the species-specific proteins. Some, such as immunoelectrophoresis, western-blotting and dot-blot methods are qualitative while others, such as enzyme linked immunoassay (ELISA), are quantitative (Sotelo et al., 1993; Huang et al., 1995b). The freshness of samples markedly affects the efficacy of these techniques, however, in some examples they are robust enough to function even in cases where the seafood products have been cooked or processed in other ways (Carrera et al., 1996; Hsieh et al., 1997). While serological methods generally require a high degree of expertise to execute properly, they do lend themselves to the development of portable non-laboratory based kits which would be helpful to compliance officials in the field. Shepard & Hartmann (1996) have developed such a technique which uses monoclonal antibody-coated glass beads that allow a technician to process up to 18 samples an hour for diagnosing the identity of sailfish species (*Istiophorus albicans*) in a non-laboratory environment. Similarly, Carrera et al. (1997) have developed a colorimetric ELISA method using species-specific rendered polyclonal antibodies bound to the paddles of immunostick tubes for identifying smoked salmon (*Salmo salar*), trout (*Oncorhynchus mykiss*) and bream (*Brama raii*).

Fixed allele differences and locus-specific features of allozymes have been most commonly used to discriminate fish species (Ward & Grewe, 1994; Shaklee & Bentzen, 1998), although these are generally being replaced by DNA based techniques. While speed and cost have been cited as the advantages of allozyme techniques (Burton, 1996), even these problems are being overcome by better technology and the proliferation of molecular facilities. Nevertheless, despite the reduction in popularity, a substantial baseline of data exists for allozymes and they can, and do, provide sufficient interspecific resolution. Allozymes require crude extracts of tissue and rely on locus-specific histochemical stains to visualise the electrophoresed alleles. The use of polyacrylamide gel electrophoresis of allozymes to identify oyster larvae (*Ostrea* and *Crassostrea* spp.) shows that allozymes can provide a high degree of sensitivity and can be used successfully with very small quantities of material (Hu et al., 1992). Other systems comprise either starch or cellulose acetate gel matrices (see Benson & Smith, 1989, for detailed methodology). Selection of appropriate loci to address any particular problem will largely depend on what already exists in the literature, since screening for species-specific loci and alleles is a time consuming exercise. A good starting point for this is Daley et al. (1997). It should be noted that one of the disadvantages of allozymes is that not all scorable enzyme loci are expressed in all tissues. Since processed seafood suitable for analysis is usually devoid of most of the internal organs, this might present a significant obstacle to success in reducing the amount of detectable polymorphisms.

DNA based techniques

The use of DNA based techniques offers many benefits over protein based ones, not the least of which is that DNA is relatively more stable (Sotelo et al., 1993). Another notable advantage is that PCR allows for analysis of minute quantities of DNA (Burton, 1996; Cary, 1996; Reischl et al., 1997; O'Ryan & Harley, 1998; Bilodeau et al., 1999, but see Hu et al., 1992). While these features allow broader application of DNA based methods, DNA is also subject to degradation, especially during processing like canning and drying of seafood, and thus has limitations in certain forensic and compliance related applications (Ram, 1996; Sweijd et al., 1998). However, by targeting small DNA fragments (<500 bp) and using carefully manipulated PCR protocols, PCR amplification from even badly preserved tissues is possible (Meyer & Candrian, 1996; Lorente et al., 1997; Carrera et al., 1998).

A wide range of analytical tools to quantify DNA polymorphisms is available (Meyer, 1994; Skibinski, 1994) and almost all of these have been applied in some way to species identification. Most of these use PCR primers that target fragments for amplification from an even wider range of genes and other DNA. One limitation of PCR is the availability of data on which to base PCR primer design, but the application of random sampling of the genome (Random Amplified Polymorphic DNA or RAPDs) to species identification has demonstrated that even this limitation can be overcome (Coffroth & Mulawka, 1995). Failing this, DNA sequencing of a targeted PCR fragment will be required, in order to actively search for species-specific characters, (e.g. Barlett & Davidson, 1991; Scholin & Anderson, 1996; Birstein et al., 1998; Quintero et al., 1998). DNA sequencing of selected loci usually provides unequivocal data for resolving forensic problems and, consequently, Bartlett & Davidson (1992) developed the system named "Forensically Informative Nucleotide Sequencing" or FINS, which might be applied where DNA sequencing is indicated. However, given the propensity of published data and DNA databases, along with the variety of extant DNA assaying methods, this step is becoming increasingly redundant. The choice of DNA marker is, however, not always so arbitrary. Palumbi & Cipriano (1998) argue that mtDNA usually provides the best separation (more fixed species-specific genetic characters), but that more than one marker might be necessary in that independent verification of species identity is sometimes required from nuclear sequences. In the case of cetaceans, natural hybridisation complicates the identity of samples and, in other instances, the use of multiple nuclear loci is desirable to control for contamination (Palumbi & Cipriano, 1998). In order to predict the utility of nuclear genes for species identification, they proposed a 'three times rule' which suggests that "phylogenetic sorting at nuclear loci is likely to produce species-specific sequences when mitochondrial alleles are reciprocally monophyletic and the branches leading to the mtDNA sequences of a species are three times longer than the average distance observed within species."

The majority of published studies have targeted known interspecific polymorphisms, mostly employing mitochondrial DNA, which has been widely used in phylogenetic reconstructions (Bermingham, 1990; Avise, 1994; Park & Moran, 1994). Some of these studies have analysed sequences of coding genes or the control region of marine vertebrate mtDNA molecules (e.g. Olson et al., 1991; Baker et al., 1996; De Salle & Birstein, 1996). A popular technique employed for species identification uses restriction endonucleases that exploit both mtDNA and nuclear gene interspecific polymorphisms, based on the consistent relative presence or absence of restriction sites in the species concerned (e.g. Ram et al., 1996; Céspedes et al., 1998; Evans et al., 1998; Quintero et al., 1998; Sweijd et al., 1998). This method is efficient in that it allows the use of a single pair of generic PCR primers that can be designed to amplify DNA from a phylogenetically wide range of species. Provided that the respective DNA sequences from the taxa in question possess differentiating recognition sites, subsequent analysis of the amplified DNA by digestion with a range of restriction endonucleases reveals species-specific fragment patterns (RFLPs) after electrophoresis on agarose or polyacrylamide gels stained with UV-fluorescent ethidium-bromide. Geller & Powers (1994) exploited sequence differences between *Mytilus trossulus* and *M. galloprovincialis* to find a distinguishing restriction site in the mitochondrial 16S ribosomal RNA gene. Other DNA based techniques use species-specific DNA probes (Wimpee et al., 1991; Medeiros-Bergen et al., 1995) and hybridisation techniques (Adachi et al., 1996). Increasingly, non-coding repetitive DNA elements (VNTRs or mini- and microsatellite DNA) are being used for discrimination of taxonomic units at both the species and sub-specific level (Wright & Bentzen, 1994; Beacham et al., 1996, Colbourne et al., 1996). Novel techniques such as single stranded conformation polymorphism assays (SSCPs, Rehbein, 1997b) and molecular beacons, which use PCR technology and fluorescent probes (Tyagi & Kramer, 1996) are sensitive to just one or two base changes in target sequences. More recent work has demonstrated the feasibility of combining probes of different colours for different alleles (Kostrikis et al., 1998), allowing rapid assaying of many samples and large segments of DNA for interspecific polymorphisms. For example, Tyrell et al. (1997) described how this technology could be used to study harmful algal blooms. This demonstrates the expanding range of applicable technology available for genetic analysis.

Some studies have employed various combinations of DNA based and protein based methods (e.g. Knowlton, 1993; Ward et al., 1995; Rehbein, 1997b; Wirgin et al., 1997) which illustrate the broad range of methodologies that are available for a given problem (see Parker et al., 1998).

Concluding remarks

In this review, we have illustrated the wide array of biochemical and molecular genetic techniques applied to species identity for conservation and management of marine organisms. Each technique has its own particular advantages and disadvantages, and each has proved its worth in various applications. The choice of technique should depend on whether the questions at hand are biological or fisheries problems as opposed to compliance related or forensic in nature. Naturally, both require absolute rigor in their execution, but there are some specific considerations that should be made in the latter case.

All published studies cited in this review have been subjected to scientific scrutiny by the process of peer review, but they may not necessarily be up to legal standards required for compliance and enforcement. While Harvey (1990) indicates that publishing is a sound basis for general acceptability (a legal requirement), it is important that laboratory standards and procedures also fulfil these standards. This might require accreditation of the specific technique in forensic laboratories since the risk of contamination of samples is high, especially where PCR is used, in routine laboratories, even when the appropriate controls are put in place (Budowle, 1995; Cunniff, 1998). Accreditation will also help in court cases where expert testimony is required, since this will help close loopholes that may exist in routine laboratories.

Another factor that should be considered is the risk of intraspecific variability that has not been adequately accounted for, as 95% confidence can transform biological fact into legal opinion in a courtroom. Thus, robust tests of monomorphism and species-specificity are a prerequisite of acceptability. In South African cases, a secondary problem has arisen in cases where consignments of abalone were confiscated. Here, in order that sub-sampling adequately represented the disputed consignment as a whole, the use of a quality control sub-sampling standard prescribed by the Codex Alimentarius (World Health Organization, 1969) was employed to obviate a defence objection (McKenzie, pers. comm.). In general, planning the development of species identification techniques should include consultation with legal and compliance enforcement officials to help account for the above issues.

It is likely that with the proliferation of molecular technology in general, and its application in fisheries in particular, (Carvalho & Pitcher, 1994; Burton, 1996) that data will become more readily available to adapt for species and sub-species level identifications. This should in turn contribute to a greater appreciation of the degree of biodiversity, and its conservation, in the marine environment.

Acknowledgements

We thank Prof. Peter Cook, Dr Nick Elliott, Dr Colleen O'Ryan and Mr Lorenzo Prendini for their helpful comments on the manuscript. We would also like to thank the editors of this edition for inviting our review and our anonymous referees for their helpful comments.

References

Adachi, M., Y. Sako & Y. Ishida, 1996. Identification of the toxic dinoflagellates *Alexandrium catenalla* and *A. tamarense* (Dinophyceae) using DNA probes and whole cell-hybridization. J. Phycol. 32: 1049–1052.

Allen, S. K. Jr. & P. M. Gaffney, 1993. Genetic confirmation of hybridization between *Crassostrea gigas* (Thunberg) and *Crassostrea rivularis* (Gould). Aquaculture 113: 291–300.

An, H., P. A. Klein, K. -K. Kao, M. R. Marshall, W. S. Otwell & C. -I. Wei, 1990. Development of monoclonal antibodies for rock shrimp identification using enzyme-linked immunosorbent assay. J. agric. Food Chem. 38: 2094–2100.

Armstrong, K. F., C. M. Cameron & E. R. Frampton, 1997. Fruit fly (Diptera: Tephritidae) species identification: a rapid molecular diagnostic technique for quarantine application. Bull. entomol. Res. 87: 111–118.

Arnason, U., E. Mitchell & B. Widegren, 1993. Cetacean mitochondrial DNA control region: sequences of all extant baleen whales and two sperm whale species. Mol. Biol. Evol. 10: 960–970.

Arnold, P. I., J. E. Serafy, M. E. Clarke & D. R. Schultz, 1996. An immunological study of predation on hatchery-reared juvenile red drum (*Sciaenops ocellatus*, Linnaeus): description of an ELISA and predator-prey studies in nature. J. exp. mar. Biol. Ecol. 199: 29–44.

Avise, J. C., 1994. Molecular Markers, Natural History and Evolution. Chapman & Hall. New York. 511 pp.

Avise, J. C., 1998. Conservation genetics in the marine realm. J. Heredity 89: 377–382.

Baker, C. S., F. Cipriano & S. R. Palumbi, 1996. Molecular genetic identification of whale and dolphin products from commercial markets in Korea and Japan. Mol. Ecol. 5: 671–685.

Baker, C. S. & S. R. Palumbi, 1994. Which whales are hunted? A molecular genetic approach to monitoring whaling. Science 265: 1538–1539.

Bartlett, S. E. & W. S. Davidson, 1991. Identification of *Thunnus* Tuna species by the polymerase chain reaction and direct sequence analysis of their mitochondrial cytochrome b genes. Can. J. Fish. aquat. Sci. 48: 309–317.

Bartlett, S. E. & W. S. Davidson, 1992. FINS (Forensically informative nucleotide sequencing): a procedure for identifying the animal origin of biological specimens. Biotechniques 12: 408–411.

Baum, D., 1992. Phylogenetic species concepts. Trends Ecol. Evol. 7: 1–2.

Beacham, T. D., R. E. Withler & T. A. Steven, 1996. Stock identification of chinook salmon (*Oncorynchus tshawytscha*) using minisatellite DNA variation. Can. J. Fish. aquat. Sci. 53: 380–394.

Benson, P. G. & P. J. Smith, 1989. A manual of techniques for electrophoretic analysis of fish and shellfish tissues. New Zealand Fish. tech. Rep.: 13, 32pp.

Bell, J. L. & J. P. Grassle, 1998. A DNA-probe for identification of larvae of the commercial surfclam (*Spisula solidissima*). Mol. mar. Biol. Biotech. 7: 127–137.

Bermingham, E., 1990. Mitochondrial DNA and the analysis of fish population structure. In Whitmore, D. H. (ed.), Electrophoretic and Isoelectric Focusing Techniques in Fisheries Management. CRC press, Boca Raton, Florida: 197–222.

Bilodeau, A. L., W. S. Lankford, T. J. Kim, D. L. Felder & J. E. Neigel, 1999. An ultrasensitive method for detection of single crab larvae (*Sesarma reticulum*) by PCR amplification of highly repetitive DNA sequence. Mol. Ecol. 8: 681–684.

Birstein, V. J., J. Betts & R. DeSalle, 1998. Molecular identification of *Acipenser sturio* specimens: a warning for recovery plans. Biol. Cons. 84: 97–101.

Bowen, B. W., A. B. Meylan & J. C. Avise, 1991. Evolutionary distinctiveness of endangered Kemp's ridley sea turtle. Nature 352: 709–711.

Brown, L. D., 1995. Genetic evidence for hybridisation between *Haliotis rubra* and *H. laevigata*. Mar. Biol. 123: 89–93.

Budowle, B., 1995. Guidelines for a quality assurance program for DNA analysis. Crime Lab. Digest 22: 22–50.

Burton, R. S., 1996. Molecular tools in marine ecology. J. exp. mar. Biol. Ecol. 200: 85–101.

Carlton, J. T. & J. B. Geller, 1993. Ecological roulette: the global transport of non-indigenous marine organisms. Science 261: 78–82.

Carrera, E., T. García, A. Céspedes, I. González, B. Sanz, P. E. Hernandez & R. Martín, 1997. Immunostick colorimetric ELISA assay for the identification of smoked salmon, trout and bream. J. Sci. Food Agric. 74: 547–550.

Carrera, E., T. García, A. Céspedes, I. González, B. Sanz, P. E. Hernandez & R. Martín, 1998. Identification of Atlantic salmon (*Salmo salar*) and Rainbow trout (*Oncorhynchus mykiss*) by using polymerase chain reaction amplification and restriction analysis of the mitochondrial cytochrome b gene. J. Food Prot. 61: 482–486.

Carrera, E., R. Martín, T. García, A. Céspedes, I. González, B. Sanz & P. E. Hernandez, 1996. Development of an enzyme-linked immunosorbent assay for the identification of smoked salmon (*Salmo salar*), trout (*Oncorhynchus mykiss*) and bream (*Brama raii*). J. Food Prot. 59: 521–524.

Carvalho, G. R. & L. Hauser, 1994. Molecular genetics and the stock concept in fisheries. Rev. Fish Biol. Fish. 4: 326–350.

Carvalho, G. R. & T. J. Pitcher, 1994. Editorial: special issue. molecular genetics in fisheries. Rev. Fish Biol. Fish. 4: 269–271.

Cary, S. C., 1996. PCR-based method for single egg and embryo identification in marine organisms. Biotechniques 21: 998–1000.

Céspedes, A., T. García, E. Carrera, I. González, B. Sanz, P. E. Hernandez & R . Martín, 1998. Identification of flatfish species using polymerase chain reaction and restriction fragment analysis of the cytochrome b gene. J. Food Sci. 63: 206–209.

Coffroth, M. A. & J. M. Mulawka III, 1995. Identification of marine invertebrate larvae by means of PCR-RAPD species-specific markers. Limnol. Oceanogr. 40: 181–198.

Colbourne, J. K., B. D. Neff, J. M. Wright & M. R. Gross, 1996. DNA fingerprinting of bluegill sunfish (*Lepomis macrochirus*) using $(GT)_n$ microsatellites and its potential for assessment of mating success. Can. J. Fish. aquat. Sci. 53: 342–349.

Cunniff, P. A. (ed.), 1998. Official Methods of Analysis of AOAC International. Vol. 1, 16th edn. 4th Revision. Association of Official Analytical Chemists, Washington, D.C. 2110 pp.

Dalebout, M. L., A. Van Helden, K. Van Waerebeek & C. S. Baker, 1998. Molecular genetic identification of Southern hemisphere beaked whales (Cetacea, Ziphiidae). Mol. Ecol. 7: 687–694.

Daley, R. K., P. R. Last, G. K. Yearsley & R. D. Ward, 1997. South east fishery quota species: an identification guide. CSIRO Division of Marine Research, Hobart, Australia. 91 pp.

De Queiroz, K. & J. Gauthier, 1992. Phylogenetic taxonomy. Ann. Rev. Ecol. Syst. 23: 449–80.

De Salle, R. & V. J. Birstein, 1996. PCR identification of black caviar. Nature 381: 187–188.

Evans, B. S., R. W. G. White & R. D. Ward, 1998. Genetic identification of asteroid larvae from Tasmania, Australia, by PCR-RFLP. Mol. Ecol. 7: 1077–1082.

Feller, R. J., G. L. Taghon, E. D. Gallagher, G. E. Kenny & P. A. Jumars, 1979. Immunological methods for food web analysis in a soft-bottom benthic community. Mar. Biol. 54: 61–74.

Ferguson, M. M., 1994. The role of genetic markers in the management of cultured fishes. Rev. Fish Biol. Fish. 4: 351–373.

Ferguson, M. M. & R. G. Danzmann, 1998. Role of genetic markers in fisheries and aquaculture: useful tools or stamp collecting? Can. J. Fish. aquat. Sci. 55: 1553–1563.

Geller, J. B., 1994. Marine biological invasions as models of dispersal: tracking secondary spread and introgressive gene flow. Calif. Coop. oceanic Fish. Invest. Rep. 35: 68–72.

Geller, J. B., J. B. Carton & D. A. Powers, 1994. PCR-based detection of mtDNA haplotypes of native and invading mussels on the northeastern Pacific coast: latitudinal pattern of invasion. Mar. Biol. 119: 243–249.

Geller, J. B. & D. A. Powers, 1994. Site directed mutagenesis with the polymerase chain reaction for the identification of sibling species of *Mytilus*. The Nautilus (supl. 2): 141–144.

Geller, J. B., E. D. Walton, E. D. Grosholz & G. M. Ruiz, 1997. Cryptic invasion of the crab *Carcinus* detected by molecular phylogeography. Mol. Ecol. 6: 901–906.

Gosling, E. M., 1994. Speciation and species concepts in the marine environment. In Beaumont, A. R. (ed.), Genetics and Evolution of Aquatic Organisms. Chapman & Hall, London: 1–14.

Haig, S. M., 1998. Molecular contributions to conservation. Ecology 79: 413–425.

Harvey, W. D., 1990. Electrophoretic techniques in forensics and law enforcement. In Whitmore, D. H. (ed.), Electrophoretic and Isoelectric Focusing Techniques in Fisheries Management. CRC press, Boca Raton, Florida: 313–322.

Heath, D. D., P. D. Rawson & T. J. Hilbish, 1995. PCR-based nuclear markers identify alien blue mussel (*Mytilus* spp.) genotypes on the west coast of Canada. Can. J. Fish. aquat. Sci. 52: 2621–2627.

Hsieh, Y.-H. P., F.-C. Chen & M. Nur, 1997. Rapid species identification of cooked red snapper using isoelectric focusing. J. Food Sci. 62: 15–19.

Hu, Y.-P., R. A. Lutz & R. C. Vrijenhoek, 1992. Electrophoretic identification and genetic analysis of bivalve larvae. Mar. Biol. 113: 227–230.

Huang, T.-S., M. R. Marshall & C.-I. Wei, 1995a. Identification of red snapper (*Lutjanus campechanus*) using electrophoretic techniques. J. Food Sci. 60: 279–283.

Huang, T.-S., M. R. Marshall, K.-J. Kao, W. S. Otwell & C.-I. Wei, 1995b. Development of monoclonal antibodies for red snapper (*Lutjanus campechanus*) identification using enzyme-linked immunosorbent assay. J. agric. Food Chem. 43: 2301–2307.

Hunt, D. J., H. D. Parkes & I. D. Lumley, 1997. Identification of the species of origin of raw and cooked meat products using oligonucleotide probes. Food Chem. 60: 437–442.

International Whaling Commission, 1998 Chairman's report of the 49th annual meeting. IWC Cambridge. Rep. int. Whal. Commn: 48.

International Whaling Commission Conference, 1946. Annex. Nomenclature of whales. IWC, Cambridge. 4 pp.

Knowlton, N., 1993. Sibling species in the sea. Ann. Rev. Ecol. Syst. 24: 189–216.

Kostrikis, L. G., S. Tyagi, M. M. Mhlanga, D. D. Ho & F. R. Kramer, 1998. Molecular beacons – spectral genotyping of human alleles. Science 279: 1228–1229.

Lorente, M., J. A. Lorente, C. J. Alvarez, B. Budowle, M. R. Wislon & E. Villaeuva, 1997. Sequential multiplex amplification: utility in forensic casework with minimal amounts of DNA and partially degraded samples. J. forens. Sci. 42: 923–925.

Lundstrom, R. C. & S. A. Roderick, 1979. Fish-species identification by thin-layer isoelectric focusing of sarcoplasmic proteins. Sci. Tools 26: 38–43.

Medeiros-Bergen, D. E., R. R. Olson, J. A. Conroy & T. D. Kocher, 1995. Distribution of holothurian larvae determined with species-specific genetic probes. Limnol. Oceanogr. 40: 1225–1235.

Meyer, A., 1994. DNA technology and the phylogeny of fish. In Beaumont, A. R. (ed.), Genetics and Evolution of Aquatic Organisms. Chapman & Hall, London: 219–248.

Meyer, R. & U. Candrian, 1996. PCR-based DNA analysis for the identification and characterisation of food components. Lebens.-Wiss. Technol. 29: 1–9.

Mendoza, H., V. López-Rodas, S. Gonzále, A. Aguilera & E. Costas, 1995. The use of polyclonal antisera and blocking of antibodies in the identification of marine dinoflagellates: species-specific and clone-specific antisera against *Gymnodinium* spp. and *Alexandrium* spp. J. exp. mar. Biol. Ecol. 186: 103–115.

Mills, J., A. Idihara, I. Sakaguchi, S. Kong, R. Parry-Jones & M. Phipps, 1997. Whale meat trade in east Asia: a review of the markets in 1997. TRAFFIC International, Cambridge.

Murray, B. W., R. A. McClymont & C. S. Strobeck, 1995. Forensic application of ungulate species using restriction digests of PCR-amplified mitochondrial DNA. J. forensic Sci. 40: 943–951.

Naganuma, T., K. Hisadome, K. Shiraishi & H. Kojima, 1998. Molecular distinction of two resemblant abalones, *Haliotis discus discus* and *Haliotis discus hannai* by 18S rDNA sequences. J. mar. Biotech. 6: 59–61.

Nixon, K. C. & Q. D. Wheeler, 1990. An amplification of the phylogenetic species concept. Cladistics 6: 211–223.

Olson, R. R., J. A. Runstadler & T. D. Kocher, 1991. Whose larvae? Nature 351: 357–358.

O'Ryan, C. O. & E. H. Harley, 1998. Identification of the species of origin in trace quantities of biological material. S. afr. J. Sci. 94: 355–356.

Palumbi, S. R. & F. Cipriano, 1998. Species identification using genetic tools: the value of nuclear and mitochondrial gene sequences in whale conservation. J. Heredity 89: 459–464.

Park, L. K. & P. Moran, 1994. Appraisal of molecular genetic techniques in fisheries. Rev. Fish Biol. Fish. 4: 272–299.

Parker, P. G., A. A. Snow, M. D. Schug, G. C. Booton & P. A. Fuerst, 1998. What molecules can tell us about populations: choosing and using a molecular marker. Ecology 79: 361–382.

Pickavance, J. R., 1970. A new approach to the immunological analysis of invertebrate diets. J. anim. Ecol. 29: 715–724.

Powell, M. L., M. E. Maede & S. A. Watts, 1995. Identification of six commercially cultured crayfish species using total protein separation by isoelectric focusing. J. Crust. Biol. 15: 763–770.

Quintero, J., C. G. Sotelo, H. Rehbein, S. E. Pryde, I. Medina, R. I. Peréz-Martín, M. Rey-Méndez & I. M. Mackie, 1998. Use of mtDNA direct polymerase chain reaction (PCR) sequencing and PCR-restriction fragment length polymorphism methodologies in species identification of canned tuna. J. agric. Food Chem. 46: 1662–1669.

Ram, J. L., M. L. Ram & F. F. Baidoun, 1996. Authentication of canned tuna and bonito by sequence and restriction site analysis of polymerase chain reaction products of mitochondrial DNA. J. agric. Food Chem. 44: 2460–2467.

Rehbein, H., 1997a. Fish as food – fish species determination for caviar by protein and DNA analysis. Anim. Res. Dev. 46: 41–46.

Rehbein, H., 1997b. Comparison of several types of precast polyacrylamide gels for fish species identification by DNA analysis (single strand conformation polymorphism and random amplified polymorphic DNA). Arch. Lebensmittelhyg. 48: 25–48.

Reischl, U., B. Haber, W. Van de Zande, T. De Baere & M. Van Eechoutte, 1997. Preparation of DNA from numerous individual microscopic organisms for PCR-based assays of environmental samples. Biotechniques 23: 585–588.

Rojas, M., 1992. The species problem and conservation: what are we protecting. Cons. Biol. 6: 170-178.

Schmidt, T. M., E. F. DeLong & N. R. Pace, 1991. Analysis of a marine picoplankton community by 16S rRNA gene cloning and sequencing. J. Bact. 173: 4371–4378.

Scholin, C. A. & D. M. Anderson, 1996. LSU rDNA-based RFLP assays for discriminating species and strains of *Alexandrium* (Dinophyceae). J. Phycol. 32: 1022–1035.

Scribner, K. T. & T. D. Bowman, 1998. Microsatellites identify depredated waterfowl remains from glaucous gull stomachs. Mol. Ecol. 7: 1401–1405.

Seeb, J. E., G. H. Kruse, L. W. Seeb & R. G. Weck, 1990. Genetic structure of red king crab populations in Alaska facilitates enforcement of fishing regulations. Alsk. Sea Grant Coll. Program Rep. 90-04: 91–102.

Seeb, L. E., J. E. Seeb & A. J. Gharrett, 1990. Genetic marking of fish populations. In Whitmore, D. H. (ed.), Electrophoretic and

Isoelectric Focusing Techniques in Fisheries Management. CRC press, Boca Raton, Florida: 223–240.

Shaklee, J. B. & P. Bentzen, 1998. Genetic identification of stocks of marine fish and shellfish. Bull. mar. Sci. 62: 589–621.

Shepard, S. & J. Hartmann, 1996. Immunologic identification of billfish carcasses. Mol. mar. biol. Biotech. 5: 215–219.

Shumway, S. E. & A. D. Cambella, 1993. The impact of toxic algae on scallop culture and fisheries. Rev. Fish. Sci. 1: 121–150.

Silberman, J. D. & P. J. Walsh, 1992. Species identification of spiny lobster phyllosome larvae via ribosome DNA analysis. Mol. mar. Biol. Biotech. 1: 195–205.

Skibinski, D. O. F., 1985. Mitochondrial DNA variation in *Mytilus edulis* L. and the Padstow mussel. J. exp. mar. Biol. Ecol. 92: 251–258.

Skibinski, D. O. F., 1994. The potential of DNA techniques in the population and evolutionary biology of aquatic invertebrates. In Beaumont, A. R. (ed.), Genetics and Evolution of Aquatic Organisms. Chapman & Hall, London: 177–198.

Smouse, P. E. & C. Chevillon, 1998. Analytical aspects of population-specific DNA fingerprinting for individuals. J. Heredity 89: 143–150.

Sotelo, C. G., C. Piñeiro, J. M. Gallardo & R. I. Pérez-Martin, 1993. Fish species identification in seafood products. Trends Food Sci. Technol. 4: 395–400.

Sweijd, N. A., R. C. K. Bowie, A. L. Lopata, A. M. Marinaki, E. H. Harley& P. A. Cook, 1998. A PCR technique for forensic species level identification of abalone tissue. J. Shellfish Res. 17: 889–896.

Taylor, D. B., A. L. Szalanski & R. D. Petersen II, 1996. Identification of screwworm species by polymerase chain reaction-restriction fragment length polymorphism. Med. vet. Entomol. 10: 63–70.

Theilacker, G. H., N. C. H. Lo & A. W. Townsend, 1993. An immunological approach to quantifying predation by euphausiids on early stages of anchovy. Mar. Ecol. Prog. Ser. 92: 35–50.

Tyagi, S. & F. R. Kramer, 1996. Molecular Beacons: probes that fluoresce upon hybridisation. Nat. Biotechnol. 14: 303–308.

Tyrrell, J. V., P. R. Bergquist, D. J. Saul, L. MacKenzie & P. L. Bergquist, 1997. Oligonucleotide probe technology as applied to the study of harmful algal blooms. New Zealand J. mar. Freshwat. Res. 31: 551–560.

Utter, F., 1991. Biochemical genetics and fishery management: an historical perspective. J. Fish. Biol. 39A: 1–20.

Van Belkum, A., P. W. M. Hermnas, L. Licciardello, S. Stefani, W. Grubb, W. Van Leeuwen & W. H. F. Goessens, 1998. Polymerase chain reaction-mediated typing of micro-organisms – tracking dissemination of genes and genomes. Electrophoresis 19: 602–607.

Van Der Bank, H. & C. Kirchner, 1997. Biochemical genetic markers to distinguish two sympatric and morphologically similar Namibian marine fish species, *Argyrosomus coronus* and *A. indodorus* (Perciformes: Sciaenidae). J. african. Zool. 11: 441–448.

Venter, J. D., S. Van Wyngaart, J. A. Verschoor, M. R. Lipinski & H. M. Verheye, 1999. Detection of zooplankton prey in squid paralarvae with immunoassay. J. Immunoassay. 20(3) 127–149.

Walsh, D., R. A. Reeves, D. J. Saul, R. D. Gray, L. MacKenzie, P. R. Bergquist & P. L. Bergquist, 1998. Heterogeneity of SSU and LSU rDNA sequences of *Alexandrium* species. Biochem. Syst. Ecol. 26: 495–509.

Ward, R. D. & P. M. Grewe, 1994. Appraisal of molecular genetic techniques in fisheries. Rev. Fish Biol. Fish. 4: 300–325.

Ward, R. D., N. G. Elliott & P. M. Grewe, 1995. Allozyme and mitochondrial DNA separation of Pacific northern bluefin tuna, *Thunnus thynnus orientalis* (Temminck and Schlegel), from southern bluefin tuna, *Thunnus maccoyii* (Castelna). Aust. J. mar. Freshwat. Res. 46: 921–930.

Whitmore, D. H., 1986. Identification of sunfish species using muscle protein isoelectric focusing. Comp. Biochem. Physiol. 84B: 177–189.

Whitmore, D. H., 1990. Isoelectric focusing of proteins. In Whitmore, D. H. (ed.), Electrophoretic and Isoelectric Focusing Techniques in Fisheries Management. CRC Press, Boca Raton, Florida: 81–106.

Wimpee, C. F., T. -L. Nadeau & K. H. Neaslon, 1991. Development of species-specific hybridisation probes for marine luminous bacteria by using in vitro DNA amplification. Apl. envir. Microbiol. 57: 1319–1324.

Wirgin, I. I., J. R. Waldman, L. Maceda, J. Stabile & V. J. Vecchio, 1997. Mixed-stock analysis of Atlantic coast striped bass (*Morone saxatilis*) using nuclear DNA and mitochondrial DNA markers. Can. J. Fish. aquat. Sci. 54: 2814–2826.

World Health Organization, 1969. Sampling plans for prepackaged foods. Codex Alimentarium Vol. 13, 2nd edn. Food and Agriculture Organization of the United Nations, New York. 15 pp.

Wright, J. M. & P. Bentzen, 1994. Microsatellites: genetic markers for the future. Rev. Fish Biol. Fish. 4: 384–388.

Yman, I. M., 1993. Identifying fish species by IEF with PhastSystem. Pharmacia LKB Biotechnology Application Note 379: 1–6.

Youn-Ho, L. & V. D. Vacquier, 1995. Evolution and systematics in Haliotidae (Mollusca: Gastropoda): inferences from DNA sequences of sperm lysin. Mar. Biol. 124: 267–278.

Hydrobiologia **420:** 165–184, 2000.
A. M. Solé-Cava, C. A. M. Russo & J. P. Thorpe (eds), Marine Genetics.

Exploited marine invertebrates: genetics and fisheries

J. P. Thorpe[1], A. M. Solé-Cava[1,2] & P. C. Watts[1]
[1]*School of Biological Sciences, University of Liverpool, Port Erin Marine Laboratory, Port Erin, Isle of Man, IM9 6JA, U.K.*
[2]*Laboratório de Biodiversidade Molecular, Departamento de Genética, Instituto de Biologia, Universidade Federal do Rio de Janeiro, Rio de Janeiro, RJ, Brazil*
E-mail: sole@biologia.ufrj.br

Key words: genetic variation, invertebrates, fisheries, stocks, larvae

Abstract

The application of genetic techniques to invertebrate fisheries is in many ways essentially similar to that in vertebrate (i.e. finfish) fisheries, for which there is already an extensive body of published data. However, there are also relative differences which lead to particular problems in the use of genetic data to study commercially important invertebrate species. The main role for genetics of both vertebrates and invertebrates has been, and is likely to continue to be, the identification of groups of interbreeding individuals as the basis for a fishery. It is in the identification of the breeding unit that the genetic differences between vertebrates and invertebrates can be of practical significance. The genetic breeding unit, usually called a 'stock' in fisheries biology, generally shows a certain uniformity of size in most marine fish which have been studied. Smaller or less mobile fish (e.g. flatfish) may only range a few tens of kilometres to their breeding grounds, whilst in more mobile, particularly migratory pelagic species (e.g. Scombridae), the area occupied by a stock is likely to be far greater and for a few (e.g. large pelagic elasmobranchs), a single unit of stock may be almost circumglobal. However, marine fish generally, particularly those large or plentiful enough to be of commercial interest, are likely to be fairly mobile and in many cases the order of mobility is likely to be in the region we might predict from our knowledge of the biology and habits of the species. In the genetic assessment of 'stocks' for invertebrate fisheries, we face a number of additional problems, mostly related to the large evolutionary range of invertebrates exploited and their widely different biology. Although in Europe and North America marine invertebrate fisheries may be thought of as being mainly for decapod crustaceans and bivalve molluscs, globally commercially important marine invertebrate fisheries range from sponges to squid and include such diverse groups as sea cucumbers, barnacles, krill, octopuses, cuttlefish, sea anemones, ascidians, polychaetes, sea urchins, gastropods and jellyfish. An obvious feature of many of these invertebrates is that the adult (i.e. commercial) stage of the life cycle is sessile (e.g. barnacles, sponges, ascidians) or of very limited mobility (e.g. sea anemones, sea urchins, bivalves, gastropods), with the result that the dispersive phase of the life cycle is the larva. Other groups (e.g. krill, jellyfish) are planktonic or nektonic and may cover very large distances, but, unlike fish, have little control over the distance or direction of travel, whilst some of the open ocean pelagic squid are more mobile than most fish and may migrate thousands or kilometres to spawning grounds. The very low mobility of both larva and adult in some invertebrates indicates that dispersal, and hence stock size, is likely to be low and that, therefore, stocks are far more vulnerable to overfishing than in most fish species. An additional difficulty is that genetic studies to date indicate a remarkably high incidence of cryptic speciation in marine invertebrates, sometimes even in comparatively well studied commercially important species. Thus, although to date marine invertebrate fisheries have not received the same level of attention from geneticist as finfish fisheries, it is clear that for invertebrate fisheries genetic data are relatively far more important if a fishery is to be exploited without being endangered.

Introduction

Background: the application of genetics to fisheries

To date, there is an extensive literature on the various uses of genetics for the study of marine fisheries (e.g. Ryman & Utter, 1987; Carvalho & Pitcher, 1995; Ward & Grewe, 1995), but the great bulk of this published work is concerned with fisheries for fish, rather than fisheries for commercially important marine invertebrates. This concentration of applied genetics on marine fish is perhaps surprising, because in the far more extensive field of the genetics of marine organisms as a whole a much larger proportion of the effort has gone into studies of invertebrates (see Ward, 1989). Indeed, in the past, marine invertebrate species have been used extensively as models to test various hypotheses concerning environmental and other parameters claimed to affect the genetic structure of populations (e.g. Ayala et al., 1973, 1974, 1975; Nevo, 1978; Valentine & Ayala, 1978; Ayala & Valentine, 1979; Lavie & Nevo, 1981; Noy et al., 1987; Mitton et al., 1989).

A wide range of 'molecular' techniques has been applied to fisheries genetics (reviews by Park & Moran, 1995; Ward & Grewe, 1995; Carvalho & Hauser, 1998), including allozyme electrophoresis (Beaumont & Pether, 1996; Gallardo & Carrasco, 1996; Stiles et al., 1996, review by Utter et al., 1987), mitochondrial DNA restriction analysis (Epifanio et al., 1996; Ward et al., 1997; review by Ferris & Berg, 1987), microsatellite DNA (e.g. Bentzen et al., 1996; Garcia de Leon et al., 1997; Heist, 1999, review by Wright & Bentzen, 1995), minisatellite DNA (e.g. Taggart & Ferguson, 1990; Taggart et al., 1995; Volpe & Ferguson, 1996; review by O'Reilly & Wright, 1995), random amplified polymorphic DNA (RAPD) (e.g. Bielawski & Pumo, 1997; Harding et al., 1997; Heipel et al., 1998), single-stranded conformational polymorphism (SSCP) (Li & Hedgecock, 1996) and DNA sequencing (e.g. Stepien, 1995). Of these several methods, allozymes are by far the longest established and remain generally the technique of choice in the first instance on grounds of cost combined with ease of use on large samples (Utter, 1995; Ward & Grewe, 1995), but their use is limited to fresh or frozen tissue samples. More 'modern' DNA based methods can give greater resolution and can also be used on very small (e.g. biopsy) or preserved tissue samples (Ward & Grewe, 1995). Of the DNA based methods, only RFLP analysis of mitochondrial DNA has been widely used in fisheries, but still far less than allozyme electrophoresis (Wright & Bentzen, 1995).

The applications for genetic techniques in relation to invertebrate fisheries are in many ways essentially similar to those in vertebrate (i.e. finfish) fisheries (reviews by e.g. Ryman & Utter, 1987; Carvalho & Pitcher, 1995; Utter, 1995; Carvalho & Hauser, 1998). However, there are also relative differences which lead to particular problems in the use of genetic data to study commercially important invertebrate species. Some of the main putative roles for genetics in marine fisheries, for either vertebrates or invertebrates, may be summarised as:

1. understanding the structuring of populations;
2. identification of stocks (breeding units) and units for conservation;
3. mixed stock analysis;
4. genetic effects on growth rate, survival, disease resistance or other important parameters;
5. development of strains for captive breeding.

All of these possible uses are potentially relevant to both vertebrates and invertebrates, and, of course, they are not all mutually exclusive; they are not completely discrete and in any practical application there is likely to be a degree of overlap between various of them. For example, the identification of breeding units is likely to entail understanding the structuring of populations and possibly also the identification and separation of mixed stocks, and in many cases it will be the aim that will differ rather than the methods or the type of data generated.

The last two roles listed are of relevance only during captive breeding and, therefore, are applicable to fisheries genetics only if captive bred fish or invertebrates are to be released back into the natural environment. In practice, this is only likely to occur either for the restocking of species which have become severely depleted in the wild (possibly through overfishing or by being wiped out by disease or pollution) or during the practice of what has been termed 'ranching' (e.g. Isaksson, 1988). In ranching, organisms are returned to the wild, usually at an early stage in the life cycle, and are then recaptured after a suitable time interval, when they are of a suitable size to be sold. Clearly ranching is only likely to be financially viable for species which are either sessile (e.g. most bivalves) or benthic and of very limited mobility (e.g. lobsters or a few fish species) or for species which will return to the same area at some later stage in the life cycle (e.g. to breed) as with anadromous fish (e.g. many salmonids). Thus, ranching is particularly suitable for 'low gene flow' species, including many commercially exploited invertebrates.

Genetics and stock identification

For both vertebrate or invertebrate fisheries, the main role for genetics has been, and is likely to continue to be, the identification of groups of interbreeding individuals as the basis for a fishery. It is in this area, the identification of the breeding unit, that the genetic differences between vertebrates and invertebrates can be of practical significance. The genetic breeding unit is usually called a 'stock' in fisheries biology (this is roughly equivalent to a 'population' to a geneticist); the concept of a fishery stock has been discussed in great detail by many authors (see, e.g. Altukov, 1981; Carvalho & Hauser, 1995) and there is little point in discussing it further here.

The assumption is that within a stock (however defined), there should be genetic homogeneity, but between stocks, if these are genetically isolated, genetic divergence may occur (through e.g. drift, mutation, possible selection) with time. Models of gene flow between populations vary greatly in their assumptions, but, almost irrespective of this, a clear conclusion is that, in the absence of extreme or abnormal selective forces, even very small levels of effective migration will be sufficient to preclude genetic divergence (for brief discussion of relevant models see e.g. Maynard-Smith, 1989; reviews of various aspects of gene flow between populations by Kimura & Weiss, 1964; Slatkin, 1981, 1985, 1994, 1995). Thus, even small levels of genetic divergence between putative stocks will provide an indication of reproductive isolation.

Much genetic work on fish stocks is concerned with freshwater species, many of these salmonids, and, given the generally very geographically restricted nature of most freshwater habitats, conclusions on freshwater stock sizes are mostly not particularly relevant to marine species. In general, many marine species, whether vertebrate or invertebrate, are more likely to be constrained by biological than by geographical barriers to dispersal, except perhaps over very large distances, when physical barriers (e.g. changes in temperature) may also be relevant. In most marine fish studied, the distance over which a species is likely to disperse, and thus the geographical size of a stock, generally shows a certain uniformity, admittedly within very broad limits. Smaller or less mobile fish (e.g. smaller flatfish, anchovies) may only range a few tens of kilometres to their breeding grounds (e.g. Bembo et al., 1996), and others (e.g. coral reef or rock pool species) will be even less mobile and stocks may have a very restricted range (e.g. Larcson et al., 1989; Doherty, 1995). For a few large and very mobile fish species like swordfish (Alvarez-Bremer et al., 1995; but see also Chow et al., 1997) and tuna (Chow & Ushiama, 1995), a single unit of stock may cover very great areas or even be circumglobal. Marine fish generally, particularly those large or plentiful enough to be of commercial interest, are likely to be fairly mobile and in many cases the order of mobility is likely to be in the region we might predict from our knowledge of the biology and habits of the species. Thus, whilst very small intertidal or inshore benthic marine fish may disperse very short distances, fish likely to be of fisheries potential are mostly those with stocks ranging over areas from tens to hundreds of kilometres, with a few of the more mobile larger or migratory pelagic species covering larger distances.

However, in attempting to make genetic assessments of 'stocks' for invertebrate fisheries, we face a number of additional problems. Firstly, relative dispersal capabilities and hence presumably stock sizes show much greater variation than is found in fish. In Europe and North America, marine invertebrate ('shellfish') fisheries may be thought of as being mainly for decapod crustaceans and bivalve molluscs, but globally there are commercial marine invertebrate fisheries for species ranging from sponges to squids and including such diverse groups as sea cucumbers, barnacles, krill, octopuses, cuttlefish, sea anemones, ascidians, polychaetes, sea urchins, gastropods and jellyfish (reviews by e.g. Caddy, 1989; Thorpe & Nash, 1993). An obvious feature of many of these invertebrates is that the adult (i.e. commercial) stage of the life cycle is sessile (e.g. barnacles, sponges, ascidians) or of very limited mobility (e.g. sea anemones, sea urchins, bivalves, gastropods), with the result that often the dispersive phase of the life cycle is the larva. Other commercially exploited invertebrates (e.g. krill, jellyfish) are planktonic or nektonic and may cover very large distances, but, unlike fish, will have little control over the distance or direction of travel. Conversely, however, some of the open ocean pelagic squids are more mobile than most fish and may migrate thousands of kilometres to spawning grounds.

A major problem in assessing invertebrate stocks is that for many species we have little or no knowledge of the biology of the larval stage; indeed for many, the larva has yet to be identified, and even where the larva has been studied and the larval life span measured (usually in the laboratory), this often does not correlate with empirical levels of genetic divergence found between populations in the natural environment (see e.g. Knowlton & Keller, 1986; Todd et al., 1988, 1998). In many marine invertebrates, genetic data indicate that there is very low mobility of both larva and adult, resulting in restricted, often only localised, dispersal. If this is the case, stock sizes are likely to be low and, therefore, stocks are far more vulnerable to overfishing than in most fish species. An additional difficulty in studying invertebrate fisheries is that genetic studies to date indicate a remarkably high incidence of cryptic speciation in marine invertebrates (Knowlton, 1993; Thorpe & Solé-Cava, 1994), sometimes even in comparatively well studied commercially important species (e.g. Smith et al., 1981; Yeatman & Benzie, 1994; Chan & Chu, 1996). Thus, although to date marine invertebrate fisheries have not received the same level of attention from geneticists as finfish fisheries, it is clear that for invertebrate fisheries it is relatively far more important to have genetic data if a fishery is to be exploited without being endangered.

Marine invertebrate fisheries

Before discussing the genetics of marine invertebrate fisheries, it is important to at least outline which types of invertebrate are commercially exploited in the sea. Since many of these fisheries, although of commercial and cultural significance where they occur, are not widely known, it is intended, here, to give greater detail of some of the more diverse invertebrates which are consumed or otherwise used commercially by man. Further detail is given in various reviews (e.g. Caddy, 1989; Thorpe & Nash, 1993; see also numerous reports by FAO).

Shellfish

As mentioned above, in western society the main marine invertebrates consumed are those groups conventionally termed 'shellfish'; that is, various crustaceans (mainly decapods; lobsters, prawns and crabs) and bivalves and, to a lesser extent, gastropod molluscs.

Throughout most of the World, the crustaceans commercially exploited are very largely decapods and fisheries for these are found in shallow continental shelf seas in most temperate or warmer areas. Of particular significance for human consumption are various species of the prawn genus *Penaeus*, which occur in shallow waters of many of the warmer countires, but which also have been extensively transported between continents (although mainly for aquaculture rather than conventional fisheries). Numerous other prawns, shrimps, crabs and lobsters are fished worldwide.

Of potentially global significance for food or for 'fishmeal' production, but to date only intermittently exploited on a large scale, are the krill. Although superficially prawn like these are euphausid crustaceans, not decapods. They exist in immense biomasses, particularly in the Antarctic, where *Euphausia superba* is the main species.

Another group of crustaceans eaten by man are the barnacles (Cirripedia), various of which are exploited commercially throughout the world. For example, the large stalked barnacle *Polliceps polliceps* is eaten throughout Iberia and the Atlantic Island groups (Canaries, Madeira and Azores), acorn barnacles are consumed in Chile (*Megabalanus psittacus*) and in the Azores (*M. tintabulum*) and the goose barnacle *Mytella mytella* in Japan.

From the other main 'shellfish' phylum, the Mollusca, bivalves are fished almost wherever they occur in shallow waters, except for close to the poles, and a very wide variety of scallops (Pectinidae), mussels (Mytilidae), Oysters (Ostreidae) and 'clams' are taken. Several species of bivalve, notably *Crassostrea gigas, C. virginica, Ostrea edulis* (Ostreidae) and *Pecten maximus* (Pectinidae), have been transported to many areas of the world and, following either deliberate introduction or escapes, now occur and are fished in many of these areas. Various gastropods are eaten in numerous countries, but the species of major commercial significance are mainly abalones (*Haliotis* spp.), which are found in warm temperate and warmer waters in various parts of the world.

'Other' invertebrate fisheries

An increasing proportion of the more usual 'shellfish' are derived from aquaculture rather than capture fisheries, which, as with finfish fisheries, are suffering from depleted wild populations as a result of overfishing. Against this background of worldwide overfishing and the decline of conventional shellfish stocks, other marine invertebrates are likely to continue to increase in commercial importance. Such species are widely, if not extensively, consumed in Europe and North America, whilst elsewhere in the world, patterns of consumption vary considerably. In various areas, particularly in the tropics, artisanal fisheries for unusual invertebrates can be locally important and in the far east there is a great demand for food species not commonly eaten in the west. One group in particular, the cephalopod molluscs (octopus, squid, cuttlefish, etc.), are currently important in terms of world catches and could become greatly more important for fisheries if the technology to exploit them can be developed (see below for further discussion).

Sponges (Porifera)

Although sponges are not used as human food, in various warmer parts of the world, particular sponge species are collected commercially, usually by divers, to be sold as bath sponges. Exploited species all have a soft elastic skeletal structure, unlike most sponges which have hard calcareous or siliceous spicules. The main species collected are *Hippospongia communis, Spongia officinalis, S. zimocca* and *S. graminea*, all of which are from the order Dictyoceratida of the class Demospongiae. In some areas, increasing prices, as supplies decline, have resulted in the virtual extinction of exploited species at diving depths.

Jellyfish and sea anemones (Cnidaria)

At least three species of jellyfish, of which *Rhopilema esculenta* is the most common, are commercially important as food in Japan. Several species of sea anemones are eaten in various countries including Japan, Samoa, France, Portugal and Italy. The intertidal *Actinia equina* and *Anemonia viridis* are eaten in Italy and France.

Marine worms (Polychaeta)

Polychaetes eaten by humans are mainly what are known as palolo worms, several species of which occur in different parts of the world. These worms have a stage in the life cycle known as a sexual epitoke, which, as part of the reproductive cycle, swarms to the surface in very large numbers at predictable times of the year. They are then collected for eating. Species consumed include the Pacific palolo (*Eunice viridis*) from Polynesia, the Wawo (*Lysidice oele*) from Indonesia, the Atlantic palolo (*Eunice schemacephala*) in the Gulf of Mexico and the Japanese palolo (*Tylorrhynchus heterochaetus*) in Japan. Also of commercial significance are various polychaetes which are dug up, or even aquacultured (Olive, 1994), for use as fishing baits (e.g. *Arenicola* spp. and *Nereis* spp. in Britain).

Ascidians (Urochordata, Ascidiacea)

All species of ascidian eaten are from the single family Cynthiidae. At least two species (*Halocynthis roretzi* and *H. aurantium*) are fished commercially by the Japanese. Ascidians of the genus *Pyura* are consumed in Chile. The French eat mainly just one species, *Microcosmus claudicans*.

Sea urchins and sea cucumbers (Echinodermata)

The sea urchins are increasingly becoming subject to commercial fisheries, primarily to satisfy the Japanese market. Two species (*Paracentrotus lividus* and *Echinus esculentus*) are traditionally eaten in some Mediterranean countries and in Ireland. Similar species are eaten in the Far East, particularly Japan, in Chile and in New Zealand. Due to increasing demand, north Atlantic species are exported to Japan, and Japanese boats have created a major sea urchin fishery in Australian waters.

The market for sea cucumbers is also mainly in Japan. The very large black species *Stichopus japonicus* is taken locally, whilst several other species are exported to Japan from various parts of the Pacific.

Cephalopods

Of the cephalopod molluscs, only species of the subclass Coleoidea are important as food for man. These include the three orders Sepioidea (cuttlefish), Teuthoidea (squids) and Octopoda (octopodes) all of which are fished commercially.

Cuttlefish Cuttlefish are almost absent from the Americas, although a few species of *Semirossia* occur along the west Atlantic coast and are fished off Argentina and in the Gulf of Mexico. Elsewhere, both large (mainly *Sepia*) and small (Sepiolidae) cuttlefish are fished in the coastal waters of most temperate and tropical latitudes. The major species fished are *Sepia officinalis* (occurs down the eastern Atlantic from northern Europe to South Africa and throughout the Mediterranean and Baltic Seas), *Sepia pharoensis* (found from the Red Sea to Japan and south to Australia), *Sepiella inermis* (occurs from the Persian Gulf to eastern Asia, fished mainly in India and Sri Lanka) and *Sepiella japonica* (Japan and China), but numerous other species are caught in smaller quantities. The tiny *Sepiola rondeleti,* which occurs from northern Europe to west Africa, is extensively eaten in many Mediterranean countries. The major markets, even for Atlantic and Mediterranean caught cuttlefish, are in Japan and Korea.

Squid There are two suborders of squids, the Myopsida which are inshore mainly demersal squids generally confined to shallower waters near shores or on continental shelves and the Oegopsida; mostly pelagic offshore squids of deep oceanic waters, which may come onto continental shelves to spawn.

All the myopsid squids currently eaten are confined to the family Loliginidae and most are from the very large genus *Loligo*, species of which are found on continental shelves worldwide. Major fishery species are *Loligo bleekeri* (Japan), *L. chinensis* and *L. edulis* (both China to Australia), *L. duvauceli* (most of Asia and north east Africa), *L. forbesi* (Sweden to Senegal, Mediterranean, Red Sea, south east Africa), *L. gahi* (southern South America, Falkland Islands), *L. japonica* (Japan, China), *L. opalescens* (Pacific coast of North America), *L. pealii* (Newfoundland to Brazil), *L. reynaudi* (South Africa), *L. vulgaris* (Europe to South Africa, Mediterranean), *Sepioteuthis lessoniana* (Red Sea to Japan and Australia) and *Alloteuthis media* (Mediterranean).

Oceanic (oegopsid) squids form well over half of the total world cephalopod catch; a few species are demersal, but most are pelagic, although this may be at great depths. For obvious reasons, the species presently exploited commercially are generally only those occurring comparatively close to the surface. Currently, only about nine (Enoploteuthidae, Octopoteuthidae, Onychoteuthidae, Gonatidae, Psychroteuthidae, Lepidoteuthidae, Histioteuthidae, Ommastrephidae, Thysanoteuthidae) of 32 recognised families have any species fished by man, although many others have potential as food. Exploited species of major commercial significance are confined to only four families:

1. Enoploteuthidae: *Watasenia scintillans* (China, Japan);
2. Onychoteuthidae: *Onychoteuthis borealijaponica* (fished off Japan and northwest U.S.A.);
3. Gonatidae: *Berryteuthis magister* (fished off Japan and northeastern Russia); and
4. Ommastrephidae: *Illex argentinus* (southeastern South America, Falkland Islands), *I. coindetii* (fished western Mediterranean and Atlantic off Spain and North Africa), *I. illecebrosus* (fished northwest Atlantic), *Todaropsis eblanae* (fished Mediterranean and northwest Africa), *Todarodes pacificus* (northern Pacific, China to Alaska, fished by Japan), *T. sagittatus* (fished Norway, Italy), *Nototodarus gouldi* (Australia), *N. sloani* (New Zealand), *Ommastrephes bartrami* (fished northern and southern Pacific) and *Dosidicus gigas* (Chile to Mexico).

Octopuses Among the octopods, only species of the suborder Incirrata are fished commercially and of these only one (Octopodidae) of the eight families is commercially important. All the main species caught are in the widely distributed genus *Octopus*: *O. briareus* (Caribbean and northern South America), *O. conispadiceus* (Japan), *O. cyaneus* (East Africa to India and Australia), *O. dofleini* (China to southwestern U.S.A.), *O. globosus* (India), *O. maya* (Gulf of Mexico), *O. membranaceus* (India to Japan and Australia, important in Japan and China), *O. variabilis* (Japan, China), *O. vulgaris* (coastal distribution from northern Europe around Africa, including Mediterranean and Red Sea, and Asia east to Japan, also western Atlantic from U.S.A. to Brazil. Fished over most of huge range, major importance Japan, northwest Africa). As with other cephalopods, much of the world market for octopus species is in Japan. For further details of species see Voss (1983) and Roper et al. (1984).

Minor food species

Many more species and other marine invertebrate groups are, or are likely to be, eaten or otherwise exploited on a limited scale somewhere in the world, but those outlined above give some indication of the diversity of marine invertebrates of use to man. An area not discussed here, but already starting to expand, is the exploitation of species of certain phyla (e.g. sponges, bryozoans, coelenterates, ascidians, nemerteans) which have allelopathic (usually anti-predator or anti-competitor) or otherwise 'useful' biologically active chemicals of value to the pharmaceutical industry (Faulkner, 1993).

Genetics of invertebrate fisheries

Outside the conventional 'shellfish' (i.e. crustaceans and shelled molluscs) fisheries, there has been remarkably little application of genetic techniques to invertebrate fisheries or fisheries related problems. Thus, the great bulk of invertebrate groups of fisheries interest have been the subject of little or no genetic study in relation to these fisheries. However, this lack of study is not indicative of some massive oversight, but merely follows from the small scale, often artisanal, nature of many of these fisheries and the geographical location of many of them; often far from major fisheries laboratories with genetic capabilities.

Among these 'other' invertebrates, the main exception is the cephalopods, particularly squids, which have now received a limited amount of attention from geneticists interested in actual or potential commercially important species. However, the reasons for this research are again largely financial; some cephalopod fisheries are on a large scale and hence are commercially important; also they are of commercial interest to economically developed countries like Japan, United States, Canada, New Zealand, Australia and Britain.

As outlined above, from the viewpoint of their population genetics, a key feature of many commercially important marine invertebrates is the mobility of the adult (i.e. commercial) stage of the life cycle. This may be sessile (e.g. barnacles, sponges, ascidians) or of very limited mobility (e.g. sea anemones, sea urchins, bivalves, gastropods). The adults may also be planktonic or nektonic (e.g. krill, jellyfish) and may cover very large distances, but have little control over the distance or direction of travel, or, like some of the open ocean pelagic squids, they may be more mobile than most fish and may migrate thousands of kilometres to spawning grounds.

High mobility exploited invertebrates: the cephalopods

Since they are among the most mobile of invertebrates, some similarity of population structuring might be expected between cephalopods, particularly squid, and fish. Genetic studies for stock assessment or other purposes have been carried out on many of the (mostly) long established inshore fisheries for benthic squids (largely *Loligo* spp.) (e.g. Ally & Keck, 1978; Christofferson et al., 1978; Augustyn & Grant , 1988; Carvalho & Pitcher, 1989; Garthwaite et al., 1989; Yeatman & Benzie, 1994; Brierley et al., 1995, 1996; Katugin, 1995; Izuka et al., 1996; Kang et al., 1996) and also for a number of the more recent, but now very important, oceanic (mainly ommastrephid) squid fisheries (e.g. *Nototodarus sloani* off New Zealand, Smith et al., 1981; *Martialia hyadesi* in the Antarctic Ocean, Brierley et al., 1993a; *Illex argentinus* around the Falkland Islands, Thorpe et al., 1986; Carvalho et al., 1992).

Genetic studies of octopus species are far fewer. Allcock (1997) and Allcock et al. (1997) studied gene flow in Antarctic populations of *Pareledone turqueti*, a species with only minor commercial potential and Levy et al. (1988) found cryptic speciation in *Eledone* off Brazil.

There are also several studies using molecular genetic methods to estimate divergence for systematic or other purposes in cephalopods (see e.g. Brierley & Thorpe, 1994; Brierley et al., 1996, 1997; Allcock, 1997). Almost all these molecular population studies on cephalopods have relied on allozyme studies for their data, although microsatellite markers have been developed at least for *Loligo forbesi* (Shaw, 1997).

A major feature common to many genetic studies of cephalopod, particularly squid, populations is the generally low levels of genetic variability found in most species. Polymorphism is often low in the myopsid (inshore) squid species (e.g. Brierley et al., 1995), but is even lower in a number of the offshore and more mobile oegopsid squid (Thorpe et al., 1986; Garthwaite et al., 1989). A further notable feature of the genetic structuring of oceanic squid populations is the tendency for samples showing genetic differentiation, perhaps collected in different areas or at different times, to exhibit fixed allelic differences with no heterozygous individuals present. For example, Thorpe et al. (1986) examined a number of samples of *Illex 'argentinus'* collected sequentially over a limited period of time in the same general area to the north of the Falkland Islands. Most samples showed no genetic variation whatsoever over 30+ individuals screened for about 40 enzyme loci, yet several of the samples were clearly genetically differentiated from all others by fixed allelic differences at a small number (2–4) of loci, with a complete absence of any heterozygotes. Similar patterns of divergence have also been found, for example, between samples of *Martialia* (Brierley et al., 1993a).

Within squid species in the open oceans, it would be reasonable to expect that single stocks of these highly mobile animals might occupy very large areas. To some extent, this is the case, but there is still often genetic differentiation with various stocks of a nominate species occurring within the same large area, in the same places, but not usually at the same time. The result is that various stocks may be found to occupy broadly the same vast area of open ocean, but the migration patterns are such that these stocks do not generally occur sympatrically and thus they are separated temporally, being found in the same place only at different times and hence maintaining spatial separation (i.e. allopatry) at any given time (see e.g. Thorpe et al., 1986; Brierley et al., 1993a).

An additional problem, again unexpected in very mobile animals occurring mainly in areas with few obvious geographical barriers, is the apparently high level of cryptic speciation in squid (e.g. Smith et al., 1981; Thorpe et al., 1986; Brierley et al., 1993a; Yeatman & Benzie, 1994). This may, however, merely reflect inadequate taxonomy in a group with few hard parts or other obvious useful taxonomic characters.

Unexpected levels of population subdivision appear to be common in squid species (see also Augustyn & Grant, 1988; Garthwaite et al., 1989; Brierley et al., 1993b, 1995, 1996; Izuka et al., 1994; Yeatman & Benzie, 1994; Katugin, 1995; Kang et al., 1996) and clearly pose problems for the management of fisheries, since, in fisheries generally, the correct identification of management units (stocks) may be considered to be the greatest single genetic problem for stock conservation (e.g. Allendorf et al., 1987; Ferguson, 1994; Ward & Grewe, 1995) and for the maintenance of evolutionarily significant units and hence the genetic diversity of species (e.g. Ryman, 1991; Bernatchez, 1994; Currens & Busack, 1995).

In squid, the correct management of stocks is essential (Smith et al., 1981), particularly since the semelparous annual life cycle of most squid leaves species highly vulnerable to population crashes, or even extinction, through overfishing. Unlike fish, squid have only one year class with the result that there are no younger cohorts developing to help a stock recover from overfishing (Bravo de Laguna, 1989). Thus, overfishing for just 1 year can exterminate a squid stock. Indeed, this may have occurred through lack of control in the early years of the Falklands fishery for *Illex argentinus*; what may have been the main genetic stock providing the fishery in 1985 was apparently absent from the fishery in 1986 (Thorpe et al., 1986). It has been suggested that for successful continuation of a fishery at least 40% of the adult population must be allowed to spawn (Patterson, 1987).

However, despite the vulnerability of single year class stocks, it is undoubtedly true that adult squid are highly mobile, particularly the oceanic species, and so are unlikely to be unduly affected by localised overfishing in a small geographic area, since most stocks will be more widely dispersed. If a stock is exploited in only part of its range, localised overfishing will probably not be serious in the longer term.

In the other, less mobile, cephalopods the scarcity of genetic data means that there is little to go on when attempting to understand how their populations may be structured. As in squid, cryptic speciation is clearly a problem in octopodids (Levy et al., 1988; Allcock, 1997). The large allozyme based study of Antarctic octopuses by Allcock (1997) shows a major complex of similar and related species of '*Pareledone*', many apparently sympatric. How these evolved in an area with few obvious barriers to gene flow is far from clear. However, some insight is given by her detailed study of genetic differentiation of *Pareledone turqueti* populations occurring on the continental shelf areas or shelf slopes of northern part of the Scotia Arc (Allcock et al., 1997). Here, the various shelves are separated by areas of water too deep (>4000 m) for the adult octopuses, which are benthic. The species maintains panmixia over distances of hundreds of kilometres where the water is less deep, but deep water only tens of kilometres wide leads to genetic differentiation of populations, although it had been previously supposed that the larvae should be easily able to cross this distance. Recent detailed assessments of the oceanographic regime of the area now indicate that apparently minor currents may act as insurmountable barriers to larvae within what was previously supposed to be an uninterrupted stretch of open water. Hence, there is probably no gene flow between the shelf areas and the octopuses on each constitute discrete stocks.

Among octopodids and other less mobile coastal demersal cephalopods the distribution of stocks or species is likely to be largely a function of the mobility of the larva (called a paralarva in octopodids). Most are probably planktonic for a few days or more and thus will undergo pelagic dispersal, but some larger octopus paralarvae may be demersal and hence probably disperse very little (Hochberg et al., 1992). For the planktonic larvae dispersal, over tens of kilometres might be expected, with stocks being distributed over distances of this sort of order or somewhat greater.

Against expectations of only moderate larval dispersal, it is notable that among the continental shelf cephalopods are two of the most widespread non planktonic invertebrate species. *Octopus vulgaris* occurs in coastal waters from South America north to Canada, down the other side of the Atlantic from northern Europe, including the Mediterranean, to South Africa, from there north to the Red Sea and across Asia to Japan and down to Australia. The continental shelf ommastrephid squid *Todaropsis eblanae* occurs further off shore, but otherwise has an essentially similar distribution except that it is absent from the coast of the western Atlantic. There is no reason to suppose that the larvae of either of these two species travel very far, but in *Todaropsis eblanae* the adult is highly mobile, although, being coastal, it probably does not disperse over large distances or undertake the very long spawning migrations of some other ommastrephids. The huge linear coastal distribution of the species is, therefore, of note and it is difficult to believe that panmixia can be maintained. *Octopus vulgaris* with an even longer and narrower coastal distribution is truly remarkable, because the adult is of low mobility and in practice largely sedentary. It is generally territorial and moves largely by pulling itself around the sea bed by the tentacles. Swimming is largely reserved for escape from predators. High gene flow is improbable and it is difficult to envisage how a genetic structure as a single species can be maintained over a linear distribution of the order of 50 000 km. However in both of these species the huge distributions may be artifacts of overconservative taxonomy.

Planktonic invertebrate species: krill and jellyfish

As outlined above, planktonic invertebrates eaten or otherwise exploited by man are essentially just krill (euphausids) and a few species of jellyfish, consumed mainly by the Japanese. As fisheries resource, the two groups are at opposite ends of the scale; the jellyfish, although of high value, are essentially only of modest and local significance, whilst *Euphausia superba* is present in huge biomasses over much of the vast Antarctic Ocean, where it has apparently been sporadically exploited on a fairly large commercial scale by the Russians and others. It is by a substantial margin the world's largest crustacean fishery.

In open ocean planktonic species, the a priori assumption would be of high mobility and hence genetic structuring only over long distances, if at all. Attempts at genetic studies of population structure in krill are confronted by major sampling problems, largely resulting from the difficulties of sampling animals which can be difficult to find over huge and inhospitable areas, but which form swarms which can be so large (up to several cubic kilometres of densely packed krill) that subsampling also becomes highly problematic. Further problems for sampling are that swarms are generally not random aggregations, often being predominantly of one sex or age group.

For krill, as with all open water planktonic invertebrates in the Antarctic, the major unknown in their population genetics is the extent to which they may or may not cross between water bodies. It is improbable that animals like krill, salps or ctenophores can move under their own power for significant horizontal distances, and, consequently, they are more or less obliged to drift with the bodies of water in which they find themselves. However, krill probably have some control over their own buoyancy and are able to regulate the depths at which they float. By doing this, they have the potential to select currents at different depths to transport them in some 'preferred' direction. Whether they actually do this is not known.

The main water mass of the Antarctic Ocean is thought to rotate in an easterly direction about the Antarctic continent, but at a speed which would take several years for a complete circumnavigation. Thus, over an extended time scale krill and other planktonic species may be unable to avoid a high degree of genetic mixing and gene flow, with the result that all Antarctic krill may constitute a single panmictic stock. Alternatively, oceanographic data indicate the separation of water masses in different areas of the Antarctic Ocean. For example, the Weddell-Scotia confluence, which closely follows the Scotia Arc, seems to separate the body of water to the east from that immediately to the west, whilst other water bodies show rotational gyres of long duration (e.g. 1 year plus in the Weddell Sea), which presumably retain, and thus genetically isolate, planktonic species occurring within them.

Against this background, currently available genetic data are difficult to interpret in terms of population structuring in *Euphausia superba*. There are several genetic studies of krill (e.g. Ayala et al., 1975; Ayala & Valentine, 1979; Fevolden, 1984) and some have led to the conclusion that there is little or no genetic structuring and possibly just a single interbreeding population in the Antarctic (e.g. Grant, 1983; MacDonald & Schneppenheim, 1983; Schneppenheim & MacDonald, 1983; Kuehl & Schneppenheim, 1986; MacDonald et al., 1986; Fevolden & Schneppenheim, 1989). Other studies have given more equivocal results, with variation between samples, but no clear pattern of stock structuring (e.g. Fevolden, 1985) or have indicated geographical differences (e.g. Patarnello et al., 1996).

Jellyfish are the other group of planktonic organisms eaten by man. Commercially exploited jellyfish are likely to be of at least moderate size and, since they live for considerable periods of time, widespread dispersal is likely. This hypothesis is supported by, for example, the occasional stranding of tropical jellyfish or siphonophores on temperate coasts. There appear to be no published studies of population genetic structuring for any jellyfish species, but intraspecific genetic differentiation in open ocean pelagic species is likely only over distances of hundreds, or more probably thousands, of kilometres. The dispersal of jellyfish can be further enhanced by the possible transport in the ballast water of ships (Greenberg et al., 1996). However, if stocks are to be assessed it should be borne in mind that genetic and other data indicate that cryptic speciation occurs (Brewer, 1991).

Low mobility exploited marine invertebrates: the importance of larval dispersal

The great bulk of species of marine invertebrates fished by man have adults of limited or zero mobility. However, most of these have planktonic larval stages, many of which are presumed to travel large distances away from the adult, thus allegedly effecting widespread gene dispersal. The identification of fishery stocks in such species is just one facet of what has become a major research area in marine biology – the extent of dispersal and the evolutionary role of marine invertebrate larvae.

For a number of years, a major use for genetic techniques in marine biology has been to attempt to understand how populations of marine species are structured. Clearly, if two or more populations of a species become geographically separated, they are likely to start to diverge genetically, mostly by genetic drift and sometimes by natural selection. There are numerous proposed models of gene flow between allopatric populations, but, as outlined above, a major conclusion from nearly all of these is that even very small levels of gene flow between populations are likely to greatly reduce or effectively eliminate divergence of gene frequencies under most natural conditions. Among marine invertebrates, a large proportion of the studies of genetic structuring have been carried out on temperate species, many of these intertidal or benthic shallow subtidal. In most of these the adults are either sessile (e.g. bivalves, barnacles, sponges, bryozoans) or of very limited mobility (e.g. snails, limpets, crabs, worms), with the result that, for most of them, the pelagic larva is the main dispersive phase of the life cycle. However, the potential for dispersal of the larvae is not always realised, and there have been many reported cases of high population structuring in species supposed to have long-lived planktotonic larvae (e.g. Knowlton & Keller, 1986; Todd et al., 1998).

Sponges

The Porifera (sponges) have received a moderate amount of attention from population geneticists, although not generally with any aim at stock identification (reviewed by Solé-Cava & Thorpe, 1994). From the available studies it appears that in general, as in sea anemones, levels of genetic polymorphism in sponges are very high (see Solé-Cava & Thorpe, 1991) and genetic structuring is frequently on a small scale, with species often showing clear differentiation over distances of only tens of kilometres (e.g. Benzie et al., 1994; Davis, 1996; Klautau et al., 1999).

The main implication of sponge population genetic data to date is that larval dispersal is limited and that gene pools and hence 'stocks' are likely to be very much localised. This is not surprising, since sponge larvae that have been studied are mostly short lived. It is, however, worrying for the future of sponge fisheries because most commercial species are thought to take a considerable time (10 years +) to reach a marketable size and, if recruitment is localised and growth is slow, fishing will need to be very carefully controlled to avoid major damage to stocks.

Sea anemones

Sea anemones are nowhere commercially very important, but they are eaten in several parts of the world. From the numerous studies of sea anemone population genetics (reviewed by, e.g. Shick, 1991; Perrin et al., 1999), it is difficult to draw many generalised conclusions about the level or geographical scale of structuring. Most species are thought to have planktonic larvae, albeit of short duration, and are therefore presumed to undergo gene dispersal. Species with little differentiation over long distances (e.g. *Oulactis mucosa* in Australia; Hunt & Ayre, 1989) are apparently rare, but the causes of the genetic variation found in most species are not always clear. Data for several species appear to show increased genetic divergence with distance and hence provide support for gene dispersal by pelagic larvae, but the rate of divergence with distance can vary markedly between closely related species or even between populations of the same species in different areas. For example, within Britain (Solé-Cava & Thorpe, 1992) or over a European scale (Monteiro et al., 1997) the edible *Actinia equina* (*sensu* Stephenson, 1935) appears to show generally much greater genetic change with distance than the Australian *A. tenebrosa* (Ayre, 1984; Ayre et al., 1991) or the South American *A. bermudensis* (Russo et al., 1994).

To define a stock in a commercially exploited sea anemone is probably straightforward in species, like *Oulactis,* which show apparent genetic homogeneity over considerable distances, but in many species, like *Actinia equina*, this must be difficult; indeed in any organism which is clonal, or which makes extensive use of self fertilisation, the whole concept of a stock is arguably inapplicable. Other complications are that, at least at allozyme loci, levels of genetic variation are often, but not invariably, very high, (Solé-Cava & Thorpe, 1991) and cryptic species are common (see, e.g. Perrin et al., 1999).

Polychaetes

Genetic homogeneity over large distances does not seem to be characteristic of polychaete populations, although several studies indicate a lack of differentiation over 10 km or more (e.g. Abbiati & Maltagliati, 1996). However, genetic differentiation over moderate to short distances is also noted (e.g. Bristow & Vadas, 1991; Abbiati & Maltagliati, 1992). Atypical patterns of differentiation are found in *Marenzellaria,* which shows little differentiation within populations in the Baltic or in the North Sea, but clear differences between these areas (Bastrop et al., 1995; Roehner et al., 1996). These results are explained because the species is not native to Europe; the Baltic and North Sea populations have apparently been introduced from different areas of the Atlantic coast of North America (Bastrop et al., 1998).

Ascidians

There are few genetic studies of population structuring in ascidians. Available data suggest that in some species the very small tadpole larvae may be amongst the least dispersive of all free swimming marine larvae. Results for the colonial ascidian *Botryllus schlosseri* indicate genetic differentiation over short distances in European populations (Sabbadin, 1978 and that larval dispersal is typically of the order of only a metre in a population from the north eastern United States (Grosberg, 1987, 1991). In British populations of the solitary species *Dendrodoa grossularia*, distances may be even smaller (Evans, 1994), although dispersal can be as high as 1.5 km in *Ascidia mentula* (Havenhand, 1991). The implication from these very few studies is that larval dispersal and gene flow in ascidians are minimal and hence that there will be little chance of overfished stocks recruiting from elsewhere. However, wide geographical distributions of some species (e.g. *Botryllus schlosseri, Ciona intestinalis*) are surprising if dispersal is so low and may indicate an ability to raft or to undergo anthropogenic dispersal (e.g. Boyd et al., 1990). Alternatively the distributions may merely reflect inadequate resolution of conventional taxonomy (Knowlton, 1993; Thorpe & Solé-Cava, 1994).

Echinoderms

Of the several major groups of echinoderms, only the echinoids (sea urchins) and holothurians (sea cucumbers) are routinely eaten by man. Population genetic studies on echinoderms are largely confined to sea urchins and many have been carried out by the Japanese (e.g. Marcus, 1977; Matsuoka & Nakamura, 1990, 1991; Watts et al., 1990; Mladenov et al., 1997).

Very long distance dispersal probably occurs in several species; most notably in *Echinothrix diadema*, which maintains gene flow across 5400 km of very deep water, between the Eastern and Central Pacific (Lessios et al., 1998) and in the edible *Evechinus chloroticus* which is considered to maintain high levels of gene flow over distances from 250 to 2200 km (Mladenov et al., 1997). However, other species show divergence over very short distances (Minokawa et al., 1992), or apparently similar species can differ greatly in levels of dispersal. Within the widely eaten genus *Strongylocentrotus, S. lividus* is considered to show panmixia from Alaska to California (Debenham, 1997), whilst *S. purpuratus* is markedly subdivided genetically between local populations in California (Edmands et al., 1996). Both of these species have apparently similar long lived larvae.

Crustaceans

As might be expected there is a considerable literature on the population genetics of marine crustaceans. The uses of genetics in assessing crustacean stocks have been reviewed by Lavery & Keenan (1995). Numerous studies indicate that in many species genetic differentiation occurs mainly over large distances (e.g. talitrids, De Matthaeis et al., 1995; Conceição et al., 1998; barnacles, Furman et al., 1989; Ford & Mitton, 1993;

Pannacciulli et al., 1997; crabs, McMillen-Jackson et al., 1994; Lavery et al., 1995; shrimps or prawns, Mattoccia et al., 1987; Benzie et al., 1992; Staton & Felder, 1995 and lobsters, Ovenden & Brasher, 1994; Thompson et al., 1996; Harding et al., 1997), although local differentiation occurs in some species (e.g. Siberman et al., 1994; Jorstad, 1999) and predictably in those without larval dispersal (e.g. Piertney & Carvalho, 1995). Most species apparently show low diversity on a local scale (Benzie et al., 1992; De Matthaeis et al., 1995; Lavery et al., 1995), but oceanographic or other barriers to larval dispersal can disrupt gene flow (Staton & Felder, 1995). Several species are considered to show particularly long distance gene flow, for example, between various Pacific islands (Lavery et al., 1996), between Australia and New Zealand (Booth et al., 1990) and between Greece and Spain (Mattoccia et al., 1987).

Bivalves and gastropods

Marine bivalves have been widely studied over several decades by population geneticists, with the result that a large body of data is now available, much of it on commercial species (e.g. Koehn et al., 1973, 1984; Ahmad et al., 1977; Buroker, 1984; Karl & Avise, 1992; Lewis & Thorpe, 1994a,b; Patwary et al., 1994; Heipel et al., 1998). In bivalves, the adults are almost invariably sessile (a few burrowing species can move over short distances) and, therefore, only the larvae are able to disperse. In broad terms, genetic differentiation occurs only over distances of tens to hundreds of kilometres (e.g. Skibinski et al., 1978; Saavedra et al., 1987; Jarne et al., 1988; Backeljau et al., 1994; King et al., 1994) and in some cases only over very large distances. For example, Moraga et al. (1994) concluded that there was gene flow between apparently isolated *Bathymodiolus* populations at deep hydrothermal vents in different ocean basins in the west Pacific and differentiation only between west and east Pacific populations. Similarly, Meehan (1985) only found substantial differentiation between *Macoma* populations in transatlantic comparisons.

There is also the possibility that apparently homogeneous population structure may show microgeographic or year class related genetic differences when very large sample sizes are used to enhance statistical resolution (see e.g. David, 1997). Another example of this effect of sample size is from studies of populations of the commercially important scallop *Aeguipecten (Chlamys) opercularis* in the northern Irish sea. This species was thought to show no genetic heterogeneity over this geographic area (Beaumont & Beveridge, 1984; Macleod et al., 1985), but the use of much larger samples revealed stable and significant, if small, allele frequency differences between stocks on a much smaller geographic scale (Lewis & Thorpe, 1994a,b).

In gastropods, few studies of population genetics are specifically directed at commercial species, although there are some data for abalone populations (e.g. Prince et al., 1987; Shepherd & Brown, 1993; Mgaya et al., 1995), whelks (Berger, 1983) and on the (locally) commercially important limpet fisheries of the Macaronesian Islands (Corte-Real et al., 1996a,b; another study of limpets by Hurst & Skibinski, 1995). Within the gastropods, the scale of genetic differentiation between populations or stocks appears generally to be related to larval type and length of larval life. Many species are thought to be very widely dispersed (e.g. Scheltema, 1986, 1989), but there are also several in which the expected larval dispersal is difficult to reconcile with the scale of genetic subdivision (e.g. Todd et al., 1988, 1998). Conversely, the littoral whelk *Bullia digitalis* shows little differentiation over large distances around the coasts of southern Africa, despite having a (lecithotrophic) larva which is apparently non planktonic (Grant & Da Silva-Tatley, 1997).

Possibilities for future exploitation of marine invertebrate stocks

Many of the world's conventional shellfish fisheries, like those for finfish, are overexploited and in decline. It is unlikely, therefore, that there is great scope for future expansion of fisheries for crabs, lobsters, prawns, bivalves or gastropods, but as fish stocks, which provide the great bulk of wild caught marine animals, appear certain to continue to decline other resources are needed.

Also, as mentioned above, there is already a certain pressure on various lower marine invertebrates, like sponges, bryozoans and ascidians, to provide biologically active natural products for the pharmaceutical industry (e.g. Higa, 1991; Higa et al., 1992; Faulkner, 1995). One such product is bryostatin, which appears to act against various cancer cells (e.g. Lilly et al., 1991; Schuchter, 1991; Berkow et al., 1993). Bryostatin is obtained from the bryozoan *Bugula neritina*, but huge quantities of the bryozoan are needed; for example Schaufelberger et al. (1991) based their work on bryostatin on a sample of 10 000 gallons (about 45 000 kg) of *Bugula*. With collection on a large scale any small invertebrate in pharmaceutical demand may become overexploited unless aquaculture can take over (see also Olson, 1996).

Future demand for invertebrate food species will presumably depend partly on the success of the aquaculture of fish and shellfish and whether untapped sources (e.g. deep sea species) can be exploited. However, whilst most of the numerous other marine invertebrate groups are eaten little or not at all in most areas, thus apparently leaving scope for the expansion of the fisheries, most of the species within these groups are either inedible or present in only relatively small biomasses. Abundant marine invertebrates for future exploitation could be Antarctic krill (*Euphausia superba*) and oceanic squid, particularly some of the deeper living species.

Krill stocks are thought to be unnaturally high, because of the near elimination of their major predators, the larger baleen whales, and offer a potentially very great fishery resource, although of low value and palatability. However, krill are also considered the 'cornerstone' species for the whole vast Antarctic marine ecosystem and, hence, major exploitation could be a serious problem and overfishing a disaster.

The squid probably offer greater potential. They currently make a useful contribution (about 2–3% by weight) to fisheries on a global scale, but with a price much higher than most fish. Squids collectively rank (by weight of catch) as the sixth most commercially important species in world fishery statistics. Since about 1970, world catches have increased annually by an average of about 6%, but there is considered to be potential to greatly increase current landings. Estimates of stocks sizes are necessarily very approximate because there has been little or no relevant scientific study of the stocks of most squid species (see Clarke, 1977; Roper et al., 1984). Indeed, it is likely that there are still many species as yet unknown (M. R. Clarke, pers. comm.).

Evidence pointing to large untapped squid populations comes from several sources (e.g. Clarke, 1977, 1987). For a few species, there are adequate scientific data for stocks to be estimated. For example, the large (up to 150 kg) cranchid squid *Mesonychoteuthis hamiltoni* is abundant in Antarctic waters and the flesh is considered to be of excellent quality and flavour (see Voss, 1983; Roper et al., 1984). This species is not caught commercially at all, but the stock is estimated at about 90 Mmt (i.e. larger than the total annual catch of fish by man). A second example is the very large squid *Architeuthis* (Architeuthidae) which is thought to be distributed throughout all the oceans apart from close to the poles. Very few have been caught by man, but its frequent consumption by whales (identified mainly from the beaks which remain undigested in whale stomachs) and occasional stranding in shallow water suggest it is not rare. As individuals weigh up to about 1000 kg, the total biomass is likely to be huge. Current direct estimates by fishery biologists of total world squid stocks vary considerably up to about 500 Mmt, but indirect estimates from other sources indicate that even this figure may be conservative (see e.g. Voss, 1973, 1983; Clarke, 1987; Thorpe & Nash, 1993).

Studies of the squid consumed by various predators suggest very large populations. In the Antarctic penguins and elephant seals, for example, are thought to take annually about 13 Mmt and 4.5 Mmt, respectively (neither of these, incidentally, eats any of the estimated 90 Mmt of *Mesonychoteuthis hamiltoni*). Sperm whales eat mainly squid found on or near the bottom in depths restricted by their diving capabilities. Sperm whale populations have been seriously depleted by over exploitation, but even so they probably eat annually a weight of squid equivalent to about twice the entire world catch of fish by man (Clarke, 1977, 1980, 1987).

Thus various sources of evidence indicate that oceanic squids are present in very large numbers and may be present in biomasses which exceed known stocks of exploited fish species. Since most squid are thought to be annual and semelparous (they grow for a year, spawn and die), they 'turn over' faster than most fish species and thus, if well managed, should be able to sustain high rates of exploitation. As long as enough adults survive to spawn, the entire stock can be exploited, because there are no juveniles which need to be preserved for future breeding (see Patterson, 1987). However, for many of the oceanic squids the depths at which they live and our lack of knowledge of their biology and behaviour present major problems for the development of commercial fisheries. Technological advances will be necessary before most species can be caught in commercial quantities.

Conclusions

It is clear that for most invertebrate fisheries, as with finfish fisheries, the key to understanding the stock structure for fisheries purposes will lie in understanding the mobility of the species. However, whereas in fish the mobility concerned is generally that of the fish that constitute the stock, in invertebrates the mobility which is critical is in most species likely to be that of the larva. This difference occurs simply because in fish the adult spawning stocks are generally likely to be the most mobile phase of the life cycle, whilst in most invertebrates the larva is far more mobile. In only a few invertebrates, like squid, is the mobile phase likely to be the adult, and even here, in some of the neritic species or the smaller benthic cuttlefish (e.g. sepiolids), it is possible that the dispersal of the pelagic larva is the main determinant of the geographical area occupied by a stock. The mobility of the adult is probably also the main factor influencing dispersal in planktonic species (e.g. krill, jellyfish), which may have little control over the magnitude or direction of their drifting.

However, in the great bulk of exploited marine invertebrate phyla the pelagic larva provides the main apparent means of gene dispersal. In various species which disperse via larvae (e.g. some molluscs, crustaceans and sea urchins; for references see above) genetic data indicate long distance dispersal (i.e. thousands of kilometres), but in many species from the same and other groups dispersal is apparently much less, and in some invertebrates with free swimming larvae (e.g. ascidians, sponges), these apparently travel short distances only. There are also, of course, numerous species without a dispersive larva. In these, the larva may be benthic, or may simply not occur, with, for example, the eggs hatching to produce a miniature adult.

When estimating the geographical area occupied by marine invertebrate 'stocks' it must be borne in mind that ideas that larvae evolved with the 'aim' of dispersal for the 'good of the species' (e.g. Hansen, 1980; Jablonski & Lutz, 1983) lead us to expect more dispersal than may occur in some (many?) species. There is still debate as to why different species have larvae and numerous putative functions have been hypothesised (reviews by many authors, e.g. Strathmann, 1980, 1985, 1990; Todd, 1985; Todd et al., 1988, 1998; Grosberg, 1992). It is not our aim to discuss in detail the 'roles' of larvae, but some of the suggestions apart from maximising dispersal are:

1. Maximising fecundity;
2. Finding more favourable habitats;
3. Reducing overcrowding and localised inbreeding;
4. Finding 'patchy' resources (e.g. space or food);
5. Tracking temporally unstable food supplies;
6. Exploiting temporally unstable substrata;
7. Exploiting the plentiful food supply in the plankton; and
8. Avoiding patchily distributed predators.

Most of these possible functions are not mutually exclusive and indeed many larvae probably achieve more than one of them, and to attempt to firmly ascribe the existence of larvae in any given species to any particular evolutionary function is possibly naive.

To further complicate the understanding of dispersal, it has become increasingly clear in recent years that in various sessile marine species long distance dispersal is achieved by means other than the regular planktonic transport of larvae. The main methods proposed are rafting on naturally occurring (e.g. wood, pumice, algae, turtles) or anthropogenic (e.g. plastics, buoys) floating substrates (e.g. Winston, 1982; Frazier et al., 1985, 1991, 1992; Highsmith, 1985; Jokiel, 1989; Smith et al., 1989; Harms, 1990; Cornelius, 1992; Helmuth et al., 1994; Worcester, 1994; Ingolfsson, 1995), or transport on the hulls or in the ballast water of ships (e.g. Carlton, 1985, 1987, 1992; Carlton & Geller, 1993; Carlton & Hodder, 1995; Pierce et al., 1997; Ruiz et al., 1997; Watts et al., 1998). In their detailed analysis of the correlates of species ranges in cheilostome Bryozoa, Watts et al. (1998) conclude that an ability to be transported by ships conveys a wider range than any likely to be achieved through the dispersive capability of the larva.

Thus, within marine invertebrates the range of dispersal strategies is vast and these can differ between closely related (e.g. congeneric) species or even between geographically separated populations of the same species. In fish estimates of genetic structuring for stock, assessment can often be made by extrapolation from earlier similar studies, but in most invertebrates stock assessments can only be made safely by treating each species and fishery as a separate entity.

Acknowledgements

The authors are thankful to Claudia Russo for many suggestions on the manuscript, and to CNPq, FAPERJ and the Linnean Society for financial support.

References

Abbiati, M. & F. Maltagliati, 1992. Genetic population structure of *Neanthes succinea* (Polychaeta: Nereididae). J. mar. biol. Ass. U.K. 72: 511–517.

Abbiati, M. & F. Maltagliati, 1996. Allozyme evidence of genetic differentiation between populations of *Hediste diversicolor* (Polychaeta: Nereididae) from the western Mediterranean. J. mar. biol. Ass. U.K. 76: 637–647.

Ahinad, M., D. 0. F. Skibinski & J. A. Beardinore, 1977. An estimate of the amount of genetic variability in the common mussel *Mytilus edulis*. Biochem. Genet. 15: 833–846.

Allcock, A. L., 1997. The genetics and taxonomy of Southern Ocean Octopodidae, with special reference to the genus *Pareledone*. Unpublished thesis, University of Liverpool, Port Erin, Isle of Man.

Allcock, A. L., A. S. Brierly, J. P. Thorpe & P. G. Rodhouse, 1997. Restricted gene flow and evolutionary divergence between geographically separated populations of the Antarctic octopus *Pareledone turgueti*. Mar. Biol. 129: 97–102.

Allendorf, F., N. Ryman & F. Utter, 1987. Genetics and fishery management: past present and future. In Ryman, N. & F. Utter (eds), Population Genetics and Fishery Management. University of Washington Press, Seattle: 1–20.

Ally, J. R. R. & S. C. Keck, 1978. A biochemical genetic population structure study of market squid, *Loligo opalescens*, along the Californian coast. Calif. Fish Game 169: 113–121.

Altukov, Y. P., 1981. The stock concept from the viewpoint of population genetics. Can. J. Fish. aquat. Sci. 38: 1523–1528.

Alvarez-Bremer, J. R., J. Mejuto & B. Ely, 1995. Global population structure of the swordfish (*Xiphias gladius*) as revealed by the analysis of the mitochondrial control region. Collect. Vol. Sci. Pap. Iccat 44: 206–216.

Augustyn, C. J. & W. S. Grant, 1988. Biochemical and morphological systematics of *Loligo vulgris vulgaris* Lamark and *Loligo vulgaris reynaudi* D'Orbigny nov. comb. (Cephalopoda: Myopsida). Malacologia 29: 215–233.

Ayala, F. J. & J. W. Valentine, 1979. Genetic variability in the pelagic environment: a paradox? Ecology 60: 24–29.

Ayala, F. J., J. W. Valentine, D. Hedgecock & L. G. Barr, 1974. Deep-sea asteroids: high genetic variability in a stable environment. Evolution 29: 203–212.

Ayala, F. J., J. W. Valentine & G. S. Zuinwalt, 1975. An electrophoretic study of the Antarctic zooplankton *Euphausia superba*. Limnol. Oceanog 20: 635–640.

Ayala, F. J., G. S. Zuinwalt, D. Hedgecock & J. W. Valentine, 1973. Genetic variation in *Tridacna maxima*, an ecological analogue of some unsuccessful evolutionary lineages. Evolution 27: 177–191.

Ayre, D. J., 1984. The effects of sexual and asexual reproduction on geographic variation in the sea anemone *Actinia tenebrosa*. Gecologia 62: 222–229.

Ayre, D. J., J. Read & J. Wishart, 1991. Genetic subdivision within the eastern Australian population of the sea anemone *Actinia tenebrosa*. Mar. Biol. 109: 379–390.

Backeljau, T., P. Boucher & F. Gofas, 1994. Genetic variation, systematics and distribution of the venerid clam *Chamelea gallina*. J. mar. biol. Ass. U.K. 74: 211–223.

Bastrop, R., K. Juerss & C. Sturmbauer, 1998. Cryptic species in a marine polychaete and their independent introduction from North America to Europe. Mol. Biol. Evol. 15: 97–103.

Bastrop, R., M. Roehner & K. Juerss, 1995. Are there two species of the polychaete genus *Marenzellaria* in Europe? Mar. Biol. 121: 509–516.

Beaumont, A. R. & C. M. Beveridge, 1984. Electrophoretic survey of genetic variation in *Pecten maximus, Chlamys opercularis, C. varia* and *C. distorta* from the Irish Sea. Mar. Biol. 81: 299–306.

Beaumont, A. R. & S. M. J. Pether, 1996. Allozyme variation and gene flow between cockle *Cerastoderma edule* populations in southern United Kingdom. Fish. Res. 28: 263–275.

Bembo, D. G., G. R. Carvalho, N. Cingolani, E. Amen, G. Giannetti & T. J. Pitcher, 1996. Allozymic and morphometric evidence for two stocks of the European anchovy *Engraulis encrasicolus* in adriatic waters. Mar. Biol. 126: 529–538

Bentzen, P., C. T. Taggart, D. E. Ruzzante & D. Cook, 1996. Microsatellite polymorphism and the population structure of Atlantic cod (*Gadus morhua*) in the northwest Atlantic. Can. J. Fish. aquat. Sci. 53: 2706–2721.

Benzie, J. A. H., S. Frusher & E. Bailment, 1992. Geographical variation in allozyme frequency in populations of *Penaeus monodon* (Crustacea: Decapoda). Aust. J. mar. Freshwat. Res. 43: 715–725.

Benzie, J. A. H., C. Sandusky & C. R. Wilkinson, 1994. Genetic structure of dictyoceratid sponge populations on the western Coral Sea reefs. Mar. Biol. 119: 335–345.

Berger, E. M., 1983. Population genetics of marine gastropods and bivalves. In Russell-Hunter, W. D. (ed.), The Mollusca. Academic Press, London: 563–596.

Berkow, R. L., L. Schlabach, R. Dodson et al., 1993. *In vivo* administration of the anti cancer agent bryostatin 1 activates platelets and neutrophils and modulates protein kinase activity. Cancer Res. 53: 2810–2815.

Bernatchez, L., 1994. Molecular biology techniques in fishery management: applications and perspectives. Bull. fr. Peche Piscic. 332: 1–9.

Bielawski, J. P. & D. E. Pumo, 1997. Randomly amplified polymorphic DNA (RAPD) analysis of Atlantic coast striped bass. Heredity 78: 32–40.

Booth, J. D., R. J. Street & P. J. Smith, 1990. Systematic status of the rock lobsters *Jasus edwardsii* from New Zealand and *J. novaehollandiae* from Australia. New Zealand J. mar. Freshwat. Res. 24: 239–249.

Boyd, H. C., I. L. Weissman & Y. Saito, 1990. Morphologic and genetic verification that Monterey *Botryllus* and Woods Hole *Botryllus* are the same species. Biol. Bull. 178: 239–250.

Bravo de Laguna, J., 1989. Managing an international multispecies fishery. The Saharan trawl fishery for cephalopods. In Caddy, J. F. (ed.), Marine Invertebrate Fisheries: Their Assessment and Management. Wiley, New York: 591–612.

Brewer, R. H., 1991. Morphological differences between and reproductive isolation of two populations of the jellyfish *Cyanea* in Long Island Sound, U.S.A. Hydrobiologia 216: 471–477.

Brierley, A. S., A. L. Allcock, J. P. Thorpe & R. D. Clarke, 1996. Biochemical genetic evidence supporting the taxonomic separation of *Loligo edulis* and *Loligo chinensis* (Cephalopoda: Teuthoidea) from the genus *Loligo*. Mar. Biol. 127: 97–104.

Brierley, A. S., M. R. Clarke & J. P. Thorpe, 1997. *Ctenopteryx sicula*, a bathypelagic loliginid squid? Am. malacol. Bull 12: 137–144.

Brierley, A. S., P. G. Rodhouse, J. P. Thorpe & M. R. Clarke, 1993a. Genetic evidence of population heterogeneity and cryptic speciation in the ommastrephid squid *Martialia hyadesi* from the Patagonian shelf and Antarctic Polar frontal zone. Mar. Biol. 116: 593–602.

Brierley, A. S. & J. P. Thorpe, 1994. Biochemical genetic evidence supporting the taxonomic separation of *Loligo gahi* from the genus *Loligo*. Antarctic Sci. 6: 143–148.

Brierley, A. S., J. P. Thorpe, M. R. Clarke & H. Martins, 1993b. A preliminary biochemical genetic investigation of the population structure of *Loligo forbesi* Steenstrup, 1856 from the British Isles and the Azores. In Okutani, T., K. O'Dor & T. Kubodera (eds), Recent Advances in Cephalopod Fisheries Biology. Tokai University Press, Shimizu City, Japan: 59–67.

Brierley, A. S., J. P. Thorpe, G. J. Pierce, M. R. Clarke & P. R. Boyle, 1995. Gene variation in the neritic squid *Loligo forbesi* (Myopsida, Loliginidae) in the Northeast Atlantic ocean. Mar. Biol. 122: 79–86.

Bristow, G. A. & R. L. Vadas, 1991. Genetic variability in bloodworm (*Glyera dibranchiata*) populations in the gulf of Maine. Mar. Biol. 109: 311–319.

Buroker, N. E., 1984. Gene flow in mainland and insular populations of *Crassostrea* (Ostreidae). Biol. Bull. 166: 550–557.

Caddy, J. F., 1989. Marine invertebrate fisheries: their assessment and management. Wiley, London.

Carlton, J. T., 1985. Transoceanic and interoceanic dispersal of coastal marine organisms: the biology of ballast water. Oceanog. mar. Biol. ann. Rev. 23: 313–371.

Canton, J. T., 1987. Patterns of transoceanic marine biological invasions in the Pacific Ocean. Bull. mar. Sci. 41: 425–465.

Carlton, J. T., 1992. Introduced marine and estuarine mollusks of North America: an end of the 20th-century perspective. J. Shellfish Res. 11: 489–505.

Carlton, J. T. & J. B. Geller, 1993. Ecological roulette: the global transport of indigenous organisms. Science 261: 78–82.

Carlton, J. T. & J. Hodder, 1995. Biogeography and dispersal of coastal marine organisms: experimental studies on a replica of a 16th-century sailing vessel. Mar. Biol. 121: 721–730.

Carvalho, G. R. & L. Hauser, 1995. Molecular genetics and the stock concept in fisheries. In Carvalho, G. R. & T. J. Pitcher (eds), Molecular Genetics in Fisheries. Chapman and Hall, London: 55–79.

Carvalho, G. R. & L. Hauser, 1998. Advances in the molecular analysis of fish population structure. Ital. J. Zool. 65: 21–33.

Carvalho, G. R. & T. J. Pitcher, 1989. Biochemical genetic studies on the Patagonian squid *Loligo gahi* d'Orbigny. II. Population structure in Falkland waters using isozymes, morphometrics and life history data. J. exp. mar. Biol. Ecol. 126: 243–258.

Carvalho, G. R. & T. J. Pitcher. 1995. Molecular Genetics in Fisheries. Chapman and Hall, London.

Carvalho, G. R., A. Thompson & A. L. Stoner, 1992. Genetic diversity and population differentiation of the shortfin squid *Illex argentinus* in the south-west Atlantic. J. exp. mar. Biol. Ecol. 158: 105–121.

Chan, T. Y. & K. H. Chu, 1996. On the different forms of *Panulirus longipes femoristriga* (Von Martens, 1872) (Crustacea: Decapoda: Palinuridae), with a description of a new species. J. nat. Hist. 30: 367–387.

Chow, S., H. Okamoto, Y. Uozumi, Y. Takeuchi & H. Takeyama, 1997. Genetic stock structure of the swordfish (*Xiphias gladius*) inferred by PCR-RFLP analysis of the mitochondrial DNA control region. Mar. Biol. 127: 359–367.

Chow, S. & H. Ushiama, 1995. Global population structure of albacore (*Thunnus alalunga*) inferred by RFLP analysis of the mitochondrial ATPase gene. Mar. Biol. 123: 39–45.

Christofferson, J. P., A. Foss, W. E. Lambert & B. Welge, 1978. An electrophoretic study of select proteins from the market squid *Loligo opalescens* along the California coast. Calif. Fish. Game 169: 123–133.

Clarke, M. R., 1977. Beaks, nets and numbers. Symp. zool. Soc. Lond. 38: 89–126.

Clarke, M. R., 1980. Cephalopods in the diet of sperm whales in the southern hemisphere and their bearing on sperm whale biology. Discovery Rep. 37: 1–324.

Clarke, M. R., 1987. Biomass of cephalopods–estimation from predation. In Boyle, P. R. (ed.), Cephalopod Life Cycles. Academic Press, London: 22 1–237.

Conceição, M. B., J. D. D. Bishop & J. P. Thorpe, 1998. Genetic relationships between ecologically divergent species of talitrid amphipod (Crustacea). Mar. Ecol. Prog. Ser. 165: 225–234.

Cornelius, P. F. S., 1992. The Azores hydroid fauna and its origin, with discussion of rafting and medusa suppression. Arquipelago 10: 75–99.

Corte-Real, H. B. S. M., S. J. Hawkins & J. P. Thorpe, 1996a. An interpretation of the taxonomic relationship between the limpets *Patella rustica* and *P. piperata*. J. mar. Biol. Ass. U.K. 76: 717–732.

Corte-Real, H. B. S. M., S. J. Hawkins & J. P. Thorpe, 1996b. Population differentiation and genetic confirmation of the taxonomic status of the exploited limpet *Patella candei* in the Macaronesian islands (Azores, Madeira, Canaries). Mar. Biol. 125: 141–152.

Currens, K. P. & C. A. Busack, 1995. A framework for assessing genetic vulnerability. Fisheries 20: 24–31.

David, P., P. Mireille-Ange, P. Anne-Françoise & J. Philippe, 1997. Fine-grained spatial and temporal population genetic structure in the marine bivalve *Spisula ovalis*. Evolution 51: 1318–1322.

Davis, A. R., D. J. Ayre, M. R. Billingham, C. A. Styan & G. A. White, 1996. The encrusting sponge *Halisarca laxus*: population genetics and association with the ascidian *Pyura spinifera*. Mar. Biol. 126: 27–33.

De Matthaeis, E., M. Cobolli, M. Mattocia & F. Scapini, 1995. Geographic variation in *Talitrus saltator* (Crustacea: Amphipoda). Boll. Zool. 62: 77–84.

Debenham, P., 1997. Molecular approaches to assessing red sea urchin (*Strongylocentrotus franciscianus*) populations: Implications of sequence variation for evolution and population genetics of the species. Unpublished thesis, University of California, Santa Barbara.

Doherty, P. J., S. Planes & P. Mather, 1995. Gene flow and larval duration in seven species of fish from the great barrier reef. Ecology 76: 2373–2391.

Edmands, S., P. E. Moberg & R. S. Burton, 1996. Allozyme and mitochondrial DNA evidence of population subdivision in the purple sea urchin *Stronglylocentrotus purpuratus*. Mar. Biol. 126: 443–450.

Epifanio, J. M., J. B. Koppelman, M. A. Nedbal & D. P. Philipp, 1996. Geographic variation of paddlefish allozymes and mitochondrial DNA. Trans. am. Fish. Soc. 125: 546–561.

Evans, R., 1994. Reproduction of the unitary larvaceous ascidian *Dendrodoa grossularia*. Unpublished thesis. University of Liverpool, Port Erin, Isle of Man.

Faulkner, D. J., 1993. Academic chemistry and the discovery of bioactive marine natural products. In Attaway, D. H. & O. R. Zaborsky (eds), Marine Biotechnology – Pharmaceutical and Bioactive Natural Products. Plenum Press, New York: 459–474.

Faulkner, D. J., 1995. Marine natural products. Natur. Product Rep. 12: 223–269.

Ferguson, A., 1994. Molecular genetics and fisheries: Current and future perspectives. Rev. Fish Biol. Fish. 4: 389–392.

Ferris, S. D. & W. J. Berg, 1987. The utility of mitochondrial DNA in fish genetics and fishery management. In Ryman, N. & F. Utter (eds), Population Genetics and Fishery Management. University of Washington Press, Seattle: 277–300.

Fevolden, S. E., 1984. Biotic and physical environmental impact on genetic variation of krill. J. Crust. Biol. 4: 206–223.

Fevolden, S. E., 1985. Genetic variation of *Euphausia superba* Dana in the Antarctic Peninsula waters. Sarsia 71: 169–175.

Fevolden, S. E. & R. Schneppenheim, 1989. Genetic homogeneity of krill (*Euphausia superba* Dana) in the Southern Ocean. Polar Biol. 9: 533–539.

Ford, M. J. & J. B. Mitton, 1993. Population structure of the pink barnacle, *Tetraclita squamosa rubescens*, along the California coast. Mol. mar. Biol. Biotechnol. 2: 147–153.

Frazier,J. G. et al., 1985. Epizoan communities on marine turtles. 1. Bivalve and gastropod mollusks. Mar. Ecol. 2: 127–140.

Frazier, J. G., I. Goodbody & C. Ruckdeschel, 1991. Epizoan communities on marine turtles. 2. Tunicates. Bull. mar. Sci. 48: 763–765.

Frazier, J. G., J. E. Winston & C. Ruckdeschel, 1992. Epizoan communities on marine turtles. 3. Bryozoa. Bull. mar. Sci. 52: 1–8.

Furman, E. R., A. B. Yule & D. J. Crisp, 1989. Gene flow between populations of *Balanus improvisus* Darwin (Cirripedia) in British estuaries. Sci. mar. 53: 465–472.

Gallardo, M. H. & J. I. Carrasco, 1996. Genetic cohesiveness among populations of *Concholepas concholepas* (Gastropoda, Muricidae) in southern Chile. J. exp. mar. Biol. Ecol. 197: 237–249.

Garcia de Leon, F. J., L. Chikhi & F. Bonhomme, 1997. Microsatellite polymorphism and population subdivision in natural populations of European sea bass *Dicentrarchus labrax* (Linnaeus, 1758). Mol. Ecol. 6: 51–62.

Garthwaite, R. L., C. J. Berg & J. Harrigan, 1989. Population genetics of the common squid *Loligo pealei* Le Seur, 1821, from Cape Cod to Cape Hatteras. Biol. Bull. 177: 287–294.

Grant, W. S., 1983. Population genetics of krill and comparison with other marine organisms. Polar Res. 4: 246–266.

Grant, W. S. & F. M. Da Silva-Tatley, 1997. Lack of genetically subdivided population structure in *Bullia digitalis* a South African gastropod with lecithotrophic development. Mar. Biol. 129: 123–137.

Greenberg, N., R. L. Garthwaite & D. C. Potts, 1996. Allozyme and morphological evidence for a newly introduced species of *Aurelia* in San Francisco Bay, California. Mar. Biol. 125: 401–410.

Grosberg, R. K., 1987. Limited dispersal and proximity dependent mating success in the colonial ascidian *Botryllus schlosseri*. Evolution 41: 130–142.

Grosberg, R. K., 1991. Sperm-mediated gene flow and the genetic structure of a population of the colonial ascidian *Botryllus schlosseri*. Evolution 45: 130–142.

Grosberg, R. K., 1992. For adults only? Supply side ecology and the history of larval biology. Trends Ecol. Evol. 7: 130–133.

Hansen, T. A., 1980. Influence of larval dispersal and geographic distribution on species longevity in neogastropods. Paleobiology 6: 193–207.

Harding, G. C., E. L. Kenchington, C. J. Bird, D. S. Pezzack & D. C. Landry, 1997. Genetic relationships among subpopulations of the American lobster (*Homarus americanus*) as revealed by random amplified polymorphic DNA. Can. J. Fish. aquat. Sci. 54: 1762–1771.

Harms, J., 1990. Marine plastic litter as an artificial hard bottom fouling ground. Helg. Meer. 44: 503–506.

Havenhand, J. N., 1991. Fertilisation and the potential for dispersal of gametes and larvae in the solitary ascidian *Ascidia mentula*. Ophelia 33: 1–15.

Heipel, D. A., J. D. D. Bishop, A. R. Brand & J. P. Thorpe, 1998. Population genetic differentiation of the great scallop *Pecten maximus* in western Britain investigated by randomly amplified polymorphic DNA. Mar. Ecol. Prog. Ser. 162: 163–171.

Heist, E. J. & J. R. Gold, 1999. Microsatellite DNA variation in sandbar sharks (*Carcharhinus plumbeus*) from the Gulf of Mexico and mid-Atlantic Bight. Copeia 1999: 182–186

Helmuth, B., R. R. Veit & R. Holberton, 1994. Long-distance dispersal of a subantartic brooding bivalve (*Gaimardia trapesina*) by kelp-rafting. Mar. Biol. 120: 421–426.

Higa, T., 1991. Bioactive phenolics and related compounds. Bioorg. mar. Chem. 4: 33–90.

Higa, T. et al., 1992. Miyakolide: a bryostain like macrolide from a sponge, *Polyfibrospongia* sp. J. am. chem. Soc. 114: 7587–7588.

Highsmith, R. C., 1985. Floating and algal rafting as potential dispersal mechanisms in brooding invertebrates. Mar. Ecol. Prog. Ser. 25: 169–175.

Hochberg, E. G., M. Nixon & R. B. Toll, 1992. Octopoda. Smiths. Contr. Zool. 513: 213–280.

Hunt, A. & D. J. Ayre, 1989. Population structure in the sexually reproducing sea anemone *Oulactis muscosa*. Mar. Biol. 102: 537–544.

Hurst, C. D. & D. O. F. Skibinski, 1995. Comparison of allozyme and mitochondrial DNA spatial differentiation in the limpet *Patella vulgata*. Mar. Biol. 122: 257–263.

Ingolfsson, A., 1995. Floating clumps of seaweed around Iceland: natural microcosms and a means of dispersal for shore fauna. Mar. Biol 122: 13–21.

Isaksson, A., 1988. Salmon ranching: a world review. Aquaculture 75: 1–33.

Izuka, T., S. Segawa & T. Okutani, 1996. Biochemical study of the population heterogeneity and distribution of the oval squid *Sepioteuthis lessoniana* complex in southwestern Japan. Am. malacol. Bull. 12: 129–135.

Izuka, T., S. Segawa, T. Okutani & K. Namuchi, 1994. Evidence on the existence of three species of the oval squid *Sepioteuthis lessoniana* complex in Ishigaki Islands, Okinawa, south west Japan. Jap. J. Malacol. 53: 217–228.

Jablonski, D. & R. A. Lutz, 1983. Larval ecology of marine benthic invertebrates: paleobiological implications. Biol. Rev. 58: 21–89.

Jane, P., P. Berrebi & 0. Guelorget, 1988. Genetic and morphological variability of five populations of the clam *Ruditapes decussatus* (Mollusca: Bivalvia). Oceanol. Acta 11: 401–407.

Jokiel, P. L., 1989. Rafting of reef corals and other organisms at Kwajalein Atoll. Mar. Biol 101: 483–493.

Jorstad, K. E. F. E., 1999. Population genetic structure of lobster (*Homarus gammarus*) in Norway, and implications for enhancement and sea-ranching operation. Aquaculture 173: 447.

Kang, Y. J., Y. H. Kim, Y. K. Hong, J. Y. Park & K. Y. Park, 1996. A population genetic analysis of the common squid, *Todarodes pacificus* Steenstrup in the Korean waters. J. korean Fish. Soc. 29: 320–331.

Karl, S. A. & J. C. Avise, 1992. Balancing selection at allozyme loci in oysters: implications from nuclear RFLPs. Science 256: 100–102.

Katugin, 0. N., 1995. Genetic differentiation in *Berryteuthis magister* from the North Pacific. In Aiken, D. E., S. L. Waddy & G. Y. Conan (eds), Shellfish Life Histories and Shell-Fishery Models. ICES, Copenhagen: 459–467.

Kimura, M. & G. H. Weiss, 1964. The stepping stone model of population structure and the decrease of genetic correlation with distance. Genetics 49: 561–576.

King, T. L., R. Ward & E. G. Zimmerman, 1994. Population structure of eastern oysters (*Crassostrea virginica*) inhabiting the Laguna Madre, Texas and adjacent bay systems. Can. J. Fish. aquat. Sci. 51: 215–222.

Klautau, M., C. A. M. Russo, C. Lazoski, N. Boury-Esnault, J. P. Thorpe & A. M. Solé-Cava, 1999. Does cosmopolitanism result from overconservative systematics? A case study using the marine sponge *Chondrilla nucula*. Evolution (in press).

Knowlton, N., 1993. Sibling species in the sea. Ann. Rev. Ecol. Syst. 24: 189–216.

Knowlton, N. & B. D. Keller, 1986. Larvae which fall far short of their potential: highly localized recruitment in an alphaeid shrimp with extended larval development. Bull. mar. Sci. 39: 213–223.

Koehn, R. K., J. G. Hall, D. J. Innes & A. J. Zera, 1984. Genetic differentiation of *Mytilus edulis* in eastern North America. Mar. Biol 79: 117-126.

Koehn, R. K., F. J. Turano & J. B. Mitton, 1973. Population genetics of marine pelecypods. II. Genetic differences in microhabitats. Evolution 27: 100–105.

Kuehl, S. & R. Schneppenheim, 1986. Electrophoretic investigation of genetic variation in two krill species *Euphausia superba* and *E. crystallorophias* (Euphausiidae). Polar Biol. 6: 17–23.

Larcson, J. M., V. M. Riccardi, S. W. Calhoun & D. C. Morizot, 1989. Genetic differentiation of bicolor damselfish (*Eupomacentrus partitus*) populations in the Florida Keys. Mar. Biol. 103: 445–451.

Lavery, S. & C. Keenan, 1995. Genetic analyses of crustacean stock structure and stock size. In Courtney, A.J. & M. G. Cosgrove (eds), Proceedings of the Workshop on Spawning Stock Recruitment Relationships. Department of Primary Industries, Brisbane, Australia: 116–121.

Lavery, S., C. Moritz & D. R. Fielder, 1995. Changing patterns of population structure and gene flow at different spatial scales in *Birgus latro* (the coconut crab). Heredity 74: 531–541.

Lavery, S., C. Moritz & D. R. Fielder, 1996. Indo-Pacific population structure and evolutionary history of the coconut crab *Birgus latro*. Mol. Ecol. 5: 557–570.

Lavie, B. & E. Nevo, 1981. Genetic diversity in marine molluscs: A test of the niche-width variation hypothesis. Mar. Ecol. 2: 335–342.

Lessios, H. A., B. D. Kessing & D. R. Robertson, 1998. Massive gene flow across the world's most potent marine biogeographical barrier. Proc. r. Soc. Lond. B 265: 583–588.

Levy, J. A., M. Haimovici & M. Conceição, 1988. Genetic evidences for two species to the genus *Eledone* (Cephalopoda: Octopodidae) in south Brazil. Comp. Biochem. Physiol. 90B: 275–277.

Lewis, R. I. & J. P. Thorpe, 1994a. Are queen scallops, *Aequipecten (Chlamys) opercularis* (L.), self recruiting? Can. tech. Rep. Fish. aquat. Sci. 1994: 214–221.

Lewis, R. I. & J. P. Thorpe, 1994b. Temporal stability of gene frequencies within genetically heterogeneous populations of the queen scallop *Aeguipecten (Chlamys) opercularis*. Mar. Biol. 121: 117–126.

Li, G. & D. Hedgecock, 1996. Mitochondrial DNA variation within and among larval cohorts of Pacific oyster, *Crassostrea gigas*, detected by PCR-SSCP analysis. J. Shellfish Res. 15: 511–512.

Lilly, M., C. Brown, G. Pettit & A. Kraft, 1991. Bryostatin 1: a potential anti leukemic agent for chronic myelomonocytic leukemia. Leukemia 5: 283–287.

MacDonald, C. M. & R. Schneppenheim, 1983. Breeding structure and stock identity in the Antarctic krill *Euphausia superba* Dana. Polar Res. 4: 240–245.

MacDonald, C. M., R. Williams & M. Adams, 1986. Genetic variation and population structure of krill (*Euphausia superba* Dana) from the Prydz Bay region of Antarctic waters. Polar Biol. 6: 233–236.

Macleod, J. A. A., J. P. Thorpe & N. A. Duggan, 1995. A biochemical genetic study of queen scallop *Chlamys opercularis*) stocks in the northern Irish Sea. Mar. Biol. 87: 77–82.

Marcus, N. H., 1977. Genetic variation within and between geographically separated populations of the sea urchin *Arabacea punctulata*. Biol. Bull. 153: 560–570.

Matsuoka, N. & Y. Nakamura, 1990. Enzyme variation within the population of the sea urchin *Glyptocidaris crenularis* from Japanese waters. Comp. Biochem. Physiol 96B: 335–338.

Matsuoka, N. & Y. Nakamura, 1991. Genetic distance and protein polymorphism in two sea urchin species of the order Arabacioidea and implications for their evolution. Comp. Biochem. Physiol. 98B: 21–27.

Mattoccia, M., G. La Rosa, E. De Mathaeis, M. Cobolli-Sbordoni & V. Sbordoni, 1987. Patterns of genetic variation and differentiation in Mediterranean populations of *Penaeus kerathurus* (Crustacea: Decapoda). In, Tiews, K. (ed.), Selection, Hybridisation and Genetic Engineering in Aquaculture. ICES, Copenhagen: 131–142.

Maynard-Smith, J., 1989. Evolutionary Genetics. Oxford University Press, New York.

McMillen-Jackson, A. L., T. M. Bert & P. Steele, 1994. Population genetics of the blue crab *Callinectes sapidus*: modest population structuring in a background of high gene flow. Mar. Biol. 118: 53–65.

Meehan, B. W., 1985. A genetic comparison of *Macoma balthica* from San Francisco Bay (California) and Coos Bay (Oregon), U.S.A. J. Shellfish Res. 7: 170.

Mgaya, Y. D., E. M. Gosling, J. P. Mercer & J. Donlon, 1995. Genetic variation at three polymorphic loci in wild and hatchery stocks of the Abalone, *Haliotis tuberculata* Linnaeus. Aquaculture 136: 71–80.

Minokawa, T., S. Amemiya & N. Matsuoka, 1992. Genetic variation and differentiation in two local Japanese populations of the sea urchin *Asthenosoma ijimai*: electrophoretic analysis of isozymes. Zool. Sci. 9: 1299.

Mitton, J. B., C. J. Berg, Jr. & K. S. Orr, 1989. Population structure, larval dispersal and gene flow in the queen conch, *Strombus gigas*, of the Caribbean. Biol. Bull. 177: 356–362.

Mladenov, P. V., R. M. Allibone & G. P. Wallis, 1997. Genetic differentiation in the New Zealand sea urchin *Evechinus chloroticus* (Echinodermata: Echinoidea). New Zealand J. mar. Freshwat. Res. 31: 261–269.

Monteiro, F. A., A. M. Solé-Cava & J. P. Thorpe, 1997. Extensive genetic divergence between populations of the common intertidal sea anemone *Actinia equina* from Britain, the Mediterranean and the Cape Verde Islands. Mar. Biol. 129: 425–433.

Moraga, D., D. Jollivet & F. Denis, 1994. Genetic differentiation across the west Pacific populations of the hydrothermal vent bivalve *Bathymodiolus* sp. and the east pacific population of *Bathymodiolus thermophilus*. Deep Sea Res. 41: 1551–1557.

Nevo, E., 1978. Genetic variation in natural populations: patterns and theory. Theor. Pop. Biol. 13: 121–177.

Noy, R., B. Lavie & E. Nevo, 1987. The niche-width variation hypothesis revisited: genetic diversity in the marine gastropods *Littorina punctata* (Ginelin) and *L. neritoides* (L.). J. exp. mar. Biol. Ecol. 109: 109–116.

Olive, P. J. W., 1994. Polychaeta as a world resource: a review of patterns of exploitation as a sea angling bait and the potential for aquaculture based production. Mem. Mus. natn. Hist. nat., Paris 162: 603–610.

Olson, S. G., 1996. Curing cancer through aquaculture. Sea Technol. 37: 89–94.

O'Reilly, P. & J. M. Wright, 1995. The evolving technology of DNA fingerprinting and its application to fisheries and aquaculture. J. Fish Biol. 47: 29–55.

Ovenden, J. R. & D. J. Brasher, 1994. Stock identity of the red (*Jasus edwardsii*) and green (*J. verreauxi*) rock lobsters inferred from mitochondrial DNA analysis. In Philips, B. F., J. S. Cobb & J. Kittaka (eds), Spiny Lobster Management. Blackwell, London: 230–249.

Pannacciulli, F. G., J. D. D. Bishop & S. J. Hawkins, 1997. Genetic structure of populations of two species of *Chthamalus* in the north east Atlantic and Mediterranean. Mar. Biol. 128: 73–82.

Park, L. K. & P. Moran, 1995. Developments in molecular genetic techniques in fisheries. In Carvalho, G. R. & T. J. Pitcher (eds), Molecular Genetics in Fisheries. Chapman and Hall, London: 1–28.

Patarnello, T., L. Bargelloni, V. Varotto & B. Battaglia, 1996. Krill evolution and the Antarctic ocean currents: evidence of vicariant specification as inferred by molecular data. Mar. Biol. 126: 603–608.

Patterson, K. R., 1987. Fishy events in the Falklands. New Scientist 1562: 44–48.

Patwary, M. U., E. L. Kenchington, C. J. Bird & E. Zouros, 1994. The use of random amplified polymorphic DNA markers in genetic studies of the sea scallop *Placopecten magellanicus* (Gmellin, 1791). J. Shellfish Res. 13: 547–553.

Perrin, M. C., J. P. Thorpe & A. M. Solé-Cava, 1999. *Actinia equina*: a genetic role model and reproductive enigma. Oceanogr. mar. Biol. ann. Rev. (in press).

Pierce, R. W., J. T. Carlton, D. A. Carlton & J. B. Geller, 1997. Ballast water as a vector for tintinnid transport. Mar. Ecol. Prog. Ser. 149: 295–297.

Piertney, S. B. & G. R. Carvalho, 1995. Microgeographical genetic differentiation in the intertidal isopod *Jaera albifrons*. 2. Temporal variation in allele frequencies. J. exp. mar. Biol. Ecol. 188: 277–288.

Prince, J. D., T. L. Sellers, W. B. Ford & S. R. Talbot, 1987. Experimental evidence or limited dispersal of haliotid larvae (genus *Haliotis*; Mollusca: Gastropoda). J. exp. mar. Biol. Ecol. 106: 243–263.

Roehner, M., R. Bastrop & K. Juerss, 1996. Colonization of Europe by two American genetic types or species of the genus *Marenzelleria* (Polychaeta: Spionidae). An electrophoretic analysis of allozymes. Mar. Biol. 127: 277–287.

Roper, C. F. E., M. J. Sweeney & C. E. Naun, 1984. FAO species catalogues. Vol 3. Cephalopods of the world. An annotated and illustrated guide to species of interest to fisheries. FAO, Rome.

Ruiz, G. M., J. T. Carlton, E. D. Grosholz & A. H. Hines, 1997. Global invasions of marine and estuarine habitats by non-indigenous species: mechanisms, extent and consequences. Am. Zool. 6: 621–632.

Russo, C. A. M., A. M. Solé-Cava & J. P. Thorpe, 1994. Population structure and genetic variation in two tropical sea anemones (Cnidaria, Actinidae) with different reproductive strategies. Mar. Biol. 119: 267–276.

Ryman, N., 1991. Conservation genetics considerations in fishery management. J. Fish Biol. 39: 211.

Ryman, N. & F. Utter, 1987. Population genetics and fishery management. Washington University Press, Seattle, Washington.

Saavedra, C., C. Zapata, A. Guerra & G. Alvarez, 1987. Genetic structure of populations of the flat oyster *Ostrea edulis*) from the north west of the Iberian Peninsula. Inv. Pesq. 51: 225–241.

Sabbadin, A., 1978. Genetics of the colonial ascidian *Botryllus schlosseri*. In Battaglia, B. & J. A. Beardinore (eds), Marine Organisms, Genetics Ecology and Evolution. Plenum Press, New York: 195–207.

Schaufelberger, D. E. et al., 1991. The large scale isolation of bryostatin 1 from *Bugula neritina* following current good manufacturing processes. J. nat. Prod. 54: 1265–1270.

Scheltema, R. S., 1986. Long-distance dispersal by planktonic larvae of shoal-water benthic invertebrates among central Pacific islands. Bull. mar. Sci. 39: 241–256.

Scheltema, R. S., 1989. Planktonic and non-planktonic development among prosobranch gastropods and its relationship to the geographic range of species. In Ryland, J. S. & P. A. Tyler (eds), Reproduction, Genetics and Distribution of Marine Organisms. Olsen & Olsen, Fredensborg, Denmark: 183–188.

Schneppenheim, R. & M. MacDonald, 1983. Population genetics of krill (*Euphausia superba*. Polar Res.4: 439.

Schuchter, R. L. et al., 1991. Successful treatment of murine melanoma with bryostatin 1. Cancer Res. 51: 682–687.

Shaw, P. W., 1997. Polymorphic microsatellite markers in a cephalopod: the veined squid *Loligo forbesi*. Mol. Ecol. 6: 297–298.

Shepherd, S. A. & L. D. Brown, 1993. What is an abalone stock–Implications for the role of refugia in conservation. Can. J. Fish. aquat. Sci. 50: 2001–2009.

Shick, J. M. 1991. A Functional Biology of Sea Anemones. Chapman & Hall, London.

Siberman, J. D., S. K. Shaver & P. J. Walsh, 1994. Mitochondrial DNA variation in seasonal cohorts of spiny lobster (*Panulirus argus*) post larvae. Mol. mar. Biol. Biotechnol. 3: 165–170.

Skibinski, D. O. F., M. Ahmad & J. A. Beardmore, 1978. Genetic evidence for naturally occurring hybrids between *Mytilus edulis* and *Mytilus galloprovincialis*. Evolution 32: 354–364.

Slatkin, M., 1981. Estimating levels of gene flow in natural populations. Genetics 99: 323–335.

Slatkin, M., 1985. Gene flow in natural populations. Ann. Rev. Ecol. Syst. 16: 393–430.

Slatkin, M., 1994. Gene flow and population structure. In Real, L. A. (ed.), Ecological Genetics. Princeton University Press, Princeton, New Jersey: 3–17.

Slatkin, M., 1995. A measure of population subdivision based on microsatellite allele frequencies. Genetics 139: 457–462.

Smith, J. M. B., P. Rudall & P. E. Keage, 1989. Driftwood on Heard Island. Polar Rec. 25: 223–228.

Smith, P. J., P. E. Roberts & R. J. Hurst, 1981. Evidence for two species of arrow squid in the New Zealand fishery. New Zealand J. mar. Freshwat. Res. 15: 247–253.

Solé-Cava, A. M. & J. P. Thorpe, 1991. High levels of genetic variation in natural populations of marine lower invertebrates. Biol. J. linn. Soc. 44: 65–80.

Solé-Cava, A. M. & J. P. Thorpe, 1992. Genetic divergence between colour morphs in populations of the common intertidal sea anemones *Actinia equina* and *A. prasina* (Anthozoa: Actiniaria) in the Isle of Man. Mar. Biol. 112: 243–252.

Solé-Cava, A. M. & J. P. Thorpe, 1994. Evolutionary genetics of marine sponges. In Van Soest, R. W. M., T. M. G. Van Kempen & J. C. Braekman (eds), Sponges in Time and Space. A. A. Balkema, Rotterdam: 55–63.

Staton, J. L. & D. L. Felder, 1995. Genetic variation in populations of the ghost shrimp genus *Callichirus* (Crustacea, Decepoda, Thalassinoidea) in the Western Atlantic and Gulf of Mexico. Bull. mar. Sci. 56: 523–536.

Stepien, C. A., 1995. Population genetic divergence and geographic patterns from DNA sequences: examples from marine and freshwater fishes. Am. Fish. Soc. Symp. 17: 263–287.

Stiles, S., J. Choromanski, D. Schweitzer & Q. Z. Xue, 1996. Preliminary investigations of genetics and breeding of the bay scallop, *Argopecten irradians*. J. Shellfish Res. 15: 461.

Strathmann, R. R., 1980. Why does a larva swim so long? Paleobiology 6: 373–376.

Strathmann, R. R., 1985. Feeding and nonfeeding larval development and life history evolution in marine invertebrates. Ann. Rev. Ecol. Syst. 16: 339–361.

Strathmann, R. R., 1990. Why life histories evolve differently in the sea. Am. Zool. 30: 197–207.

Taggart, J. B. & A. Ferguson, 1990. Minisatellite DNA fingerprints of salmonid fishes. Anim. Genet. 21: 377–389.

Taggart, J. B., E. Verspoor, P. T. Galvin, P. Moran & A. Ferguson, 1995. A minisatellite DNA marker for discriminating between European and North-American Atlantic salmon (*Salmo salar*). Can. J. Fish. aquat. Sci. 52: 2305–2311.

Thompson, A. P., J. R. Hanley & M. S. Johnson, 1996. Genetic-structure of the western rock lobster *Panulirus cygnus*, with the benefit of hindsight. Aust. J. mar. freshwat. Res. 47: 889–896.

Thorpe, J. P., J. N. Havenhand & K. R. Patterson, 1986. Report of the University of Liverpool (Department of Marine Biology) to the Falkland Islands Development Corporation on stock and species identities of Patagonian Shelf *Illex*. Falkland Islands Development Corporation, Port Stanley, Falkland Islands.

Thorpe, J. P. & R. D. M. Nash, 1993. Other invertebrates used as food. In Macrae, R. R. Robinson & M. Sadler (eds), Encyclopaedia of Food Science, Food Technology and Nutrition. Academic Press, London: 2898–2902.

Thorpe, J. P. & A. M. Solé-Cava, 1994. The use of allozyme electrophoresis in invertebrate systematics. Zool. Scr. 23: 3–18.

Todd, C. D., 1985. Settlement timing hypothesis: a reply to Grant and Williamson. Mar. Ecol. Prog. Ser. 23: 197–202.

Todd, C. D., J. N. Havenhand & J. P. Thorpe, 1988. Genetic differentiation, pelagic larval transport and gene flow between local populations of the intertidal marine mollusc *Adalaria proxima* (Alder & Hancock). Funct. Ecol. 2: 441–451.

Todd, C. D., W. J. Lambert & J. P. Thorpe, 1998. The genetic structure of intertidal populations of two species of nudibranch molluscs with planktotrophic and pelagic lecithotrophic larval stages: are pelagic larvae 'for' dispersal? J. exp. mar. Biol. Ecol. 228: 1–28.

Utter, F. M., 1995. Perspective of molecular genetics and fisheries into the 21st century. In Carvalho, G. R. & T. J. I. Pitcher (eds), Molecular Genetics in Fisheries. Chapman and Hall, London: 105–109.

Utter, F. M., P. Aebersold & G. Winans, 1987. Interpreting genetic variation detected by electrophoresis. In Ryman, N. & F. Utter (eds), Population Genetics and Fishery Management. University of Washington Press, Seattle: 21–46.

Valentine, J. W. & F. J. Ayala, 1978. Adaptive strategies in the sea. In Battaglia, B. & J. A. Beardmore (eds), Marine Organisms: Genetics, Ecology and Evolution. Plenum Press, New York: 323–345.

Volpe, J. P. & M. M. Ferguson, 1996. Molecular genetic examination of the polymorphic Arctic charr *Salvelinus alpinus* of Thingvallavatn, Iceland. Mol. Ecol. 5: 763–772.

Voss, G. L. 1973. Cephalopod resources of the world. FAO, Rome.

Voss, G. L., 1983. Review of cephalopod fishery biology. Mem. natl. Mus. Vic. 44: 229–241.

Ward, R. D., 1989. Molecular population genetics of marine organisms. In Ryland, J. S. & P. A. Tyler (eds), Reproduction, Genetics and Distribution of Marine Organisms. Olsen and Olsen, Fredensborg, Denmark: 235–249.

Ward, R. D., N. G. Elliott, B. H. Innes, A. J. Smolenski & P. M. Grewe, 1997. Global population structure of yellowfin tuna, *Thunnus albacares*, inferred from allozyme and mitochondrial DNA variation. Fish. Bull. 95: 566–575.

Ward, R. D. & P. M. Grewe, 1995. Appraisal of molecular genetic techniques in fisheries. In Carvalho, R. & T. J. Pitcher (eds), Molecular Genetics in Fisheries. Chapman and Hall, London: 29–54.

Watts, P. C., J. P. Thorpe & P. D. Taylor, 1998. Natural and anthropogeneic dispersal mechanisms in cheilostomatid Bryozoa. Phil. Trans. r. Soc., Lond. B 353: 453–464.

Watts, R. J., M. S. Johnson & R. Black, 1990. Effects of recruitment on genetic patchiness in the urchin *Echinometra mathaei* in Western Australia. Mar. Biol. 105: 145–151.

Winston, J. E., 1982. Drift plastics–an expanding niche for a marine invertebrate? Mar. Pollut. Bull. 13: 348–351.

Worcester, S. E., 1994. Adult rafting versus larval swimming: dispersal and recruitment of a botryllid ascidian on eelgrass. Mar. Biol. 121: 309–317.

Wright, J. M. & P. Bentzen, 1995. Microsatellites: genetic markers for the future. In Carvalho, G. R. (ed.), Molecular Genetics in Fisheries. Chapman & Hall, London: 117–121.

Yeatman, J. & J. A. H. Benzie, 1994. Genetic structure and distribution of *Photololigo* spp. in Australia. Mar. Biol. 118: 79–87.

Hydrobiologia **420:** 185–189, 2000.
A. M. Solé-Cava, C. A. M. Russo & J. P. Thorpe (eds), Marine Genetics.

Effects of copper on the fitness of the Chilean scallop *Argopecten purpuratus* (Mollusca: Bivalvia)*

L. Troncoso[1,3], R. Galleguillos[2] & A. Larrain[1]
[1]*Laboratory of Bioassays, Department of Zoology, Faculty of Natural Sciences & Oceanography, University of Concepción. P.O. Box 160-C, Concepción, Chile*
E-mail: ltronco@udec.cl
[2]*Department of Oceanography. Faculty of Natural Sciences & Oceanography, University of Concepción, P.O. Box 160-C, Concepción, Chile*
E-mail: rgalleg@udec.cl
[3]*Centre EULA. University of Concepción. P.O. Box 160-C, Concepción, Chile*

Key words: allozymes, toxicity, mollusks, pollution, heterozygosity

Abstract

Exposure to high concentrations of metals can affect populations and individuals at morphological, physiological, biochemical or genetic levels. Metal pollution is a source of environmental stress that can have deleterious effects on organisms and generate selective pressure upon populations. This work attempts to establish whether concentrations of copper, above physiological requirements, can affect the genetic structure of a cohort of Chilean scallop *Argopecten purpuratus*. Following the determination of LC_{50} for 96 h experiments, other acute toxicity tests were carried out, exposing the juveniles to solutions of 150 ppb copper for 120 h. Dead, surviving and control individuals from these bioassays were genetically characterized for five polymorphic loci, Isocitrate dehydrogenase (*Idh*), Octopine dehydrogenase (*Ocdh*), Phosphogluconate dehydrogenase (*Pgd*), Leucine aminopeptidase (*Lap*) and Phosphoglucose isomerase (*Pgi*). Results showed no significant differences in allele and genotypic frequencies between surviving and dead individuals, although for the same groups significant differences were found in estimated mean heterozygosity. *Pgi* and *Ocdh* showed significantly different heterozygosity values for dead and surviving individuals. A positive relationship between multilocus heterozygosity and survival was found for young *A. purpuratus* exposed to high copper concentrations; thus, we found evidence of a differential response to exposure to high copper concentrations, related to degree of heterozygosity.

Introduction

Metal pollution is a source of environmental stress that can exert significant selective pressure on organisms (Posthuma et al., 1993). Reactions to pollutant exposure can be a continuum of biochemical and physiological changes both at population and community levels (Moriarty, 1988).

Tolerance to metals, which can be defined as the ability to prevent, reduce or repair damage by assimilation of metals, can develop during individual exposure (acclimation) or by the action of selection on genetic variability (adaptation), or by a combination of both possibilities (Levitt, 1980).

Studies of enzyme polymorphism have shown that certain ecological events can result in the replacement of particular genotypes or, alternatively, can maintain stability of some observed variability. The adaptive importance of genetic variability has been suggested (Koehn & Hilbish, 1987). If the genetic structure of some specific organisms is sensitive and subject to change under exposure to different levels and kinds of pollutants, then those changes could be analyzed and used in a biological monitoring system (Beardmore, 1980; Nevo et al., 1981, 1984; Lavie & Nevo, 1986).

This study tests whether exposure to copper salts produces significant differences in patterns of genetic

* This work was supported by FONDEF Grant 2-72.

variability in the Chilean scallop *Argopecten purpuratus*. The main objective is to analyze the effects of exposure to high concentrations of copper salts on fitness, estimated as differential survival, of a cohort of juveniles of *A. purpuratus*.

Materials and methods

Juveniles of *A. purpuratus* were obtained from a commercial hatchery. They were part of a 3 month old cohort and had an average length of 18.5 (± 2.1) mm.

After pilot toxicity tests, in which the LC_{50} (for 96 h acute experiments) was estimated using the Probit program (U.S. EPA, 1993), a copper concentration of 150 ppb was selected for the experiments. A 10 l glass jar was set up with 6 l of the experimental copper solution; another jar was set up as a control, with 6 l of filtered, sterilized sea water. Eighty scallop juveniles were placed in the control jar, and 160 juveniles in the treatment jar. The copper solution was prepared from $CuSO_4$ dissolved in deionized water and was changed every 48 h for 120 h. Juveniles were fed with microalgae half an hour before each water change. The seawater used was free of copper, filtered, sterilized and saturated with oxygen at over 8 mgO_2 l^{-1}. The pH and temperature were registered at each change of solution.

The scallops were inspected daily and were considered dead when they were found gaping and showed no response to tactile stimuli. Dead animals were removed, numbered and stored at −20 °C. At the end of the 120 h period, all remaining scallops were also frozen and stored for analysis by protein gel electrophoresis. The experiment was replicated three times, using identical materials and methods.

Electrophoresis of extracts from the adductor muscle of each scallop was carried out for 4 h on horizontal starch gels. Five polymorphic enzyme systems were used to characterize the juveniles of *A. purpuratus*: Isocitrate dehydrogenase, 1.1.1.42 (*Idh*), Phosphogluconate dehydrogenase 1.1.1.44 (*Pgd*); Octopine dehydrogenase 1.1.1.46 (*Ocdh*), Leucine amine peptidase 3.4.11.1 (*Lap*) and Phosphoglucose isomerase 5.3.1.9 (*Pgi*). Three buffers were used: Lithium hydroxide, pH 8.0 (Ridgway et al., 1970), Tris-citric, pH 7.0 (Siciliano & Shaw, 1976) and Tris-citric, pH 5.1–6.0 (Galleguillos & Troncoso, 1991).

Genetic analysis was carried out using the program BIOSYS 1.7 (Swofford & Selander, 1989). The relationship between juvenile survival and multilocus heterozygosity was estimated for the three pooled replicates. Five heterozygosity levels (0, 1, 2, 3, 4) were considered and the numbers of dead and alive scallops in each category were counted. Data were arcsine-transformed to apply linear regression analysis. The homogeneity test (G-test) was used to compare allele frequencies, while between-group heterozygosity values were compared using a heterogeneity *t*-test (Zar, 1984).

Results

The acute toxicity tests with *A. purpuratus* juveniles generated a LC_{50} of 117 ppb, but to ensure more extreme environmental conditions, the copper concentration actually used in the experiments was equivalent to a LC_{80} (150 ppb).

The homogeneity test showed no differences in allele frequencies for the five polymorphic loci analyzed both among the replicate experiments and among the surviving, dead and control groups ($P>0.05$). Because of these results, the three replicates of the experiment were pooled into a single set. Table 1 shows allele frequencies per locus for each group (surviving, dead and control individuals) for the pooled set. The χ^2 values for deviations from expected Hardy-Weinberg equilibrium were not significant.

Mean observed heterozygosity for survivors was 0.355 ± 0.076, while for dead specimens it was 0.328 ± 0.067. The comparison of heterozygosity per locus for all three groups showed significant differences between surviving and dead scallops for the *Ocdh* and *Pgi* loci (t=4.22; P=0.05 and t=6.6; $P<0.05$, respectively). The analysis of the D index, which shows excess or deficiency of heterozygotes, showed significant values for survivors at the loci *Ocdh* (f=5.25; $P<0.05$), *Idh* (f=3.1; P=0.05) and *Pgi* (f=10.25; $P<0,05$) loci, compared with dead individuals (Table 1).

The comparison of genotypes showed significant differences between survivors and dead at the *Pgi* locus for the 100/100 and 115/100 genotypes (t=7.0; P=0.02 and t=5.0; P= 0.038, respectively). Differences were also found for the *Idh* locus between surviving and dead scallops for the 100/100 genotype (t=5.77; P=0.03) and between the dead and control groups for the 110/110 genotype. The *Lap* locus showed the most significant difference between the groups of survivors and dead for the 100/100 genotype (t=15.59; P=0.004).

Table 1. Argopecten purpuratus. Allelic frequencies for five loci, χ^2Hardy-Weinberg equilibrium, observed heterozygosity per locus and excess or deficiency of heterozygotes (D)

		Control	Surviving	Dead
Ocdh	N	174	153	158
	100	0.833	0.820	0.835
	107	0.167	0.180	0.165
	χ^2	3.15 ns	0.005 ns	2.63 ns
	Ho	0.293	0.293	0.256
	D	−0.149	−0.009	−0.120
Idh	N	192	176	155
	110	0.190	0.142	0.152
	100	0.799	0.855	0.848
	90	0.001	0.003	0.000
	χ^2	2.97 ns	1.02 ns	4.84 ns
	Ho	0.252	0.225	0.227
	D	−0.223	−0.062	−0.168
Pgd	N	150	142	146
	100	0.843	0.782	0.805
	90	0.157	0.218	0.195
	χ^2	0.74 ns	0.42 ns	1.63 ns
	Ho	0.237	0.308	0.266
	D	−0.089	−0.085	−0.136
Pgi	N	178	159	157
	120	0.098	0.129	0.111
	115	0.365	0.302	0.306
	100	0.503	0.553	0.561
	85	0.034	0.016	0.022
	χ^2	8.68 ns	7.45 ns	3.98 ns
	Ho	0.601	0.654	0.585
	D	−0.007	0.108	0.003
Lap	N	182	175	150
	105	0.201	0.154	0.197
	100	0.791	0.837	0.800
	95	0.008	0.009	0.003
	χ^2	1.36 ns	0.80 ns	0.26 ns
	Ho	0.323	0.265	0.319
	D	−0.034	−0.064	−0.017

Table 2 shows the range of heterozygosity categories considered (0, 1, 2, 3 and 4) and the number of surviving and dead animals for each category. A positive correlation between survival and heterozygosity was observed. Linear regression analysis applied to these data showed a positive slope with a good fit (r^2=0.860).

Table 2. Argopecten purpuratus. Number of individuals for category of heterozygosity

Category	*N*	Surviving	Dead
0	71	34 (48%)	37 (52%)
1	101	55 (54%)	46 (46%)
2	103	54 (52%)	49 (48%)
3	45	26 (58%)	19 (42%)
4	10	6 (60%)	4 (40%)

Discussion

The effects of trace metals on natural ecosystems are complex and often unpredictable (Depledge et al., 1995). Thus, it is necessary to develop methods for *in situ* toxicity assessments. Current ecotoxicology assessments are mainly based on toxicity assays under laboratory conditions. However, this approach limits the applicability of the results to prediction of pollution events due to metals or other xenobiotics in natural environments (Depledge et al., 1995). Another approach, that deserves more attention, is the identification of genetic differences that affect the intra- and inter-population sensitivity to metal traces (Depledge, 1996).

Earlier work on acute toxicity for *A. purpuratus* with an average length of 4.6 cm estimated a mean lethal copper concentration of 130 ppb (Sánchez & Tupayachi, 1988). This value is comparable with the results for the 2 cm average length scallops used in this study (96 h; 117 ppb of copper). Other studies have found similar results for several bivalve species from the Chilean coast (i.e. *Perumytilus purpuratus*, 122 ppb; and *Aulacomya ater*, 147 ppb). A higher tolerance value was found for *Semimytilus algosus* (479 ppb; Zuñiga pers. comm.).

The analysis of the genetic variability suggests that individuals with higher heterozygosity at *Ocdh* and *Pgi* loci have higher tolerance to copper concentrations, since the more heterozygous individuals showed higher capacity for survival. In general, surviving scallops had higher heterozygosity than the dead or control groups. For the *Pgi* locus, survivors showed a significant excess of heterozygotes.

Exposure to stress can generate significant changes in metabolism of an organism as it adapts to the new condition (Calow, 1991). Metabolic adaptation requires energy, thus an increase in basal metabolism can be observed. Clark & Koehn (1993) suggested that the differential rate of protein replacement, which is the main component of basal metabolism, could be genotype-dependent. Koehn & Bayne (1989) suggest that the energetic cost of maintenance is also genotype-dependent and that heterozygotes require less energy. Thus, heterozygotes may have more energy available for other metabolic requirements (Volckaert & Zouros, 1989).

Under conditions of stress, the energy balance may favor heterozygotes (Koehn & Shumway, 1982; Koehn & Bayne, 1989) and differences in fitness might be predicted by assaying loci involved in energy production or in the most important metabolic pathways (Blackstock, 1980). The *Pgi* enzyme has been much studied both for its participation in metabolism and for its high polymorphism.

Nevo et al. (1984) described also a positive relationship between degree of heterozygosity and survival for marine species exposed to mercury. On the other hand, a similar analysis of *Mytilus edulis* larvae exposed to copper salts showed a negative relationship (Hoare et al., 1994).

For *A. purpuratus*, no clear relationship was found between any particular genotype and survival at high copper concentrations. In other bivalves, such as *Mytilus edulis*, Hvilson (1983) found significant differences in allele frequencies for the *Pgi* and *Pgm* loci between surviving and dead individuals and established copper-induced differential survival for the *Pgi* locus.

Nevo et al. (1981) studied *Pgi* genotypic frequency and survivorship in the gastropods *Monodonta turbinata* and *M. turbiformis* exposed to Cu^{2+} and Zn^{2+} salts and found that certain genotypes appeared more tolerant to pollution and were more abundant in populations that lived in contaminated areas. Similar results were found by Lavie & Nevo (1982) under laboratory conditions.

References

Beardmore, J. A., 1980. Genetical considerations in monitoring effects of pollution. Rapp. P. verb. Reun. Cons. int. Expl. Mer 179: 258–266.

Blackstock, J., 1980. Estimation of activities of some enzymes associated with energy-yielding metabolism in the polychaete, *Glycera alba* (Muller), and application of the methods to the study of the effect of organic pollution. J. exp. mar. Biol. Ecol. 46: 197–217.

Calow, P., 1991. Physiological costs of combating chemical toxicants: ecological implications. Comp. Biochem. Physiol. 100C: 3–6.

Clark, A. & R. K. Koehn, 1993. Enzymes and adaptation. In Berry, R. J. (ed.), Genes in Ecology. Oxford University Press, London.

Depledge, M. H., 1996. Genetic ecotoxicology: an overview. J. exp. mar. Biol. Ecol. 200: 57–66.

Depledge, M. H., A. Agard & P. Györkos, 1995. Assessment of trace metal toxicity using molecular physiological and behavioral biomarkers. Mar. Pollut. Bull. 31: 19–27.

Galleguillos, R. & L. Troncoso, 1991. Protein variation in the Chilean-Peruvian scallop *Argopecten purpuratus*. In Shumway, S. E. & P. A. Sandifer (eds), International Compendium of Scallop Biology and Culture vol. 1. World Aquaculture Workshops: 146–150

Hoare, K., A. Beaumont & J. Davenport, 1994. Effects of copper exposure during early life stages on heterozygosity in laboratory-reared mussel (*Mytilus edulis*) populations. In Beaumont, A. R. (ed.), Genetics and Evolution of Aquatic Organisms. Chapman & Hall, London: 425–434.

Hvilson, M. M., 1983. Copper-induced differential mortality in the mussel *Mytilus edulis*. Mar. Biol. 76: 291–295.

Koehn, R. K. & B. L. Bayne, 1989. Towards a physiological and genetical understanding of the energetics of the stress response. Biol. J. linn. Soc. 37: 157–171.

Koehn, R. K. & T. J. Hilbish, 1987. The adaptative importance of genetic variation. Am. Sci. 75: 134–141.

Koehn, R. K. & S. E. Shumway, 1982. A genetic/physiological explanation for differential growth rate among individuals of the American oyster, *Crassostrea virginica* (Gmelin). Mar. Biol. Letters 3: 35–42.

Lavie, B. & E. Nevo, 1982. Heavy metal selection of phosphoglucose isomerase allozymes in marine gastropods. Mar. Biol. 71: 17–22.

Lavie, B. & E. Nevo, 1986. Genetic selection of homozygote allozyme genotypes in marine gastropods exposed to cadmium pollution. Sci. Tot. Environ. 57: 91–98.

Levitt, J., 1980. Responses of Plants to Environmental Stress. Academic Press. New York.

Moriarty, F., 1988. Ecotoxicology: The Study of Pollutants in Ecosystems. Academic Press. London. 347 pp.

Nevo, E., R. Ben-Shlomo & B. Lavie, 1984. Mercury selection of allozymes in marine organisms: prediction and verification in nature. Proc. natn. Acad. Sci. U.S.A. 81: 1258–1259.

Nevo, E., T. Perl, A. Beiles & D. Wool, 1981. Mercury selection of allozyme genotypes in shrimp. Experientia 37: 1152–1154.

Posthuma, L., R. F. Hogervorst, E. N. G. Joose & N. M. Van Straalen, 1993. Genetic variation and covariation for characteristics associated with cadmium tolerance in natural populations of the springtail *Orchesella cincta* (L.). Evolution 47: 619–631.

Ridgway, G. J., S. W. Sherbourne & R. D. Lewis, 1970. Polymorphism in the esterases of Atlantic herring. Trans. am. Fish. Soc. 99: 145–151.

Sánchez, G. & M. Tupayachi, 1988. Pruebas preliminares sobre toxicidad aguda del cobre en la concha abanico *Argopecten purpuratus*. In Salzwedel, H. & A. Landa (eds), Recursos y Dinámica del Ecosistema de Afloramiento Peruano. Bol. Inst. Mar Perú-Callao: 191–194.

Siciliano, M. J. & C. R. Shaw, 1976. Separation and visualization of enzymes on gels. In Smith, Y. (ed.), Chromatographic and Electrophoretic Techniques. Heinemann, London 2: 185–209.

wofford, D. L. & R. B. Selander, 1989. BIOSYS-1. A computer program for the analysis of allelic variation in population genetics and biochemical systematics. Smithsonian Illinois Natural History Survey.

.S. EPA, 1993. Methods for measuring the acute toxicity of effluents and receiving waters to freshwater and marine organisms. Office of Research and Development, U.S. Environmental Protection Agency, Washington, D.C. 206460.

Volckaert, F. & E. Zouros, 1989. Allozyme and physiological variation in the scallop *Placopecten magellanicus* and a general model for the effects of heterozygosity on fitness in marine molluscs. Mar. Biol. 103: 51–61.

Zar, J. H., 1984. Biostatistical Analysis. Prentice Hall, Englewood Cliffs, New Jersey.

Hydrobiologia **420:** 191–201, 2000.
A. M. Solé-Cava, C. A. M. Russo & J. P. Thorpe (eds), Marine Genetics.

Genetics in fisheries management

R. D. Ward
CSIRO Marine Research, GPO Box 1538, Hobart, Tasmania 7001, Australia
E-mail: Bob.Ward@marine.csiro.au

Key words: resource, sustainability, stocks, genetic structure, forensics

Abstract

Genetic analyses have much to offer fisheries managers, especially in the provision of tools enabling unequivocal specimen identification and assessment of stock structure. The three commonly-used genetic tools – allozymes, mitochondrial DNA and microsatellites – differ in their properties. These differences must be born in mind, especially when interpreting gene frequency data collected for stock structure research. Examples where genetic approaches have been used to identify specimens are given, with special attention being given to compliance and labelling issues. Treatment of stock structure focuses on Atlantic cod and on yellowfin, bigeye and albacore tunas. The different resolving powers of the various techniques are discussed. Marine fish typically show low levels of population genetic differentiation, and in such species the use of large sample sizes and, preferably, multiple types of markers are desirable to resolve stock structure issues.

Introduction

The days of free and unfettered access to wild fisheries are long gone. Many important commercial stocks are severely depleted, and there has been a gradual replacement of high value 'top predators' by species that were once discarded (FAO, 1993; Pauly et al., 1998). Management of wild fisheries is now seen as necessary, both to ensure that present fisheries can continue to be exploited in perpetuity and to assist in the recovery of depleted stocks.

Fisheries managers have as one of their primary objectives the need to ensure resource sustainability and to avoid population crashes. This is not easy. Biological communities are naturally highly variable in space and time, and the additional impacts of fishing practices are often very poorly understood. Decisions cannot be reached without full consideration of their implications for the societies that depend for their livelihood on fishing. Thus, fisheries managers need to take into account scientific, economic, social and sometimes political factors in determining the controls that need to be exercised in any particular fishery.

The scientific data required for fisheries management include, primarily, stock abundance and productivity estimates and information on stock dynamics. Immediately, then, the question of 'what is a stock' is posed. Indeed, a recent review of stock assessment methods (Deriso & Quinn, 1998) included as number one in a recommended check list of items, information on spatial distribution or stock structure. It suggested that such information could be garnered from tagging, microconstituents, genetics and/or morphometrics.

Genetic technologies have been long recognised for their usefulness in defining stock structures of fish and some of the earliest attempts date back to the 1930s. However, this is not the only use of genetics in fisheries management; other important applications include specimen identification, monitoring fisheries enhancement (re-stocking), and aquaculture operations. In this paper, I will briefly outline some of the genetic tools that are available, and then give examples of their use in specimen identification and in stock structure analysis.

The genetic tools

The potential of genetic approaches for identifying fish stocks has been recognised for a long time. L. Von Toth, in the early 1930s (see de Ligny, 1969), stated that his attempts to find blood group variation in fish were to differentiate, for example, herring populations. In the 1950s and 1960s, several laboratories were engaged in the analysis and description of blood group variation in fish. This early work demonstrated the genetic nature of the serological variants, but serology failed to gain widespread adoption as a tool for stock separation; blood group variation is rarely studied in fish populations today.

The next genetic tool to be used, and one which made a far greater impact and which continues in widespread use today, was the analysis of genetically determined electrophoretic variation of proteins. One of the earliest such fish studies was that of Sick (1961), who described haemoglobin variants in whiting (*Gadus melangus*) and cod (*Gadus morhua*). The extension of these methods to the examination of enzymes using histochemical stains led to an explosion of population genetic studies of fish and other animals. This was triggered by the seminal studies of *Drosophila* and man by Lewontin & Hubby (1966) and Harris (1966), respectively, which revealed the ubiquity of such variation.

Protein electrophoretic variation is a somewhat indirect indicator of genetic variation; generally being caused by amino-acid substitutions which in turn reflect DNA changes. With the recognition and purification of restriction enzymes, which cut DNA molecules at particular base sequences, came the ability to look at DNA variation more directly. Initially, these studies tended to focus on mitochondrial DNA, largely because this molecule was much smaller and easier to work with than nuclear DNA. Studies of nuclear DNA were largely restricted to examination of minisatellite variability (minisatellites being tandem arrays of short oligonucleotide repeats, with the repeat unit being less than 65 base pairs), a technique often referred to as DNA fingerprinting.

The next, and perhaps most significant major scientific advance, was the development of polymerase chain reaction (PCR) technology. This enables the rapid amplification of DNA between two primers, and this amplified DNA can then be assessed for sequence or, more commonly, for size variability. There is an ever-increasing number of ways in which PCR approaches are used in the genetic analysis of fish, including RAPDs (randomly amplified polymorphic DNA), microsatellites (one to four nucleotide base repeats that are generally less than 300 bp in total), EPICs (exon-primed intron-crossing amplification) and AFLP (amplified fragment length polymorphisms).

These techniques and their advantages and disadvantages for fisheries analysis are described elsewhere (e.g. Park & Moran, 1994; Ward & Grewe, 1994). Some major advantages of protein electrophoretic methods include their speed and simplicity; disadvantages include a requirement for fresh or frozen tissue, relatively low levels of variation, and uncertainty over whether the variability recorded is neutral or subject to selective pressures. Major advantages of PCR-based procedures include ease of sample collection and storage (and samples can be extremely small), access to either high or low variability DNA sequences by judicious selection of genomic regions, and use of non-coding regions to render data less liable to the influence of natural selection. Disadvantages include longer development times and slower collection of data, the need for greater technical expertise and, often, complex and expensive equipment. Perhaps the major interpretational issue with microsatellite data concerns uncertainty regarding the underlying evolutionary model to be used for these highly mutable loci (see e.g. Estoup & Angers, 1998)

Specimen and species identification

Accurate fish identification is a prerequisite for effective fisheries management, as catch data are used to estimate biomass sizes and, for those species managed under quota systems, to monitor quota uptake throughout the year. Sometimes, species are morphologically very similar and mistakes in identification may be quite common. Misidentification of fish, whether intentional or accidental, greatly reduces the value of log-book data to managers and can contribute to species depletions.

It is also necessary that species are correctly labelled in the market-place, in order to maintain consumer confidence. Sadly, cases of fish substitution do occur, generally (and unsurprisingly) by substituting cheaper fish for more expensive fish. Such practices are likely to be more common when fish are marketed with heads removed or as fillets. Compliance officers often require a tool that enables unequivocal identification of species in disputed cases, and which will be acceptable to a court of law. The accurate identification of threatened or endangered species is required for conservation purposes. Finally, eggs and larvae may be virtually impossible to identify morphologically, yet accurate identification may be essential for particular research (especially ecological studies of early life history stages) or monitoring purposes.

Molecular genetic methods have proved invaluable in distinguishing between morphologically very similar species (and in revealing previously unrecognised cryptic sibling species), between fish fillets and in enabling the identification of eggs and larvae. Allozyme analysis is a very satisfactory method for resolving taxonomic issues and for identifying species, provided that the material either fresh or frozen (preferably at −80 °C). Some examples of the discovery of cryptic species in fisheries using allozyme electrophoresis include tropical sharks from Australia (Lavery & Shaklee, 1991) and sprats and arrow squid off New Zealand (Smith & Robertson, 1981 and Smith et al., 1987, respectively). Allozyme methods have been used for fish fillet identification (Shaklee & Keenan, 1986) and have been used in law courts in the US (Harvey, 1990) and Australia (Keenan, pers. comm.). Seeb et al. (1990) used allozyme analysis to locate a king crab catch (*Paralithodes camtschatica*) to an area closed to fishing rather than to the claimed open area. The methods are sufficiently sensitive to be able to identify fish eggs (Mork et al., 1983), fish larvae (Smith & Crossland, 1977; Sidell et al., 1978; Seeb & Kendall, 1991) and even bivalve larvae (Hu et al., 1992).

My laboratory has been using protein fingerprinting to discriminate among commercial species of fish from Australian waters. The technique is based on the Coomassie blue staining of the major proteins of fish muscle following electrophoretic separation on a cellulose acetate gel matrix. We opted for this approach as it is very quick (less than one hour) and inexpensive, needs a minimum of equipment and training, and could if necessary be carried out on a bench in a fish market.

Our initial project was the production of an identification guide to the fish of the trawl sector of the South East Fishery of Australia (Daley et al., 1997). This fishery consists of 21 quota species and about 65 non-quota relatives of commercial size. These 85 or so species are among the most commonly sold at fresh fish markets in Melbourne and Sydney, and some are exported. The guide includes the usual type of information used for identification, including photographs, maps and notes on the quota and non-quota species, but more unusually carries protein fingerprint figures of all depicted species (alongside a reference ladder of proteins). Usually the protein fingerprints alone are sufficient to separate species, but for sibling species where the protein fingerprints are identical, we give allozyme information which enables discrimination. Normally eight specimens of each species were examined so that at least the commonly occurring protein polymorphisms could be identified, but for most species such variation was not observed. Reference samples are maintained at −80° C.

Following the launch of the guide, samples bought from various fish outlets were tested, and frequently these samples proved to be incorrectly labeled. Similar techniques were used by Huang et al. (1995) to show that sought-after fillets of red snapper (*Lutjanus campechanus*) were substituted with congeneric species in 70% of tested samples from Florida between 1988 and 1992. However, we have not carried out a scientifically rigorous assessment of the scale of the Australian misidentification problem. In Australia, and perhaps other parts of the world too, the problem of labeling is compounded by varying local names for the same species; there is now a list of marketing names to promote uniformity of nomenclature across Australia (Department of Primary Industries and Energy, 1995).

This work has been expanded to cover most of Australia's domestic seafood species (about 450 species of fish and invertebrates), and a guide to all these species, including protein fingerprints, is now available (Yearsley et al., 1999); a second volume dealing with imported species will be available later.

Sometimes, identification of disputed specimens using protein or allozyme methods may not be possible. The species may be so similar that they cannot be allozymically unambiguously separated, but can be separated by DNA-based methods. There may be insufficient material available for allozyme analysis (although, as noted above, bivalve larvae have been successfully identified). Specimens may have been preserved in such a way as to destroy protein integrity or enzyme activity. Another circumstance may be that of a recreational angler who is uncertain about the identification of a specimen but may be unwilling to sacrifice it or to take a muscle biopsy sample for allozyme analysis. This is especially likely in big-game fisheries where the specimen is often released soon after capture. In all these situations, PCR methods allow the amplification and characterisation of DNA, usually mitochondrial DNA (mtDNA) from minute alcohol-stored samples, including fin clips.

Examples of the use of PCR amplification of mtDNA (followed by sequencing or RFLP analysis) for identification purposes are: billfish (Chow, 1994;

Innes et al., 1998), snappers, Lutjanidae (Chow et al., 1993), tunas (Ward et al., 1995), wrasse of the genus *Xyrichtys* (including larvae) (Hare et al., 1994, 1998), fish eggs (caviar) (DeSalle & Birstein, 1996), sea cucumber larvae (Olson et al., 1991), starfish larvae (Evans et al., 1998) and calanoid copepods (Bucklin & Kann, 1991). Quinteiroet al. (1998) showed that DNA was somewhat degraded during canning, but the use of mtDNA-PCR enabled identification of canned tuna samples. A recent guide to Australian billfish incorporates a section on PCR-mtDNA identification (Pepperell & Grewe, 1998).

Nuclear DNA has been less commonly used for identification purposes. A few examples include RAPD–PCR to identify fish species, including Tilapia and barbels (Dinesh et al., 1993; Bardakci & Skibinski, 1994; Callejas & Ochando, 1998), PCR amplification of 28S RNA from nuclear DNA to identify spiny lobster phyllosome larvae (Silberman & Walsh, 1992).

An interesting example of using DNA analysis to identify not just species but individuals was recently published (Cipriano & Palumbi, 1999). Here, mtDNA and nDNA approaches were combined to identify a sample of whale meat bought in Japan in 1993 as the offspring of a mating between a blue whale mother and a fin whale father, and this hybrid individual was thought to have been harpooned near Iceland in 1989 under a scientific whaling permit issued by Iceland. Thus, it is now possible to genetically track individual whales through international commercial channels and make it possible to distinguish legally caught animals from all others. Although not used in this example, multi-locus typing of hypervariable microsatellite loci lends itself particularly well to identification and tracking of individuals.

Stock structure

How to define a stock? The most commonly quoted biological definition of a stock is that proposed by Ihssen et al. (1981): "a stock is an intraspecific group of randomly mating individuals with temporal and spatial integrity". Such a stock may then be assessed and attempts made to set sustainable catch levels. If a fishery consists of more than one fishing ground, as of course most do, the question is whether the fishery consists of more than one stock. If so, separate assessments of sustainable fishing levels should be made for each stock. In practice, stock assessments for management purposes often do not incorporate information on biological stock structure, because 1. genetic and other biological data on population structure are not available or are ambiguous, 2. the stock may straddle political and/or management boundaries and there is no joint management authority, and 3. it may not be economically feasible to monitor and set catch levels for each biological stock. Fisheries managers may then define as a stock the fish that occur in a specified area at a specified time, without regard to genetic integrity. Stocks thus defined cannot be as effectively managed as true biological stocks. These issues are considered further by Carvalho & Hauser (1994).

Determining whether two samples are from a single panmictic population or from two separate stocks is not always straight forward. If there are genetic differences between samples of fish from two areas, and if it is assumed that these differences arise from restricted gene flow rather than local selection, then two stocks can be inferred. If no genetic differences are detected, then either there is a single stock or there are two stocks which could not be resolved by the test. Sample homogeneity does not necessarily equate to population homogeneity. The null hypothesis, in this case panmixia, can be rejected if heterogeneity is detected, but a failure to reject the null hypothesis does not mean the null hypothesis is true. A finding of sample heterogeneity allows more powerful conclusions concerning stock structure than a finding of sample homogeneity. Gene flow rates of 1%, 5%, 20% and 50% will give genetic homogeneity among samples and thus cannot be distinguished, yet each of these cases should have different consequences for stock assessment models. Findings of sample homogeneity are thus of little assistance to fishery managers.

Determining stock structure for marine fish is particularly difficult as differentiation among samples of marine fish (mean F_{ST} estimated at 0.062) is much less than that between samples of freshwater fish (0.222), with anadromous species being intermediate (0.108) (Ward et al., 1994a). The estimated mean F_{ST} value for marine fish is similar to that estimated for other highly vagile groups such as birds (0.076) and insects (0.097) (Ward et al., 1992). In fact, a few marine fish had relatively high F_{ST} values, inflating the mean value; the median F_{ST} for marine fish is only 0.020 (Waples, 1998). The low F_{ST} no doubt reflects the lack of obvious barriers to dispersal in the marine environment compared with the isolated lakes and catchment areas of freshwater species. In addition, many marine fish have long larval periods allowing widespread dispersal in currents (although gyres may enable local retention) and long-lived migratory adults, further increasing levels of gene flow. Because genetic differentiation among marine populations is both expected and observed to be low, it becomes more difficult to discriminate this signal from the genetic noise associated with limited sampling. These issues are dealt with in some depth by Waples (1998), who *inter alia* recommends sample replication over time to discern signal from noise.

Rather than provide a review of the many, probably hundreds, of genetic studies of stock structure of marine fish, I have chosen to focus on just a couple of species. From these examples, I hope to be able to make some general points about the utility of the different methods and issues that must be considered in interpreting allele frequency data.

Cod in the North Atlantic

The Atlantic cod (*Gadus morhua*) was in 1996 the ninth largest single species fishery, at 1.33×10^6 mt (FAO, 1998) and was probably more valuable than several of the larger fisheries of small pelagic fish harvested mainly for agricultural feed. This demersal species has been the focus of numerous genetically-based examinations of population structure (see Shaklee & Bentzen, 1998), commencing as early as 1960 with a survey of haemoglobin variability (Sick, 1961; Frydenberg et al., 1965). Most interest has centered on stock structure within the North Atlantic. Several surveys of allozyme variability found evidence for significant spatial variation in allele frequencies, but the largest such survey, that of Mork et al. (1985), found only very limited differentiation between east and west Atlantic cod. From the data presented, Pogson et al. (1995) estimated an F_{ST} value of only 0.014 for Atlantic regions. Mork et al. concluded that gene flow was high enough to preclude the formation of genetically different cod stocks below the scale of ocean basins and suggested that selection accounted for some of the more marked cases of heterogeneity recorded in earlier studies.

The conclusions from mtDNA studies of cod have ranged from lack of genetic differentiation across the Atlantic (Smith et al., 1989) to discrete eastern Atlantic stocks (Dahle, 1991). Sequencing studies of the mtDNA cytochrome b gene showed appreciably more haplotype diversity than earlier RFLP studies, and the heightened diversity in the eastern Atlantic compared with western Atlantic (h.d. = 0.88 vs. 0.40, Carr & Marshall, 1991a,b) suggested a restriction in gene flow across the Atlantic.

Pogson et al. (1995) examined Atlantic cod samples using anonymous nuclear DNA markers and RFLP analysis. They found a higher basin-scale F_{ST} value (0.069) than that (0.014) from the allozyme analysis of Mork et al. (1985), with significant differences among all locations. The relative homogeneity of allozymes was attributed to balancing selection and the heterogeneity of the RFLP loci to restriction on gene flow. This interpretation is of course the opposite of that of Mork et al. (1985) who suggested that the homogeneity of most allozymes reflected extensive gene flow and the heterogeneity of a few allozymes reflected selection. These conflicting interpretations illustrate one of the difficulties of using allozymes: allozymes are important gene products and the extent to which their variation reflects deterministic (selective) processes or stochastic (drift) processes remains unknown and controversial. On the other hand, much DNA work concentrates on non-coding regions (Pogson et al. (1995) markers were from the flanking regions of single-copy genes) where neutrality is expected and sample differentiation is likely to reflect stochastic processes alone.

Recent cod studies have concentrated on the analysis of mini and microsatellite variation. Galvin et al. (1995) examined a single minisatellite locus, finding highly significant allele frequency differences across the Atlantic and significant differences between Newfoundland and Nova Scotia. Two eastern regions, the Irish Sea and north Norway, were just non-significantly different. The overall F_{ST} was 0.03.

Most cod microsatellite studies have focused on the western Atlantic, where fine-scale differentiation has been detected. Bentzen et al. (1996) examined winter aggregations of northwestern Atlantic populations, and found genetic differences separating four regions, from Labrador in the north to Nova Scotia in the south. Inshore overwintering cod from Trinity Bay, Newfoundland, showed small, but significant, genetic differences from offshore overwintering cod from the Grand Bank region (Ruzzante et al., 1996a), with subsequent evidence of temporal stability of these differences over the period 1992–1995 (Ruzzante et al., 1997). These differences appear to be maintained even though individuals from the two regions intermingle during summer and fall. Analysis of cod larvae off Nova Scotia showed that the larval aggregation as a whole showed heterozygote deficiencies for microsatellite loci, while larvae from a single water mass and a single cohort did not, suggesting that mixing of larvae from distinct spawning events led to a Wahlund effect in the aggregation (Ruzzanteet al., 1996b). The fine-scale genetic differentiation observed in northwestern Atlantic cod for microsatellite loci are consistent with postdispersal spawning fidelity to natal areas, perhaps facilitated by topographically-induced gyre-like oceanic circulations that act as retention mechanisms (Ruzzante et al., 1998). These findings of multiple genetically discrete stocks, most likely reflecting genetic isolation, at a level finer than that currently recognised by management, should be considered carefully in future management plans aimed at rebuilding these depleted stocks (Ruzzante et al., 1997).

Interestingly, and perhaps unexpectedly, these fine-scale microsatellite differences are not reflected in increased large-scale differentiation. The single region examined from the eastern Atlantic, the Barents Sea, was only distinguishable from the northwest Atlantic samples at one of six microsatellite loci (Bentzen et al., 1996). It was suggested that convergent mutations at these highly mutable loci may have erased the effects of genetic drift that are more obvious from the allozyme, mtDNA and nuclear RFLP analyses, and that therefore microsatellites may not be the markers of choice for broad-scale surveys.

This very brief overview of cod population genetics shows clearly that relying on the results of a single survey using a single technique may produce misleading conclusions. Failure to reject the null hypothesis of sample homogeneity does not necessarily mean that the samples are homogeneous, just that the samples and methods followed have failed to show heterogeneity. Increasing sample size or using a different class of marker with different population dynamics may reveal heterogeneity and restrictions on gene flow among regions once classed as homogeneous.

Yellowfin tuna

The yellowfin tuna (*Thunnus albacares*) is another very valuable fishery. In 1996, it ranked 13th in terms of gross tonnage (0.99×10^6 mt, FAO, 1998). It is globally distributed in warm waters, and is recognised as a highly migratory species. As such, genetic differentiation is expected to be very limited, at least within oceans.

Some genetic diversity parameters are summarised in Table 1. The major impression is of very little differentiation within or among oceans, whether allozymes, mtDNA or microsatellites are used as markers. The second overall impression is that where differentiation is found, it is attributable to differences within oceans rather than between oceans ($F_{SO} > F_{OT}$, , where O = oceans, T = total and S = sample diversity). In fact, this within-ocean differentiation is almost entirely due to differentiation within the Pacific Ocean; samples are fewer in number from the Atlantic and Indian Oceans and the existence of within-ocean differentiation in those oceans cannot be well tested.

The largest F_{ST} values come from allozyme analysis, and this reflects differentiation within the Pacific Ocean at only one of the four polymorphic allozyme loci. For this locus, *Gpi*-A, the F_{ST} value within the Pacific is high at about 10% (with eastern samples being separable from western-central samples), while it is not significantly greater than zero for the other three allozymes (Ward et al., 1994b, 1997), and is no more than about 1% for mtDNA nor for each of six microsatellite loci.

Interpreting these data is not straightforward. The finding of no differentiation for either mtDNA or allozymes between oceans (Scoles & Graves, 1993; Ward et al., 1997) suggests either that there is significant gene flow between the Atlantic and IndoPacific Oceans (yellowfin tuna are found around the Cape of Good Hope), or that the sampling regime has failed to detect the existence of presumably low amounts of inter-oceanic differentiation.

With respect to within-ocean differentiation, at present most of the data came from the Pacific Ocean. Coverage of samples within the Pacific with these different markers is similar, except that, at the time of writing, only four of the six microsatellite loci have been examined in eastern populations. The marked *Gpi*-A differentiation, being an order of magnitude higher than that shown by three other allozymes, mtDNA, and six microsatellite loci, may reflect selective forces specific to this locus. If this locus is ignored, then the $F/_{ST}$ values for genetic markers in the Pacific are all 1% or less. Still, there is evidence of some significant, but very limited, spatial heterogeneity. For example, the overall F_{ST} value for the six microsatellite loci, while less than 1%, attains statistical significance as total sample sizes are quite large, averaging about 700 per locus, and four of the six loci show individual probability values of 0.10 or less. In addition, there is also evidence of weak, but

Table 1. Genetic diversity estimates for yellowfin tuna

Oceans	Marker	H_1 H_2	N_1 N_2	F_{ST} sig	F_{SO} sig	F_{OT} sig	Ref
P, A	mtDNA (whole: 12)	0.85 0.82–0.86	6 19–20	−0.021 ns	−0.015 ns	−0.006 ns	a
P, I, A	mtDNA (whole: 2)	0.68 0.63–0.73	9 21–176	0.010 **	0.012 *	−0.002 ns	b
P, I, A	allozyme (4 polymorphic)	0.36 0.34–0.39	8 21–175	0.025 **	0.027 **	−0.002 ns	b
P	mtDNA (whole: 12)	0.85 0.82–0.86	5 19–20	−0.015 ns			a
P	mtDNA (whole: 2)	0.68 0.63–0.71	6 40–176	0.012 *			b, c
P	allozyme (4 polymorphic)	0.36 0.34–0.38	6 39–175	0.027 **			b
P	microsatellites (6 polymorphic)	0.78 0.77–0.81	5–8 34–285	0.001 *			c
I	mtDNA (whole: 2)	0.71 0.70–0.73	2 21, 91	0.009 ns			b

Oceans: P = Pacific Ocean, A = Atlantic Ocean, I = Indian Ocean. Marker: mtDNA (whole = whole genome assayed by RFLP methods, with the given number of restriction enzymes), allozyme and microsatellites (number of polymorphic loci examined). H_1: overall heterozygosity (genetic diversity for mtDNA), H_2: range of sample heterozygosity, N_1: number of samples, N_2: range of sample sizes, F_{ST}: proportion of total genetic variation attributable to sample differences, F_{SO} and F_{OT}: proportion of within ocean genetic variation attributable to sample differences and proportion of total genetic variation attributable to ocean differences respectively. Note that the F_{ST} parameter for the single ocean studies is analogous to the F_{SO} parameter for the multiple-ocean studies. F estimates are from Hauser & Ward (1998) and used the AMOVA package. Sig: significance of F_{ST}; ns = not significant, $* P < 0.05$, $** P < 0.01$. Refs: a, Scoles & Graves, 1993; b, Ward et al., 1994b; c, Grewe & Hampton, 1998.

significant, mtDNA differentiation. Note that this latter differentiation was not significant in earlier studies (Scoles & Graves, 1993; Ward et al., 1994b). In fact, it only became detectable as sample size increased, from a Pacific-wide 434 (Ward et al., 1994b), through 561 (Ward et al., 1997), to the present-day 649.

An important question for Australian managers of the yellowfin tuna resource is whether the Australian Coral Sea and east coast fishery is independent of a larger western Pacific population or whether there is a single panmictic Australian-western Pacific population. We are currently analysing more samples from this region of the Pacific, using microsatellite loci, together with microchemical techniques (J. Gunn, unpublished data), to try to resolve this question.

Yellowfin tuna, bigeye tuna and albacore tuna

Bigeye tuna (*Thunnus obesus*) and albacore tuna (*T. alalunga*) are other important commercial tuna species, with catches of around one third and one fifth, respectively, that of yellowfin tuna. All three species are found circumglobally in tropical and temperate waters, and all are highly migratory. Interestingly, their global population structures appear to be quite different from yellowfin tuna.

In bigeye tuna, very pronounced differentiation of the Atlantic Ocean population from that of the Pacific Ocean was observed following mtDNA analysis (Alvarado Bremer et al., 1998). A large and highly significant F_{ST} of 0.260 was very largely due to Atlantic-Pacific partitioning (F_{OT} = 0.218 and F_{SO} = 0.042). Within the Pacific Ocean, no differentiation was observed for either mtDNA (Alvarado Bremer et al., 1998; Grewe & Hampton, 1998) or four microsatellite loci (Grewe & Hampton, 1998).

With respect to albacore tuna, pronounced differentiation of the Atlantic Ocean population from that of the Pacific Ocean was also observed, again following mtDNA examination (Chow & Ushiama, 1995). This was entirely due to Pacific/Atlantic partitioning (F_{ST} = 0.092, F_{OT} = 0.096, and F_{SO} = −0.004; see Hauser & Ward, 1998). As yet, no non-mtDNA studies have been carried out on this species.

Thus, while all three of these tunas occupy a similar global range, one of them (yellowfin) shows no evidence of significant Atlantic-IndoPacific isolation, while two (bigeye and albacore) show high (albacore) to very high (bigeye) levels of Atlantic-IndoPacific differentiation. These differences among species occur despite all three being highly migratory, with similar ecologies and life-histories, and all three being found off the Cape of Good Hope, the mixing zone between the Atlantic and the Indo-Pacific regions. It can be seen that extrapolating population structures from one species to another, even between those with ostensibly similar ecologies, can be dangerous.

Discussion

It is not just wishful thinking to state that genetic tools and genetic data are becoming increasingly important to fisheries managers. This brief review has focused on uses of genetics for species identification and population structure analyses; other major uses in stock enhancement projects and aquaculture production have not been covered.

Accurate and unequivocal specimen identification is essential for compliance officers engaged in legal disputes with alleged illicit fishing operators, and fish substitutions in the market place can be readily detected. In addition, the ability to identify fish carcasses from which external diagnostic features, such as heads and fins, have been removed makes it possible to monitor more precisely the accuracy of log-book details, thus improving the efficacy of management schemes based on catch levels and quotas. Eggs and larvae can be identified, improving studies of life-history and ecology. Most of these identifications are based on molecular genetic examination using protein electrophoresis or PCR amplification of DNA sequences, the particular approach to be followed varying from case to case, but in some instances serological methods have been deployed (see Sweijd et al., 1998). Species identification using genetic approaches is, in general, non-controversial and unequivocal.

The same cannot be said of stock structure analyses. Marine fish constitute the bulk of commercial fisheries, and levels of genetic population differentiation are low (Ward et al., 1994a). Discerning signal from noise can be difficult (Waples, 1998). If differentiation is present, it is expected to be low, and sample sizes are often too small to detect low levels of differentiation. Findings of no detectable genetic differentiation between samples are common-place, but are of little value to managers. Such results are consistent with any model of population structure ranging from complete panmixia to the exchange of as little as 1% or fewer of individuals.

Identification of species is simple, identification of stocks is difficult. Essentially, three types of markers are used in genetic assessments of stock structure – allozymes, mitochondrial DNA and microsatellites. The examples given in this review show that inferred population structure may vary from study to study. Of course, the underlying population structure for any particular species will be, essentially, a constant. So how can different studies infer different structures?

Part of the problem lies in our lack of knowledge regarding the ubiquity or otherwise of natural selection at allozyme loci. Generally, population differentiation is assumed, for any marker, to reflect reproductive isolation, but in the presence of selection this assumption becomes invalid. While the null hypothesis of neutral alleles seems appropriate for all classes of marker, allozymes, representing expressed products, are more likely to be subject to differential selection than microsatellite alleles. Uniformity of allozyme allele frequencies is generally ascribed to gene flow but, where this uniformity is associated with heterogeneity at non-coding marker loci, has been ascribed to balancing selection. Heterogeneity between samples for microsatellite alleles is likely to reflect genetic isolation, heterogeneity for allozymes might reflect genetic isolation or disruptive selection within a panmictic population. Interpretation of allozyme data is thus more difficult, or more subject to different views, than microsatellite data. Perhaps the major issue with microsatellite data concerns some uncertainty that high mutation rates have on evolutionary models of population structure.

The effect of sampling size should also be considered. Small samples taken from two regions may be statistically homogeneous, but the power of small samples to reveal low level differentiation is very limited. Subsequent surveys with increased sample size may well reveal significant differentiation. It is thus wrong to claim from the former survey that the two regions form a single panmictic unit, all that can be formally claimed is that the null hypothesis of panmixia has not been rejected. A second and larger survey may well reject panmixia. Sample sizes should be reasonably large; a realistic target would be at least 100 per sample. Ideally, temporally separated samples would also be analysed.

A related issue concerns the power of the loci, assuming neutrality, to reveal population structure. More highly polymorphic loci apparently have higher asymptotic power for exact tests of population differentiation, and thus microsatellites may permit the detection of genetic differences that are not apparent from allozymes (Estoup & Angers, 1998). Microsatellite analysis, for several reasons, would therefore seem a better approach to the examination of fine-scale stock structure than allozyme analysis. For broad-scale surveys, it has been suggested that hypervariable microsatellites might be less appropriate than less variable loci, as there is the possibility of convergent mutation erasing the effects of drift (Bentzen et al., 1996). These issues have not been fully resolved and the recommendation must be to use as many markers of different types as possible and as large a sample size as possible. This will assist in teasing out the effects of drift, mutation and selection, and will result in a better understanding of stock structure than any single genetic approach.

It is perhaps also worth repeating that these analyses have to be carried out on each species of interest separately. Extrapolation of population structure from a well-studied species to any other species, even to a closely-related species with similar range, ecology and life-history attributes, is fraught with danger. Graves (1998) makes this point in comparing the population structures of cosmopolitan marine fish.

This paper has focused on genetic approaches to resolving the stock structures of commercially important fish. It should be clear from the above that genetic analysis is not without its problems and limitations, but despite these difficulties it may well be the single most informative tool. There are, however, many other tools that can be applied, including the analysis of life-history traits, tagging returns, parasite loads and microchemical variation of hard parts. All these non-genetic approaches come with their attendant advantages and disadvantages, but all provide valuable and complementary information. Genetic and non-genetic data can then be considered along with oceanographic data, such as bathymetry and the direction and strength of eddies and currents, to synthesise the most likely population structure model for the species under study. Such a co-ordinated multi-disciplinary approach is the optimal strategy, but may be too resource-demanding for any other than the largest fisheries. Consequently, most fisheries will continue to be managed with rather little understanding of underlying population structures, making it difficult to maximise returns while ensuring sustainability.

References

Alvarado Bremer, J. R., B. Stequert, N. W. Robertson & B. Ely, 1998. Genetic evidence for inter-oceanic subdivision of bigeye tuna (*Thunnus obesus* Lowe) populations. Mar. Biol. 132: 547–557.

Bardakci, F. & D. O. F. Skibinski, 1994. Application of the RAPD technique in *Tilapia* fish – species and subspecies identification. Heredity 73: 117–123.

Bentzen, P., C. T. Taggart, D. E. Ruzzante & D. Cook, 1996. Microsatellite polymorphism and the population structure of Atlantic cod (*Gadus morhua*) in the northwest Atlantic. Can. J. Fish. aquat. Sci. 53: 2706–2721.

Bucklin, A. & L. Kann, 1991. Mitochondrial DNA variation of copepods: markers of species identity and population differentiation in *Calanus*. Biol. Bull. 181: 357.

Callejas, C. & M. D. Ochando, 1998. Identification of Spanish barbel species using the RAPD technique. J. Fish Biol. 53: 208–215.

Carr, S. M. & H. D. Marshall, 1991a. Detection of intraspecific DNA sequence variation in the mitochondrial cytochrome b gene of Atlantic cod (*Gadus morhua*) by the polymerase chain reaction. Can. J. Fish. aquat. Sci. 48: 48–52.

Carr, S. M. & H. D. Marshall, 1991b. A direct approach to the measurement of genetic variation in fish populations: applications of the polymerase chain reaction to studies of Atlantic cod, *Gadus morhua*. J. Fish Biol. 39 (Suppl. A): 101–107.

Carvalho, G. R. & L. Hauser, 1994. Molecular genetics and the stock concept in fisheries. Rev. Fish Biol. Fish. 4: 326–350.

Chow, S., 1994. Identification of billfish species using mitochondrial cytochrome b gene fragment amplified by polymerase chain reaction. Collect. Vol. Sci. Pap. ICCAT 42: 549–556.

Chow, S., M. E. Clarke & P. J. Walsh, 1993. PCR-RFLP analysis on thirteen western Atlantic snappers (subfamily Lutjaninae): a simple method for species and stock identification. Fish. Bull. U.S.A. 91: 619–627.

Chow, S. & H. Ushiama, 1995. Global population structure of albacore (*Thunnus alalunga*) inferred by RFLP analysis of the mitochondrial ATPase gene. Mar. Biol. 123: 39–45.

Cipriano, F. & S. R. Palumbi, 1999. Genetic tracking of a protected whale. Nature 397: 307–308.

Dahle, G., 1991. Cod, *Gadus morhua* L., populations identified by mitochondrial DNA. J. Fish Biol. 38: 295–303.

Daley, R. K., P. R. Last, G. K. Yearsley & R. D. Ward, 1997. South east fishery quota species: an identification guide. CSIRO Division of Marine Research, Hobart, Australia.

Deriso R. & T. Quinn (eds), 1998. Improving Fish Stock Assessments. National Academy Press, Washington, D.C.

De Ligny, W., 1969. Serological and biochemical studies in fish populations. Oceanogr. mar. Biol. ann. Rev. 7: 411–513.

De Salle, R. & V. J. Birstein, 1996. PCR identification of black caviar. Nature 381: 197–198.

Dinesh, K. R., T. M. Lim, K. L. Chua, W. K. Chan & V. P. E. Phang, 1993. RAPD analysis: an efficient method of DNA fingerprinting in fishes. Zool. Sci. 10: 849–854.

Department of Primary Industries and Energy, 1995. Marketing Names for Fish and Seafood in Australia. Department of Primary Industries and Energy, Canberra, Australia.

Estoup, A. & B. Angers, 1998. Microsatellites and minisatellites for molecular ecology, theoretical and empirical considerations. In Carvalho, G. R. (ed.), Advances in Molecular Ecology. IOS Press, Amsterdam: 55–86.

Evans, B. S., R. W. G. White & R. D. Ward, 1998. Genetic identification of asteroid larvae from Tasmania, Australia, by PCR-RFLP. Mol. Ecol. 7: 1077–1082.

FAO, 1993. Review of the state of world marine fishery resources. FAO Fisheries Technical Paper No. 335. FAO, Rome.

FAO, 1998. FAO Yearbook 1996. Fishery Statistics: Catches and Landings, Vol. 82. FAO, Rome.

Frydenberg, O., D. Moller, G. Naevdal & K. Sick, 1965. Haemoglobin polymorphism in Norwegian cod populations. Hereditas 53: 257–271.

Galvin, P., T. Sadusky, D. McGregor & T. Cross, 1995. Population genetics of Atlantic cod using amplified single locus minisatellite VNTR analysis. J. Fish Biol. 47 (Suppl. A): 200–208.

Graves, J. E., 1998. Molecular insights into the population structures of cosmopolitan marine fishes. J. Heredity 89: 427–437.

Grewe, P. M. & J. Hampton, 1998. An assessment of bigeye (*Thunnus obesus*) population structure in the Pacific Ocean, based on mitochondrial DNA and DNA microsatellite analysis. CSIRO Marine Research, Hobart, Australia.

Hare, J. A., R. K. Cowen, J. P. Zehr, F. Juanes & K. H. Day, 1994. Biological and oceanographic insights from larval labrid (Pisces, Labridae) identification using mtDNA sequences. Mar. Biol. 118: 17–24.

Hare, J. A., R. K. Cowen, J. P. Zehr, F. Juanes & K. H. Day, 1998. A correction to biological and oceanographic insights from larval labrid (Pisces, Labridae) identification using mtDNA sequences. Mar. Biol. 130: 589–592.

Harris, H., 1966. Enzyme polymorphisms in man. Proc. r. Soc. Lond. 164B: 298–310.

Harvey, W. D., 1990. Electrophoretic techniques in forensics and law enforcement. In Whitmore, D. H. (ed.), Electrophoretic and Isoelectric Focusing Techniques in Fisheries Management. CRC Press, Boca Raton, Florida: 314–321.

Hauser, L. & R. D. Ward, 1998. Population identification in pelagic fish: the limits of molecular markers. In Carvalho, G. R. (ed.), Advances in Molecular Ecology. IOS Press, Amsterdam: 191–224.

Hu, Y.P., R. A. Lutz & R. C. Vrijenhoek, 1992. Electrophoretic identification and genetic analysis of bivalve larvae. Mar. Biol. 113: 227–230.

Huang, T. -S., M. R. Marshall & C. -I. Wei, 1995. Identification of red snapper (*Lutjanus campechanus*) using electrophoretic techniques. J. Food Sci. 60: 279–283.

Ihssen, P. E., H. E. Booke, J. M. Casselman, J. M. McGlade, N. R. Payne & F. M. Utter, 1981. Stock identification: materials and methods. Can. J. Fish. aquat. Sci. 38: 1838–1855.

Innes, B. H., P. M. Grewe & R. D. Ward, 1998. PCR-based genetic identification of marlin and other billfish. Aust. J. mar. Freshwat. Res. 49: 383–388.

Lavery, S. & J. B. Shaklee, 1991. Genetic evidence for separation of two sharks, *Carcharhinus limbatus* and *C. tilstoni*, from northern Australia. Mar. Biol. 108: 1–4.

Lewontin, R. C. & J. L. Hubby, 1966. A molecular approach to the study of genic variation in natural populations. II. Amount of variation and degree of heterozygosity in natural populations of *Drosophila pseudoobscura*. Genetics 54: 595–609.

Mork, J., P. Solemdal & G. Sundnes, 1983. Identification of marine fish eggs: a biochemical genetics approach. Can. J. Fish. aquat. Sci. 40: 361–369.

Mork, J. P., N. Ryman, G. Ståhl, F. Utter & G. Sundnes, 1985. Genetic variation in Atlantic cod (*Gadus morhua*) throughout its range. Can. J. Fish. aquat. Sci. 42: 1580–1587.

Olson, R. R., J. A. Runstadler & T. D. Kocher, 1991. Whose larvae? Nature 351: 357–358.

Park, L. K. & P. Moran, 1994. Developments in molecular genetic techniques in fisheries. Rev. Fish Biol. Fish. 4: 272–299.

Pauly, D., V. Christensen, J. Dalsgaard, R. Froese & F. Torres, Jr., 1998. Fishing down marine food webs. Science 279: 860–863.

Pepperell, J. & P. Grewe, 1998. A field guide to the Indo-Pacific billfishes. CSIRO Division of Marine Research, Hobart, Australia.

Pogson, G. H., K. A. Mesa & R. G. Boutilier, 1995. Genetic population structure and gene flow in the Atlantic cod *Gadus morhua*: a comparison of allozyme and nuclear RFLP loci. Genetics 139: 375–385.

Quinteiro, J., C. G. Sotelo, H. Rehbein, S. E. Pryde, I. Medina, R. I. Peréz-Martin, M. Rey-Méndez & I. M. Mackie, 1998. Use of mtDNA direct polymerase chain reaction (PCR) sequencing and PCR-restriction fragment length polymorphism methodologies in species identification of canned tuna. J. agric. Food Chem. 46: 1662–1669.

Ruzzante, D. E., C. T. Taggart & D. Cook, 1996b. Spatial and temporal variation in the genetic composition of a larval cod (*Gadus morhua*) aggregation: cohort contribution and genetic stability. Can. J. Fish. aquat. Sci. 53: 2695–2705.

Ruzzante, D. E., C. T. Taggart & D. Cook, 1998. A nuclear DNA basis for shelf- and bank-scale population structure in northwest Atlantic cod (*Gadus morhua*): Labrador to Georges Bank. Mol. Ecol. 7: 1663–1680.

Ruzzante, D. E., C. T. Taggart, D. Cook & S. V. Goddard, 1996a. Genetic differentiation between inshore and offshore Atlantic cod (*Gadus morhua*) off Newfoundland: microsatellite DNA variation and antifreeze level. Can. J. Fish. aquat. Sci. 53: 634–645.

Ruzzante, D. E., C. T. Taggart, D. Cook & S. V. Goddard, 1997. Genetic differentiation between inshore and offshore Atlantic cod (*Gadus morhua* L.) off Newfoundland: a test and evidence of temporal stability. Can. J. Fish. aquat. Sci. 54: 2700–2708.

Scoles, D. R. & J. E. Graves, 1993. Genetic analysis of the population structure of yellowfin tuna, *Thunnus albacares*, from the Pacific Ocean. Fish. Bull. U.S.A. 91: 690–698.

Seeb, L. W. & A. W. Kendall, Jr., 1991. Allozyme polymorphisms permit the identification of larval and juvenile rockfishes of the genus *Sebastes*. Envir. Biol. Fishes 30: 191–201.

Seeb, L. W., J. E. Seeb & J. J. Polovina, 1990. Genetic variation in highly exploited spiny lobster *Panulirus marginatus* populations from the Hawaiian archipelago. Fish. Bull. U.S.A. 88: 713–718.

Shaklee, J. B. & P. Bentzen, 1998. Genetic identification of stocks of marine fish and shellfish. Bull. mar. Sci. 62: 589–621.

Shaklee, J. B. & C. P. Keenan, 1986. A practical laboratory guide to the techniques and methodology of electrophoresis and its application to fish fillet identification. CSIRO Mar. Lab. Report 177, CSIRO, Hobart, Australia.

Sick, K., 1961. Haemoglobin polymorphism in fishes. Nature 192: 894–896.

Sidell, B. D., R. G. Otto & D. A. Powers, 1978. A biochemical method for distinction of striped bass and white perch larvae. Copeia 1978: 340–343.

Silberman, J. D. & P. J. Walsh, 1992. Species identification of spiny lobster phyllosome larvae via ribosomal DNA analysis. Mol. mar. Biol. Biotech. 1: 195–205.

Smith, P. J. & J. Crossland, 1977. Identification of larvae of snapper *Chrysophrys auratus* Forster by electrophoretic separation of tissue enzymes. New Zealand J. mar. Freshwat. Res. 11: 795–798.

Smith, P. J., A. Jamieson & C. A. Bishop, 1989. Mitochondrial DNA in the Atlantic cod, *Gadus morhua*: lack of genetic divergence between eastern and western populations. J. Fish Biol. 34: 369–373.

Smith, P. J., R. H. Mattlin, M. A. Roeleveld & T. Okutani, 1987. Arrow squids of the genus *Nototodarus* in New Zealand waters: systematics, biology and fisheries. New Zealand J. mar. Freshwat. Res. 21: 315–326.

Smith, P. J. & D. A. Robertson, 1981. Genetic evidence for two species of sprat (*Sprattus*) in New Zealand waters. Mar. Biol. 62: 227–233.

Sweijd, N. A., R. C. K. Bowie, A. L. Lopata, A. M. Marinaki, E. H. Harley & P. A. Cook, 1998. A PCR technique for forensic species level identification of abalone tissues. J. Shellfish Res. 17: 889–896.

Waples, R. S., 1998. Separating the wheat from the chaff: patterns of genetic differentiation in high gene flow species. J. Heredity 98: 438–450.

Ward, R. D., N. G. Elliott, P. M. Grewe & A. J. Smolenski, 1994b. Allozyme and mitochondrial DNA variation in yellowfin tuna (*Thunnus albacares*) from the Pacific Ocean. Mar. Biol. 118: 531–539.

Ward, R. D., N. G. Elliott & P. M. Grewe, 1995. Allozyme and mitochondrial DNA separation of Pacific northern bluefin tuna, *Thunnus thynnus orientalis* (Temminck and Schlegel), from southern bluefin tuna, *Thunnus maccoyii* (Castelnau). Aust. J. mar. Freshwat. Res. 46: 921–930.

Ward, R. D., N. G. Elliott, B. H. Innes, A. J. Smolenski & P. M. Grewe, 1997. Global population structure of yellowfin tuna (*Thunnus albacares*) inferred from allozyme and mitochondrial DNA variation. Fish. Bull. U.S.A. 95: 566–575.

Ward, R. D. & P. M. Grewe, 1994. Appraisal of molecular genetic techniques in fisheries. Rev. Fish Biol. Fish. 4: 300–325.

Ward, R. D., D. O. F. Skibinski & M. Woodwark, 1992. Protein heterozygosity, protein structure and taxonomic differentiation. Evol. Biol. 26: 73–159.

Ward, R. D., M. Woodwark & D. O. F. Skibinski, 1994a. A comparison of genetic diversity levels in marine, freshwater and anadromous fish. J. Fish Biol. 44: 213–232.

Yearsley, G. K., P. R. Last & R. D. Ward, 1999. Australian seafood handbook: domestic species. CSIRO Division of Marine Research, CSIRO, Hobart, Australia.

Hydrobiologia **420:** 203–210, 2000.
A. M. Solé-Cava, C. A. M. Russo & J. P. Thorpe (eds), Marine Genetics.

Genetic population structure of the swimming crab *Callinectes danae* (Crustacea: Decapoda) in southern Brazil

L. I. Weber & J. A. Levy
Fundação Universidade Federal do Rio Grande, Departamento de Química, Laboratório de Bioquímica Marinha, Caixa Postal 474, CEP 96201-900, Rio Grande, RS, Brazil

Key words: Portunidae, population genetics, gene flow, allozymes, estuarine retention

Abstract

The population genetic structure of the swimming crab *Callinectes danae* (Crustacea, Portunidae) was studied by allozyme electrophoresis along the southern coast of Brazil (Santa Catarina and Rio Grande do Sul States). The biology of *C. danae* is poorly known, but some studies suggest that this species depends on estuaries for reproduction and for completing its life cycle, using them as recruitment sites and nursery grounds. If estuarine retention is an important process acting in favour of the recruitment of local populations, we should expect restriction of gene flow among populations inhabiting different estuaries. Therefore, our aim was to establish whether gene flow between populations of *C. danae* from different estuaries was restricted. Samples were collected in four estuaries: São Francisco do Sul, Laguna (Santa Catarina), Patos Lagoon and Chuí Stream (Rio Grande do Sul). Eleven loci were resolved. Contingency table tests (χ^2 and G) showed significant differences ($P \leq 0.05$) between pairwise subpopulation comparisons. The estimated F_{ST} was θ=0.065±0.019, suggesting a moderate structuring of *C. danae* populations. No relationship was found for the number of migrants between pairs of subpopulations and the geographic distance separating them (P= 0.292; r^2= 0.269). Nevertheless, UPGMA analysis clustered together those estuaries separated by less than 250 km. Low adult migration, the requirement of estuaries as reproductive areas, recruitment sites and nursery grounds for juveniles, together with larval and post-larval retention processes, are discussed in the context of their importance in preventing panmixia among subpopulations inhabiting different estuaries.

Introduction

In southern Brazil, the swimming crab *Callinectes danae* Smith (1869) is the second most abundant species of the genus after *C. sapidus*. The distribution of *C. danae* extends from Florida (Rathbun, 1900) through Central America (Rathbun, 1930; Powers, 1977) to Brazil (Rathbun, 1895, 1930; Coelho & Ramos, 1972, 1980; Williams, 1974; Abreu, 1980; Medeiros, 1982; Moreira et al., 1988). Prior to this study, the southernmost record for *C. danae* was from Patos Lagoon in southern Brazil (Rio Grande do Sul State) (Garcia et al., 1996), we present a new record from close to the Brazil-Uruguay border (Chuí Stream). *C. danae* is a euryhaline species (Shumway, 1983), living in estuaries, bays and adjacent coasts (Coelho & Ramos, 1972; Williams, 1974; Abreu, 1980). From Santa Catarina State towards northern Brazil, *C. danae* becomes more abundant than *C. sapidus* (Sawaya & Pereira, 1946; Farias, 1980; Medeiros, 1982; Pita et al., 1985a,b; Moreira et al., 1988). In Santa Catarina and Rio Grande do Sul States, both species are caught together by artisanal fishermen, most of whom are unable to distinguish between them.

The life cycle of *C. danae* is not well known and it is not yet clear how much it depends on estuaries for reproduction, as recruitment sites for its post-larvae, and as nursery grounds for its juveniles. Juveniles are found on *Spartina* in estuarine areas of São Paulo State (Abreu, 1980) and a few have been found on *Ruppia maritima* in the estuary of the Patos Lagoon (Garcia et al., 1996). Ovigerous females are found in estuaries during autumn and/or winter (Negreiros-Fransozo & Fransozo, 1995; Garcia et al., 1996), but it is not known whether males are restricted to low salinity areas of estuaries, as is the case in *C. sapidus*.

If *C. danae* depends on estuaries for reproduction

and for completing its life cycle, estuarine retention, including larval retention and postlarval reinvasion (also referred to in this study as postlarval retention) (Sulkin & Van Heukelem, 1981, 1982) may be an important process acting in favour of the recruitment of local populations. If this is the case, restricted gene flow between populations inhabiting different estuaries is to be expected (Hedgecock, 1981). Therefore, our aim was to investigate whether populations of *C. danae* are genetically structured along the coast of southern Brazil.

Materials and methods

Samples of *C. danae* were collected (Figure 1) by beam trawl at two localities in Rio Grande do Sul State (RS): Chuí Stream (33° 33 S, 53° 24 W), March-1997 (N=75) and Saco do Mendanha, Patos Lagoon estuary (32° 30 S, 52° 10 W), April-1997 (N=80); and at two localities in Santa Catarina State (SC): Laguna estuary (27° 28 S, 48° 51 W), July-1996 (N=80) and São Francisco do Sul estuary (26° 15 S, 48° 45 W), July-1996 (N=57). Samples from Laguna and São Francisco were frozen in dry ice upon collection and samples from Patos Lagoon and Chuí stream were brought live to the laboratory. All samples were kept at −24 °C in a freezer until required for electrophoresis.

Standard allozyme electrophoresis was used for studying the genetic structure of *C. danae*. Homogenates were prepared by macerating muscle tissue from pereopods in distilled water. Homogenates were centrifuged (Eppendorf centrifuge 5415-C) at 14 000 rpm and supernatants were used for electrophoresis. Horizontal 14% starch-gel electrophoresis was carried out for the following enzymes: aspartate aminotransferase (*Aat*) (2.6.1.1), malic enzyme (*Mep*) (1.1.1.40), di-peptidase (*di-Pep*) (Leu-Phe) (3.4.13.11), phosphogluconate dehydrogenase (*Pgdh*) (1.1.1.44), isocitrate dehydrogenase (*Idhp*) (1.1.1.42). Vertical 10% polyacrylamide electrophoresis was carried out for esterases (*Est*) (3.1.1.-). The following buffer systems were used: I. Tris-citrate-borate-LiOH, pH 8.15 (Shaklee & Keenan, 1986) for *Est*; II. Tris-citrate-borate-LiOH, pH 8.26–8.31 (Redfield & Salini, 1980) for *di-Pep*; III. Discontinuous Tris-citrate-borate, pH 8.2–8.7 (Poulik, 1957) for *Mep* and *Aat*; and IV. Tris-citrate-EDTA, pH 7.0 (Ayala et al., 1972) for *Idhp* and *Pgdh*. Staining procedures followed the protocols compiled by Murphy et al. (1990) with modifications. Genotypic data were analysed using the programmes BIOSYS-1 (Swofford & Selander, 1981) and FSTAT v. 1.2 (Goudet, 1995).

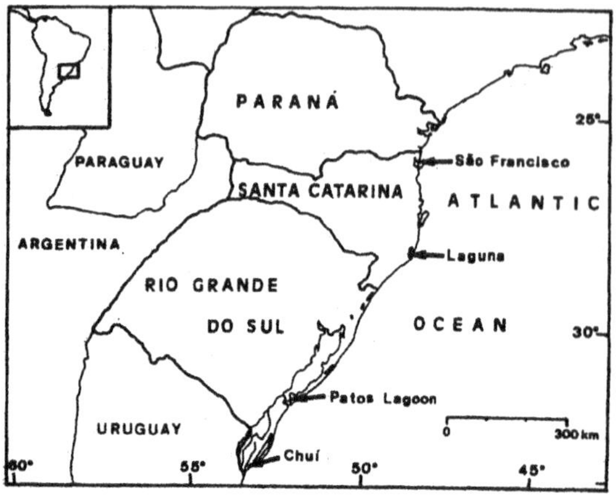

Figure 1. Callinectes danae. Sampling sites in southern Brazil, indicated by arrows.

The following information and analyses were obtained from BIOSYS-1: allele frequencies; variability measures (e.g. polymorphism (P), mean observed heterozygosity (H_o), unbiased mean expected under Hardy-Weinberg heterozygosity (H_e) (Nei, 1978)); χ^2 tests for the deviation from Hardy-Weinberg equilibrium; contingency χ^2 tests for the independence of allele frequencies between estuaries; Nei's (1978) unbiased genetic distance (D) between pairs of populations ($D = -\ln I$, where I is the Nei's (1978) genetic identity, a measure of genetic similarity); and UPGMA (unweighted pair-group method with arithmetic means) cluster analysis. The FSTAT program was used to obtain an estimator (θ) of the F statistic F_{ST} (Weir & Cockerham, 1984), which can be defined as the heterozygote deficiency among populations due to a Wahlund effect and represents a measure of genetic differentiation. This program also gives the standard error and the 99% confidence intervals of θ by Jackknife and Bootstrap procedures, respectively. The number of migrants per generation between pairs of populations (N_em) was also obtained by this program and by Slatkin's (1993) formula $N_e m = 1/(4F_{ST}) - 1/4$.

All significant tests ($P \leq 0.05$) obtained after a common Pearson χ^2 contingency table test for the independence of allele frequencies between localities (BIOSYS-1) were redone. Pooling was applied for the cases of expected frequencies less than five, and Yates correction (Cochran, 1954) was applied for all 2 × 2 tables. Then, χ^2_c and G_c were calculated (Sokal & Rohlf, 1995).

Results

Eleven loci gave useful results. The most common allele for each locus was the same in all localities. Allele frequencies and variability measures for all *Callinectes danae* samples are given in Table 1. The most variable samples were those from São Francisco (H_0=0.166 ± 0.037) and Chuí (H_0=0.180 ± 0.035). The least variable one was Laguna (H_0=0.071 ± 0.071). Genotype frequencies in all samples did not differ significantly from Hardy-Weinberg equilibrium with the exception of the Patos Lagoon sample, which showed disequilibrium at *Mep-1* and *Est-4* loci, due to a deficiency and an excess of heterozygotes, respectively. Contingency Chi-square tests showed significant independence of allele frequencies between all pairwise sample comparisons (Table 2). The most divergent subpopulations were São Francisco (SC) and Patos Lagoon (RS), and the most closely related were Patos Lagoon and Chuí (RS). UPGMA analysis based on Nei's (1978) genetic distance distinguished two groups of subpopulations, separating those of Santa Catarina State-1996 (São Francisco and Laguna) from those of Rio Grande do Sul-1997 (Patos Lagoon and Chuí) by a genetic distance, *D*, of 0.012 (see Figure 2). Genetic identities and distances between all pairs of populations are given in Table 3. Four out of eleven loci (*Pgdh*, *Idhp-1*, *Idhp-2*, *Est-3*) analysed did not contribute to differentiation between subpopulations. Populations of *C. danae* were moderately structured (θ=0.065±0.019; 99% confidence interval: 0.025–0.106) along Santa Catarina and Rio Grande do Sul States. The number of migrants (N_em) between pairs of subpopulations ranged from 2 to 7 individuals per generation (as shown in Table 4). No significant relationship was observed for the number of migrants per generation between pairs of subpopulations and the geographical distance separating them (P=0.292; r^2=0.269).

Discussion

The degree of population genetic structure found in *C. danae* suggests that the individuals from the four sites sampled represent distinct subpopulations (or genetic units) (Hastings & Harrison, 1994). *C. danae* depends on estuarine habitats for reproductive success, and the distribution of its subpopulations along the coast of southern Brazil appears to reflect this dependence.

Table 1. Allele frequencies and variability measures for all *Callinectes danae* samples

Locus	Populations			
	São Francisco	Laguna	Patos Lagoon	Chuí
Aat-2				
(*N*)	(56)	(78)	(80)	(75)
A	0.125	0.000	0.000	0.007
B	0.875	1.000	1.000	0.993
Mep-1				
(*N*)	(56)	(77)	(77)	(75)
A	0.071	0.045	0.162	0.200
B	0.027	0.000	0.000	0.000
C	0.902	0.955	0.838	0.800
Mep-2				
(*N*)	(56)	(79)	(80)	(48)
A	0.000	0.000	0.000	0.000
B	1.000	1.000	0.981	0.833
C	0.000	0.000	0.019	0.156
di-Pep-2				
(*N*)	(57)	(80)	(80)	(75)
A	0.096	0.006	0.000	0.100
B	0.009	0.000	0.000	0.000
C	0.895	0.994	0.994	0.867
D	0.000	0.000	0.006	0.033
Pgdh				
(*N*)	(57)	(78)	(79)	(72)
A	0.018	0.019	0.000	0.007
B	0.982	0.981	0.981	0.965
C	0.000	0.000	0.019	0.028
Idhp-1				
(*N*)	(53)	(80)	(80)	(75)
A	0.962	0.988	1.000	1.000
B	0.038	0.000	0.000	0.000
C	0.000	0.013	0.000	0.000
Idhp-2				
(*N*)	(53)	(80)	(80)	(73)
A	0.972	1.000	0.994	0.966
B	0.028	0.000	0.006	0.034
Est-2				
(*N*)	(57)	(80)	(80)	(75)
A	0.798	0.925	0.950	0.920
B	0.202	0.075	0.050	0.080
Est-3				
(*N*)	(57)	(80)	(80)	(75)
A	0.000	0.013	0.031	0.000
B	0.886	0.913	0.913	0.900
C	0.114	0.075	0.056	0.100

Continued on p. 206

Table 1. continued

Locus	Populations			
	São Francisco	Laguna	Patos Lagoon	Chuí
Est-4				
(*N*)	(57)	(80)	(80)	(75)
A	0.026	0.038	0.338	0.187
B	0.930	0.950	0.663	0.813
C	0.044	0.013	0.000	0.000
Est-5				
(*N*)	(57)	(80)	(63)	(35)
A	0.000	0.013	0.016	0.000
B	0.000	0.006	0.000	0.000
C	0.298	0.100	0.103	0.100
D	0.667	0.881	0.865	0.871
E	0.035	0.000	0.016	0.029
$Ho^{(a)}\pm$ Standard error	0.166± 0.037	0.071± 0.024	0.126± 0.052	0.180± 0.039
$He^{(b)}\pm$ Standard error	0.172± 0.041	0.071± 0.023	0.121± 0.044	0.169± 0.035

[a] Direct-count mean heterozygosity.
[b] Unbiased estimate of Hardy-Weinberg expected mean heterozygosity (Nei, 1978).

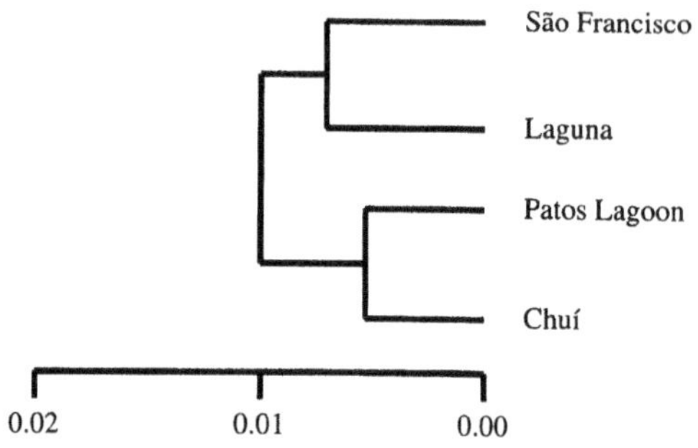

Figure 2. UPGMA cluster analysis of *Callinectes danae* subpopulations.

There is evidence indicating that *C. danae* has a life cycle similar to that of *C. sapidus*, developing in river inlets, with ovigerous females moving to spawn in the adjacent coast of estuaries (Pita et al., 1985b). There the larvae hatch and afterwards develop in the open sea, returning to nearby estuaries prior to metamorphosis to first crab (Schemy, 1980; Gaspar, 1981; Negreiros-Fransozo & Fransozo, 1995). The constant presence of ovigerous females of *C. danae* during certain seasons of the year inside estuaries (Negreiros-Fransozo & Fransozo, 1995; Garcia et al., 1996; pers. obs.) suggests that larval retention (not only post-larval retention) might be more frequent in this species than in *C. sapidus*.

Larval and post-larval retention may be important processes in preventing high levels of gene flow between estuaries in a species with a long larval dispersal phase and with low adult migration. Estuarine retention has been well studied for *C. sapidus* populations from the Atlantic coast of the United States and the Gulf of Mexico (Dittel & Epifanio, 1981; Wolf, 1981; McConnaughey & Sulkin, 1984; Sulkin & Epifanio, 1986; Sulkin & Van Heukelem, 1986; Epifanio et al., 1989; Johnson & Hester, 1989; Johnson & Hess, 1990; Brumbaugh & McConaughey, 1995; Johnson, 1995; Jones & Epifanio, 1995; Morgan et al., 1996), but unfortunately genetic data on the interchange of Atlantic U.S.A and Gulf of Mexico estuaries and adjacent coasts are contradictory (Cole & Morgan, 1978; Kordos & Burton, 1993; McMillen et al, 1994; Berthelemy-Okazaki & Okazaki, 1997).

Along the southern coast of Brazil, studies of invertebrate species with pelagic larvae have shown high homogeneity in allele frequencies (e.g. Weber et al., 1993; Delevedove, 1996; Levy, 1997). However, the species studied by those authors showed high levels of adult migration, which is clearly not the case for *C. sapidus* (Cargo, 1958; Tagatz, 1968; Norse, 1977) or *C. danae*. We should expect, as for molluscs (which tend to have a larval dispersal phase and low adult migration) that the degree of population differentiation along the geographical distribution of *C. danae* should be inversely related to the length of planktonic life. However, this is not always the case for crustaceans, which have shown high levels of genetic heterogeneity within a few kilometers (Tracey et al., 1975; Burton & Feldman, 1981; Knowlton & Keller, 1986; Weber et al., 2000). Unexpected levels of differentiation among crustacean populations have been attributed in some cases to the active role of larval and adult behavioral responses in their dispersal process (Burton & Feldman, 1981; Burton, 1983, 1986), to unexpected rapid larval settlement (Knowlton & Keller, 1986) and to hydrographic barriers (Tracey et al., 1975).

The duration of the planktonic larval phase in *C. danae* is not yet known. However, other closely related species of *Callinectes*, for which the timing of larval development has been determined, exhibit a general similarity in morphology and development (e.g. *C. similis sensu* Williams, 1974; and *C. ar-*

Table 2. Callinectes danae. Contingency table tests (χ^2, χ^2_c, and G_c) for the heterogeneity in allele frequencies for all pairwise comparisons. A sequential Bonferroni procedure for multiple test was applied, and results are shown for all loci that were significant ($P \leq 0.05$) for the common χ^2 test

Pairwise comparisons	Locus	Common Test			Tests with corrections					
		χ^2	df	*P*	χ^2_c	*P*	df	G_c	*P*	α'
S. Francisco & Laguna	*Aat-2*	20.58	1	0.0000	18.13	0.000	1	20.74	0.0000*	0.0050
	Est-5	25.96	4	0.0000	17.33	0.0000	1	17.18	0.0000*	0.0055
	DiPep-2	14.47	2	0.0007	12.33	0.0004	1	13.07	0.0003*	0.0062
	Est-2	9.60	1	0.0020	8.50	0.0036	1	8.38	0.0038*	0.0071
	Idhp-2	4.58	1	0.0324	2.39	0.1218	1	2.43	0.1189	0.0083
	Mep-1	5.09	2	0.0783	2.09	0.1487	1	2.05	0.1519	0.0100
	Idhp-1	7.40	2	0.0247	0.87	0.3496	1	0.85	0.3562	0.0125
S. Francisco & Patos Lagoon	*Est-4*	44.15	2	0.0000	25.67	0.0000	1	28.64	0.0000*	0.0045
	Aat-2	21.08	1	0.0000	18.60	0.0000	1	21.14	0.0000*	0.0050
	Est-2	15.28	1	0.0001	13.80	0.0002	1	13.77	0.0002*	0.0055
	Dipep-2	18.24	3	0.0004	12.33	0.0004	1	13.07	0.0003*	0.0062
	Est-5	17.38	3	0.0006	12.24	0.0005	1	12.40	0.0004*	0.0071
	Idhp-1	6.13	1	0.0133	3.85	0.0498	1	3.99	0.0457	0.0083
	Mep-1	8.75	2	0.0126	1.76	0.1842	1	1.81	0.1786	0.0100
	Est-3	6.38	2	0.0411	0.27	0.6025	1	0.27	0.6040	0.0125
S. Francisco & Chuí	*Mep-2*	20.22	2	0.0000	17.94	0.0000	1	21.53	0.0000*	0.0045
	Aat-2	16.64	1	0.0000	14.52	0.0001	1	15.70	0.0001*	0.0050
	Est-5	10.15	2	0.0063	8.51	0.0035	1	9.09	0.0026*	0.0055
	Est-2	8.35	1	0.0039	7.32	0.0068	1	7.28	0.0070	0.0062
	Est-4	21.78	2	0.0000	6.51	0.0107	1	6.90	0.0086	0.0071
	Idhp-1	5.75	1	0.0165	3.56	0.0593	1	3.75	0.0529	0.0083
	Mep-1	12.11	2	0.0023	3.56	0.0593	1	3.67	0.0554	0.0100
Laguna & Patos Lagoon	*Est-4*	48.60	2	0.0000	40.51	0.0000	1	44.28	0.0000*	0.0050
	Mep-1	11.30	1	0.0008	10.08	0.0015	1	10.56	0.0012*	0.0055
	Pgdh	6.00	2	0.0498	0.16	0.6918	1	0.16	0.6911	0.0062
Laguna & Chuí	*Mep-2*	28.10	2	0.0000	25.35	0.0000	1	27.81	0.0000*	0.0045
	DiPep-2	19.86	2	0.0000	17.84	0.0000	1	20.70	0.0000*	0.0050
	Mep-1	16.98	1	0.0000	15.56	0.0001	1	16.47	0.0000*	0.0055
	Est-4	19.22	2	0.0001	12.79	0.0003	1	13.31	0.0003*	0.0062
	Idhp-2	5.57	1	0.0183	3.64	0.0563	1	4.12	0.0425	0.0071
Patos Lagoon & Chuí	*DiPep-2*	20.28	2	0.0000	17.84	0.0000	1	20.70	0.0000*	0.0050
	Mep-2	19.22	2	0.0001	17.01	0.0000	1	16.82	0.0000*	0.0055
	Est-4	9.05	1	0.0026	8.29	0.0040	1	8.41	0.0037*	0.0062
	Est-3	6.62	2	0.0366	0.03	0.8551	1	0.03	0.8552	0.0071

$\alpha' = \alpha/(1+k\text{-i})$; *k*: total number of tests; *i*: rank of the values of *p* when ordered from the smallest (p_1) to the largest (p_k); α=0.05. When any $p_i \leq \alpha'$, then the corresponding test indicates significance at the 'table-wide' α level (Rice, 1989).

*Significant test.

Table 3. Callinectes danae. Nei (1978) unbiased genetic identity (above the diagonal) and genetic distances (below the diagonal) between pair of subpopulations

	S. Francisco	Laguna	Patos Lagoon	Chuí
S. Francisco	*****	0.993	0.982	0.987
Laguna	0.007	*****	0.991	0.994
Patos Lagoon	0.018	0.009	*****	0.995
Chuí	0.013	0.006	0.005	*****

Table 4. Pairwise number of migrants per generation for four *Callinectes danae* subpopulations

	Santa Catarina		Rio Grande do Sul	
	S. Francisco	Laguna	Patos Lagoon	Chuí
S. Francisco	****	3.87	2.32	4.09
Laguna		****	2.74	4.08
Patos Lagoon			****	7.64
Chuí				****

cuatus, Dittel & Epifanio, 1984). These have seven to eight zoeal stages and a megalopa, that may take from 1 to 3 months before reaching metamorphosis. We should expect, therefore, *C. danae* to show larval and post-larval development similar to its congeners. Recent studies have shown that the time from hatching to metamorphosis in *C. sapidus* depends on environmental conditions (e.g. salinity and temperature; Sulkin & Van Heukelen, 1986) and on the concentration in the water of chemical cues associated with estuaries (Forward et al., 1997). The absence of chemical cues associated with estuaries delays the time to metamorphosis in *C. sapidus* (e.g. when megalopae are still offshore; Wolcott & De Vries, 1994). Metamorphosis in *C. sapidus* is shortened by the presence of low salinity water (Wolcott & De Vries, 1994); by chemical cues from some species of aquatic vegetation (Forward et al., 1994, 1996; Brumbaugh & McConaughey, 1995); and by the presence of humic acids common in rivers and estuaries (Forward et al., 1997).

From the apparent potential for long distance larval dispersal of *Callinectes* species, a high level of gene flow might be expected between populations of *C. danae* inhabiting different estuaries separated by less than 550 km (Laguna– São Francisco do Sul; and Patos Lagoon – Chuí are less than 250 km; Laguna–Patos Lagoon is less than 550 km). The oceanographic conditions in the coast of Santa Catarina are mainly determined by the influence of the warm saline southward Brazil Current, while the oceanographic conditions along the coast of Rio Grande do Sul are mainly determined by the Brazil–Malvinas confluence, which represents the convergence of the Brazil Current with the cold and relatively fresh northward Malvinas (or Falkland) Current. The northern limit of this confluence moves seasonally from 35° –30° S during the winter to 40° –46° S during the summer (see Zavialov et al., 1998). Although the Brazil Current is greater during the Summer in Rio Grande do Sul (RS), Zavialov et al. (1998) found that during all seasons northward near surface currents dominate the continental shelf (200 km wide on the southernmost coast of RS to 100 km at nearly 31° S) with a maximum speed of 0.3 m s^{-1}, while offshore (off the continental shelf) the Brazil Current predominates with southward near-surface flow.

The transport of postlarvae (megalopae) from offshore back to the entrance of estuaries has been studied in *C. sapidus* and seems to be mainly accomplished by drift in surface and near-surface waters (Epifanio et al., 1984; Goodrich et al., 1989; Johnson & Hess, 1990). Considering the speed of near-surface currents over the shelf of Rio Grande do Sul and the distances separating the estuaries studied, we expect that larvae of *C. danae* should spend less than ten days travelling between adjacent estuaries (e.g. Patos Lagoon and Chuí in Rio Grande do Sul; Laguna and São Francisco in Santa Catarina) and less than 21 d between the furthest estuaries (e.g. Laguna and Patos Lagoon). Travelling these distances may be possible given the planktonic period of *Callinectes* species.

However, the genetic heterogeneity found in *C. danae* from different estuaries suggests that something associated with estuaries is restricting gene flow. The results are consistent with the assumption that estuaries are important areas for reproduction of*C. danae*, and may provide evidence of the importance of larval and postlarval retention processes in preventing panmixia between populations inhabiting different estuaries, particularly for those species that depend on estuarine conditions for completing their life cycles and where dispersal is mainly during the larval phase.

Although it is not yet clear whether estuarine retention is a passive or active process (Wolf, 1981), larval behavior or local recruitment appear to be the most plausible explanations for the observed levels of genetic heterogeneity in *C. danae*. Further studies of allele frequencies in *C. danae* recruits will be necessary to determine whether post-settlement selection acts to enhance local divergence (Hedgecock, 1981, 1986).

Acknowledgements

The authors thank CAPES (Coordenação de Aperfeiçoamento de Pessoal de Nível Superior) and FURG (Fundação Universidade do Rio Grande) for funding this research.

References

Abreu, J., 1980. Distribuição e ecologia dos decápoda numa área estuarina de Ubatuba (SP). Bol. Inst. oceanogr. USP 29: 1–3.

Ayala, F. J., J. R. Powell, M. L. Tracey, C. A. Mourão & S. Pérez-Salas, 1972. Enzyme variability in *Drosophila willistoni* group. IV. Genetic variation in natural populations of *Drosophila willistoni*. Genetica 70: 113–139.

Berthelemy-Okazaki, N. J. & R. K. Okazaki, 1977. Population Genetics of the blue crab *Callinectes sapidus* from northwestern Gulf of Mexico. Gulf Mex. Sci. 15: 35–39.

Brumbaugh, R. D. & J. R. McConaughey, 1995. Time to metamorphosis of the blue crab *Callinectes sapidus* megalopae: effects of benthic macroalgae. Mar. Ecol. Progr. Ser. 129: 113–118.

Burton, R. S., 1983. Protein polymorphisms and genetic differentiation of marine invertebrate populations. Mar. Biol. Lett. 4: 193–206.

Burton, R. S., 1986. Evolutionary consequences of restricted gene flow among natural populations of the copepod *Tigriopus californicus*. Bull. mar. Sci. 39: 526–535.

Burton, R. S. & M. W. Feldman, 1981. Population genetics of coastal and estuarine invertebrates: are behavioral differences among species reflected in population structure? Estuaries 4: 239.

Cargo, D. G., 1958. The migration of adult female of blue crabs, *Callinectes sapidus* Rathbun, in Chincoteague bay and adjacent waters. J. mar. Res. 16: 180–191.

Cochran, W. G., 1954. Some methods for strengthening the common χ^2 tests. Biometrics 10: 417–451.

Coelho, P. A. & M. A. Ramos, 1972. A constituição e a distribuição da fauna de decápodos do litoral leste da América do sul entre as latitudes 5° N e, 39° S. Trab. oceanogr. Univ. fed. Pe. 13: 133–236.

Coelho, P. A. & M. A. Ramos, 1980. Crustáceos decápodos da costa do Maranhão, Brasil. Bol. Inst. oceanogr. USP 29: 135–138.

Cole, M. A. & R. P. Morgan II, 1978. Genetic variation in two populations of blue crab, *Callinectes sapidus*. Estuaries 1: 202–205.

Delevedove, G. C., 1996. Genética populacional do camarão-rosa *Penaeus paulensis* (Decapoda: Penaeidae) na região estuarina da Lagoa dos Patos, Rio Grande do Sul, e no litoral de Santa Catarina. Unpublished M. Sc. Thesis. Fundação Universidade do Rio Grande, Rio Grande, Brazil.

Dittel, A. I. & C. E. Epifanio, 1981. Dispersal and recruitment of crab larvae in tropical estuaries. Estuaries 4: 238.

Dittel, A. I. & C. E. Epifanio, 1984. Growth and development of the portunid crab *Callinectes arcuatus* Ordway: zoeae, megalopae, and juveniles. J. crust. Biol. 4: 491–494.

Epifanio, C. E., A. K. Masse & R. W. Garvine, 1989. Transport of blue crab larvae by surface currents off Delaware Bay, U.S.A. Mar. Ecol. Progr. Ser. 54: 35–41.

Epifanio, C. E., C. C. Valenti & A. E. Pembroke, 1984. Dispersal and recruitment of blue crab larvae in the Delaware Bay, U.S.A. Estuar. coast. shelf. Sci. 18: 1–12.

Farias, M. C. Q., 1980. Crustáceos decápodes da Ilha da Restinga. Bol. Inst. oceanogr. USP. 29: 169–172.

Forward, R. B. Jr., M. C. DeVries, D. Rittschof, D. A. Z. Frankel, J. P. Bischoff, C. M. Fisher & J. M. Welch, 1996. Effects of environmental cues on metamorphosis of the blue crab *Callinectes sapidus*. Mar. Ecol. Prog. Ser. 131: 165–177.

Forward, R. B. Jr., D. A. Z. Frankel & D. Rittschof, 1994. Molting of megalopae from the blue crab *Callinectes sapidus*: effects of offshore and estuarine cues. Mar. Ecol. Prog. Ser. 113: 55–59.

Forward, R. B. Jr., R. A. Tankersley, D. Blondel & D. Ritschof, 1997. Metamorphosis of the blue crab *Callinectes sapidus*: effects of humic acids and ammonium. Mar. Ecol. Prog. Ser. 157: 277–286.

Garcia, A. M., J. P. Viera, C. E. Bemvenuti & R. M. Geraldi, 1996. Abundância e diversidade da assembléia de Crustáceos Decápodos dentro e fora de uma pradaria de *Rupia maritima* L., no estuário da Lagoa dos Patos (RS-Brasil). Nauplius 4: 113–128.

Gaspar, M. H., 1981. Contribuição ao estudo biológico do siri *Callinectes danae* Smith, 1869 (Decapoda, Portunidae) do rio Itiberê (Paranaguá-Paraná). Unpublished M.Sc. Thesis, Federal University of Paraná, Curitiba, Brazil.

Goodrich, D. M., J. Van Montfrans & R. J. Orth, 1989. Blue crab megalopal influx to the Chesapeake Bay: evidence for a wind-driven mechanism. Estuar. coast. shelf. sci. 29: 2447–260.

Goudet, J., 1995. FSTATv-1.2. A computer program to calculate F-statistics. J. Hered. 86: 485–486.

Hastings, A. & S. Harrison, 1994. Metapopulation dynamics and genetics. Ann. Rev. Ecol. Syst. 25: 167–188.

Hedgecock, D., 1981. Genetic consequences of larval retention: theoretical and methodological aspects. Estuaries 4: 239.

Hedgecock, D., 1986. Is gene flow from pelagic larval dispersal important in the adaptation and evolution of marine invertebrates? Bull. mar. Sci. 39: 550–564.

Johnson, D. F. & K. W. Hess, 1990. Numerical simulations of blue crab larvae dispersal and recruitment. Bull. mar. Sci. 46: 195–213.

Johnson, D. R., 1995. Wind forced surface currents at the entrance to Chesapeake Bay: their effect on blue crab larval dispersion and post-larval recruitment. Bull. mar. Sci. 57: 726–738.

Johnson, D. R. & B. S. Hester, 1989. Larval transport and its association with recruitment of blue crab to Chesapeake Bay. Estuar. coast. shelf. Sci. 28: 459–472.

Jones, M. B. & C. E. Epifanio, 1995. Settlement of brachyuran megalopae in Delaware Bay: an analysis of time series data. Mar. Ecol. Progr. Ser. 125: 67–76.

Knowlton, N. & B. D. Keller, 1986. Larvae which fall far short of their potential: highly localized recruitment in an alpheid shrimp with extended larval development. Bull. mar Sci. 39: 213–223.

Kordos, L. M. & R. S. Burton, 1993. Genetic differentiation of Texas Gulf populations of the blue crab *Callinectes sapidus*. Mar. Biol. 117: 227–233.

Levy, J. A., 1997. Genetic biochemical differentiation among marine organisms. In Seeliger, U., C. Odebrecht & J. P. Castello (eds), Subtropical Convergence Environments. The Coast and Sea in the Southwestern Atlantic Springer-Verlag, Berlin: 159–160.

McConnaughey, R. A. & S. D. Sulkin, 1984. Measuring the effects of thermoclines on the verical migration of larvae of *Callinectes sapidus* (Brachyura: Portunidae) in the laboratory. Mar. Biol. 81: 139–145.

McMillen-Jackson, A. L., T. M. Bert & P. Steele, 1994. Population genetics of the blue crab *Callinectes sapidus*: modest population structuring in a background of high gene flow. Mar. Biol. 118: 53–65.

Medeiros, M. F. M., 1982. Ocorrência do gênero *Callinectes* (Crustacea-Portunidae) com especial referência à *Callinectes* no estuário do Rio Sergipe, Brasil. Atlântica 5: 78–79.

Moreira, P. S., A. M. Paiva-Filho, C. M. Okida, J. M. M. Schmiegelow & R. Giannini, 1988. Bioecologia de crustáceos decápodos, braquiúros, no sistema baía-estuário de Santos e São Vicente, S.P. 1. Ocorrência e composição. Bol. Inst. oceanogr. USP 36: 55–62.

Morgan, S. G., R. K. Zimmer-Faust, K. L. Heck Jr. & L. D. Coen, 1996. Population regulation of blue crabs *Callinectes sapidus* in the northern Gulf of Mexico: postlarval supply. Mar. Ecol. Prog. Ser. 133: 73–88.

Murphy, R. W., J. W. Sites, D. G. Buth & C. H. Haufler, 1990. Proteins I: Isozymes Electrophoresis. In Hillis, D. M. & C. Moritz (eds), Molecular Systematics. Sinauer Associates, Sunderland, Massachusetts: 45–126.

Negreiros-Fransoze, M. L. & A. Fransozo, 1995. On the distribution of *Callinectes ornatus* ordway, 1863 and *Callinectes danae* Smith, 1869 (Brachyura, Portunidae) in the Fortaleza Bay, Ubatuba, Brazil. Iheringia Ser. Zool. P.A. 79: 13–25.

Nei, M., 1978. Estimation of average heterozygosity and genetic distance from a small number of individuals. Genetics 89: 583–590.

Norse, E. A., 1977. Aspects of the zoogeographic distribution of *Callinectes* (Brachyura: Portunidae). Bull. mar. Sci. 27: 440–447.

Pita, J. B., E. S. Rodrigues, G. Lopes & J. A. P. Coelho, 1985a. Levantamento da familia Portunidae (Crustacea, Decapoda, Bachyura) no complexo baía-estuário de Santos, São Paulo, Brasil. Bol. Inst. Pesca S.P. 12: 153–162.

Pita, J. B., E. S. Rodrigues, G. Lopes & J. A. P. Coelho, 1985b. Observaciones bioecológicas sobre o siri *Callinectes danae* Smith, 1869 (Crustacea, Portunidae), no complexo baía-estuário de Santos, Estado de São Paulo, Brasil. Bol. Inst. Pesca S.P. 12: 35–43.

Poulik, M. D., 1957. Starch gel electrophoresis in a discontinuous system of buffers. Nature 180: 1477–1479.

Powers, L. W., 1977. A catalogue and bibliography to the crabs (Brachyura) of the Gulf of México. Contrib. mar. Sci. 20: 1–190.

Rathbun, M., 1895. The genus *Callinectes*. Proc. U.S. nat. Mus. 18: 349–375.

Rathbun, M., 1900. The Brachyura and Macrura of Porto Rico. Bull. U.S. Fish. Comm. 20: 1–127.

Rathbun, M., 1930. The cancroid crabs of America of the families Euryalidae, Portunidae, Atelecyclidae, Cancridae and Xanthidae. U.S. natl. Mus. Bull. 152: 1–609.

Redfield, J. A. & J. P. Salini, 1980. Techniques of Starch-gel electrophoresis of penaeid prawn enzymes (*Penaeus* spp. and *Metapenaeus* spp.). CSIRO Aust. Div. Fish. Oceanogr. Rep. 166: 1–19.

Rice, W. R., 1989. Analysing tables of statistical tests. Evolution 43: 223–225.

Sawaya, P. & R. S. Pereira, 1946. Nota sobre a ecologia de alguns crustáceos decápodos marinhos de São Paulo. Bol. Fac. Filos. Ciênc. Univ. S.P., Zool. 11: 383–392.

Shaklee, J. B. & C. P. Keenan, 1986. A practical guide to the techniques and methodology and its application to fish fillet identification. CSIRO mar. Labs. Rep. 177: 1–59.

Schemy, R. A., 1980. Aspectos da biologia de *Callinectes danae* (Smith, 1869) da região de Santos, São Paulo. Unpublished M. Sc. Thesis. São Paulo University, São Paulo, Brazil.

Shumway, S. E., 1983. Oxygen consumption and salinity tolerance in four Brazilian crabs. Crustaceana 44: 76–82.

Slatkin, M., 1993. Isolation by distance in equilibrium and non-equilibrium populations. Evolution 47: 264–279.

Sokal, R. R. & F. J. Rohlf, 1995. Biometry. W.H. Freeman, San Francisco.

Sulkin, S. D. & C. E. Epifanio, 1986. Natural regulation of juvenile recruitment in the blue crab (*Callinectes sapidus* Rathbun) and its consequences for sampling and management strategy. Can. sp. Publ. Fish aquat. Sci. 92: 117–123.

Sulkin, S. D. & W. Van Heukelem, 1981. Redefining estuarine retention: factors regulating retention within the estuary versus off-shore recruitment of larvae of the blue crab *Callinectes sapidus*. Estuaries 4: 239.

Sulkin, S. D. & W. Van Heukelem, 1982. Larval retention in the crab *Callinectes sapidus* Rathbun: an amendment to the concept of larval retention in estuaries. In Kennedy, V. S. (ed.), Estuarine Comparisons Academic Press, New York: 459–475.

Sulkin, S. D. & W. Van Heukelem, 1986. Variability in the length of the megalopal stage and its consequence to dispersal and recruitment in the portunid crab *Callinectes sapidus* Rathbun. Bull. mar. Sci. 39: 269–278.

Swofford, D. L. & R. B. Selander, 1981. BIOSYS-I. A program for the comprehensive analysis of electrophoretic data in population genetics and systematics. J. Hered. 72: 281–283.

Tagatz, M. E., 1968. Biology of the blue crab *Callinectes sapidus* Rathbun in the St. John's River, Florida. Fish. Bull. U.S. 67: 17–33.

Tracey, M. L., K. Nelson, D. Hedgecock, R. A. Schleser & M. L. Pressick, 1975. Biochemical genetics of lobsters: genetic variation and the structure of American lobster (*Homarus americanus*) populations. J. Fish Res. Bd Can. 32: 2091–2101.

Weber, L. I., M. B. Conceição, M. L. Ruffino & J. A. Levy, 1993. Population genetics of the shrimp *Artemesia longinaris* (Crustacea: Penaeidae) on the south-west Atlantic coast. Comp. Biochem. Physiol. 106B: 1015–1020.

Weber, L. I., R. G. Hartnoll & J. P. Thorpe, 2000. Genetic divergence and larval dispersal in two spider crabs (Crustacea: Decapoda). In Solé-Cava, A. M., C. A. M. Russo & J. P. Thorpe (eds), Marine Genetics. Developments in Hydrobiology 144. Kluwer Academic Publishers, Dordrecht: 211–219. Reprinted from Hydrobiologia 420.

Weir, B. S. & C. C. Cockerham, 1984. Estimating F-statistics for the analysis of population structure. Evolution 38: 1358–1370.

Williams, A. B., 1974. The swimming crabs of the genus *Callinectes* (Decapoda, Portunidae). Fish. Bull. U.S.A. 72: 685–798.

Wolcott, D. L. & M. C. De Vries, 1994. Offshore megalopae of *Callinectes sapidus*: depth of collection, molt stage and response to estuarine cues. Mar. Ecol. Prog. Ser. 109: 157–163.

Wolf, P., 1981. Is retention the result of active or passive phenomena? Estuaries 4: 239.

Zavialov, P. O., R. D. Ghisolfi & C. A. E. Garcia, 1998. An inverse model for seasonal circulation over the Southern Brazilian Shelf: near-surface velocity from the heat budget. J. phys. Oceanogr. 28: 545–562.

Hydrobiologia **420:** 211–219, 2000.
A. M. Solé-Cava, C. A. M. Russo & J. P. Thorpe (eds), Marine Genetics.

Genetic divergence and larval dispersal in two spider crabs (Crustacea: Decapoda)

L. I. Weber[1,2], R. G. Hartnoll[1] & J. P. Thorpe[1]
[1]*School of Biological Sciences, The University of Liverpool, Port Erin Marine Laboratory, Port Erin, Isle of Man, IM9 6JA, U.K.*
[2]*Present Address: Fundação Universidade Federal do Rio Grande, Dpto, Fisiologia, C.P. 474, CEP 96201-900, Rio Grande, RS, Brazil*

Key words: Crustacea, Majidae, allozymes, gene flow, larval dispersal, *Inachus*, *Hyas*

Abstract

The spider crabs *Inachus dorsettensis* (Pennant) and *Hyas coarctatus* Leach are widespread in subtidal areas of muddy sand or gravel around western Europe. Both species have a life cycle with an obligatory planktonic larval phase of several weeks, which might be expected to cause widespread larval dispersal and consequent genetic homogeneity over considerable distances. However, earlier work on both taxa has indicated differences in growth pattern between populations separated by tens of kilometres. This study was undertaken to determine whether these differences were purely environmental or whether, despite the short distances involved, differences may have a genetic basis. A study of gene frequencies, as indicated by allozymes in samples of adults collected off the Isle of Man (northern Irish Sea), indicates significant genetic differentiation between populations over a geographical distance of only about 40 km in both *Inachus dorsettensis* (θ = 0.086 ± 0.048) and *Hyas coarctatus* (θ = 0.023 ± 0.017). Variability measures differed between species, showing *I. dorsettensis* to have a mean number of alleles per locus of 2.5–2.6 and a range of gene diversity of 0.216–0.241, while *H. coarctatus* showed lower values of mean number of alleles (1.9–2.0) and a range of gene diversity from 0.122 to 0.124. Given the high expected larval mobility of the two species the results are most surprising. Possible explanations are discussed in relation to population discontinuities and patterns of larval drift.

Introduction

Inachus dorsettensis (Pennant) and *Hyas coarctatus* Leach are two of the most common spider crabs (Majidae) in the north east Atlantic. In British waters, both species are widespread from the Channel Islands and Scillies north to the Shetlands (Ingle, 1983), but they differ in overall geographical distribution. *H. coarctatus* is an arctic-boreal species near its southern limit around Britain, whereas *I. dorsettensis* is a warm temperate species which extends little further north than Britain (Christiansen, 1969). Both occur on a wide range of substrata from the shallow subtidal to about 200 m. Because they are common, they are of ecological significance and have featured in numerous studies of subtidal faunas over many years (e.g. Beaumont, 1900; Sinel, 1907; Jones, 1956; Allen, 1967; Thurston, 1970; Hartley & Bishop, 1986). There are differences in ecological preference between the two species. *H. coarctatus* is widespread in areas of muddy sand and on coarser substrata, whilst *I. dorsettensis* is the commonest species on muddy sand, but is scarce on harder bottoms (Hartnoll, 1963).

The two species are found in many areas around the Isle of Man (Bruce et al., 1963), where they have been the subject of major investigations of their general biology (Hartnoll, 1963) and of their breeding strategies (Bryant & Hartnoll, 1995). Hartnoll et al. (1993) made a detailed study of size distributions in populations of each species from two different areas. This work confirmed earlier indications that in each species there were marked differences in growth patterns, and specifically in size range at the terminal moult, between populations separated by geographical distance of 40 km or less. The authors hypothesised that these differences could either be under genetic control, or could result from differing environments.

There are strong tidal currents around the Isle of Man, and since the larval life is thought to be at least 80 d in *H. coarctatus* (Anger, 1984) and 30 d in *I. dorsettensis* (in summer, longer in winter, see Lebour, 1927), genetic separation over tens of kilometres would seem intuitively improbable. However, if the size differences are environmental in origin the factor responsible is not obvious. An alternative, if improbable, possibility which needs to be eliminated is that sibling species (now known from many genetic studies of marine invertebrates; Knowlton, 1993; Thorpe & Solé-Cava, 1994) may be involved. The aim of the present work was to assess levels of genetic differentiation between two populations of the two spider crab species studied by Hartnoll et al. (1993). This was done partly to establish whether growth differences might be genetic, but also because the two species provided a good opportunity to study the genetic consequences of larval dispersal in species with known larval durations and with adults having negligible powers of dispersal.

Previous similar work on a range of invertebrates has shown that actual genetic dispersal is often poorly correlated with apparent dispersal capability. In particular, several studies have demonstrated genetic differences over far shorter distances than might be expected from a knowledge of the larval planktonic period. This in turn has caused some authors (e.g. Strathmann, 1985; Todd et al., 1988) to question the previously widely held assumption (reviews by Jablonski & Lutz, 1983; Todd, 1985) that most marine invertebrate larvae have evolved for purposes of dispersal.

The technique chosen was enzyme electrophoresis, a method which has been commonly used to investigate the genetic structure of natural populations in a wide range of plant and animal species (reviewed by e.g. Nei, 1987); including quantifying genetic divergence between populations of many marine invertebrates (e.g. Macleod et al., 1985; Todd et al., 1988). Genetic divergence in Crustacea has been reviewed by Hedgecock et al. (1982; see also Hedgecock, 1986).

Materials and methods

Samples of the spider crabs *Inachus dorsettensis* and *Hyas coarctatus* were collected by dredging from the

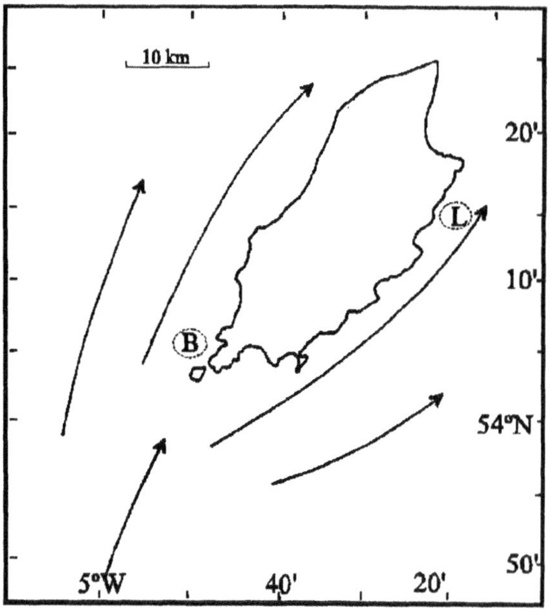

Figure 1. Map of the Isle of Man to show sampling sites at Bradda (B) and Laxey (L) from which spider crabs were collected. The general flow of residual surface water currents is also shown (based on Ramster & Hill, 1969).

University of Liverpool research vessel *Roagan* from two areas off the coast of the Isle of Man (see Figure 1). The areas selected were about 2 km east of Laxey (approximate position 54° 14′ N, 4° 24′ W) and about 4 km west of Bradda Head (approximate position 54° 5′ N, 4° 45′ W). Depths were about 20 m and about 30 m, respectively. These collection areas are subsequently referred to as 'Laxey' and 'Bradda'. Collections were made on various dates between July 1994 and January 1995. Samples were stored in an ultra deep freeze (−75°C) until required for electrophoresis.

Horizontal 12.5% starch-gel electrophoresis was carried out using standard procedures (Harris & Hopkinson, 1976; Murphy et al., 1990 with modifications). Homogenates were obtained by macerating whole chelipeds in 0.02 M Tris–HCl, pH 8.0 buffer and centrifuging for 4 min at 10 000 rpm. Buffer systems used were discontinuous Tris-Citrate-Borate (Poulik, 1957), Tris-Citrate, pH 8.0 (Siciliano & Shaw, 1976), and Tris-Citrate-Borate-LiOH (Redfield & Salini, 1980). The enzymes, their E.C. numbers, buffer systems, and number of loci scored are shown in Table 1.

Most statistical analyses of the data were carried out using the programme BIOSYS-1 (Swofford &

Table 1. Inachus dorsettensis and *Hyas coarctatus*. Enzymes analysed, their E.C. number, buffer system used, and number of loci scored for both species. Buffer systems used are: 1. discontinuous Tris-Citrate-Borate (Poulik, 1957), 2. Tris-Citrate, pH 8.0 (Siciliano & Shaw, 1976), and 3. Tris-Citrate-Borate-LiOH (Redfield & Salini, 1980)

(Abbreviation)	Enzyme Number	Buffer System	Number of loci	
			Inachus dorsettensis	*Hyas coarctatus*
Aspartate aminotransferase (*Aat*)	2.6.1.1	1	2	2
Esterase (*Est*)	3.1.1.-	1	2	2
Glucose-6-phosphate isomerase (*Gpi*)	5.3.1.9	2	1	1
Isocitrate dehydrogenase (*Idhp*)	1.1.1.42	2	1	1
l-Lactate dehydrogenase (*Ldh*)	1.1.1.27	2	1	1
Malate dehydrogenase(*Mdh*)	1.1.1.37	2	1	2
Malic Enzyme (*Mep*)	1.1.1.40	1	1	2
Mannose-6-phosphate isomerase (*Mpi*)	5.3.1.8	3	1	1
Phosphogluconate dehydrogenase (*Pgdh*)	1.1.1.44	2	1	1
Phosphoglucomutase (*Pgm*)	5.4.2.2	2	2	1
Superoxide dismutase (*Sod*)	1.15.1.1	1	2	2

Selander, 1981). Mean observed heterozygosity per locus (H_o) was estimated by direct count and mean expected heterozygosity per locus (H_e) was from Nei's (1978) unbiased estimate of the mean heterozygosity expected under Hardy-Weinberg equilibrium. Tests for the deviation from Hardy-Weinberg equilibrium at each locus and sample were performed using Levene's correction for small sample size and using exact probabilities (Swofford & Selander, 1981). For each species, a contingency χ^2 test for independence in allele frequencies between Laxey and Bradda populations was also done.

Nei's (1978) unbiased genetic identity, *I*, and another measure of the degree of genetic differentiation between populations, the standardised variance across populations, F_{ST} (Wright, 1978; Nei, 1987) were also calculated using the programme BIOSYS-1 (Swofford & Selander, 1981). An estimator of F_{ST} (θ) over all loci and its standard deviation were obtained by FSTAT v.1.2 (Goudet, 1995) from the formula given by Weir & Cockerham (1984); and estimations of the number of migrants per generation were calculated from Slatkin's (1993) formula $N_e m = 1/(4F_{ST}) - 1/4$, where F_{ST} and its estimator Theta were used.

F_{ST} values, obtained by BIOSYS-1, were tested by $\chi^2 = 2NF_{ST}\ (k-1)$ (Workman & Niswander, 1970) for H_0: $F_{ST} = 0$, with $(k-1)(s-1)$ degrees of freedom, where *k* is the number of alleles, *N* is the sample size, and *S* is the number of populations in comparison.

Sequential Bonferroni procedure for multiple tests (Rice, 1989) was used for the analysis of the deviation from Hardy-Weinberg equilibrium, for testing independence in allele frequencies between populations, and for the significance of F_{ST} over all loci.

Results

Eleven enzymes were successfully resolved in each species; these gave a total of 15 loci for *I. dorsettensis* and 16 loci for *H. coarctatus*. Allele frequencies, sample sizes and variability measures for Bradda and Laxey samples of both species are given in Table 2. *I. dorsettensis* showed higher values of genetic variability than *H. coarctatus* (Table 2). The only significant deviation (after Bonferroni procedure) from Hardy-Weinberg equilibrium was observed in the Laxey population of *I. dorsettensis,* which showed a significant deficiency of heterozygotes at the *Sod-2* locus. Variability measures differed between species, showing a mean number of alleles per locus of 2.5–2.6 *I. dorsettensis* and a range of gene diversity of 0.216–0.241, while *H. coarctatus* showed lower values of mean number of alleles (1.9–2.0) and a range of gene diversity from 0.122 to 0.124.

Contingency χ^2 analysis for both species showed that Bradda and Laxey populations differed significantly in their allele frequencies (Table 3). Only one locus (*Pgm-2*) contributed to this differentiation in *H. coarctatus*, while in *I. dorsettensis* five loci (*Est-1, Mep-1, Pgdh, Pgm-2, Sod-2*) showed significant

Table 2. Allele frequencies and variability for Bradda and Laxey populations of *Inachus dorsettensis* and *Hyas coarctatus*. Note that mobilities of allozymes (designated from A to G) were not compared between species

Locus	*Inachus dorsettensis*		*Hyas coarctatus*	
	Bradda	Laxey	Bradda	Laxey
Aat-1				
(N)	(57)	(57)	(42)	(35)
A	1.000	1.000	0.024	0.014
B	—	—	0.952	0.986
C	—	—	0.024	0.000
Aat-2				
(N)	(57)	(57)	(42)	(35)
A	0.044	0.018	1.000	1.000
B	0.956	0.974	—	—
C	0.000	0.009	—	—
Est-1				
(N)	(57)	(57)	(42)	(35)
A	0.000	0.070	0.000	0.057
B	0.781	0.904	0.940	0.800
C	0.184	0.026	0.048	0.143
D	0.035	0.000	0.012	0.000
Est-3				
(N)	(57)	(57)	(42)	(35)
A	1.000	1.000	1.000	1.000
Gpi				
(N)	(57)	(57)	(42)	(35)
A	0.044	0.018	0.012	0.014
B	0.044	0.018	0.988	0.986
C	0.289	0.307	—	—
D	0.000	0.018	—	—
E	0.263	0.289	—	—
F	0.342	0.351	—	—
G	0.018	0.000	—	—
Idhp-1				
(N)			(42)	(35)
A	—	—	0.012	0.000
B	—	—	0.060	0.014
C	—	—	0.917	0.971
D	—	—	0.012	0.000
E	—	—	0.000	0.014
Idhp-2				
(N)	(33)	(35)		
A	1.000	1.000	—	—

Table 2. contd.

Locus	*Inachus dorsettensis*		*Hyas coarctatus*	
	Bradda	Laxey	Bradda	Laxey
Ldh				
(N)	(33)	(56)	(42)	(35)
A	0.727	0.554	1.000	1.000
B	0.273	0.438	—	—
C	0.000	0.009	—	—
Mdh-1				
(N)	(57)	(57)	(42)	(35)
A	0.009	0.000	1.000	0.986
B	0.009	0.018	0.000	0.014
C	0.026	0.000	—	—
D	0.956	0.974	—	—
E	0.000	0.009	—	—
Mdh-2				
(N)			(42)	(35)
A	—	—	0.345	0.157
B	—	—	0.655	0.843
Mep-1				
(N)	(49)	(57)	(42)	(35)
A	0.112	0.412	1.000	1.000
B	0.888	0.588	—	—
Mep-2				
(N)			(42)	(35)
A	—	—	0.667	0.743
B	—	—	0.333	0.257
Mpi				
(N)	(57)	(57)	(42)	(35)
A	0.974	1.000	1.000	1.000
B	0.026	0.000	—	—
Pgdh				
(N)	(42)	(55)	(42)	(35)
A	0.262	0.091	0.000	0.043
B	0.738	0.900	1.000	0.957
C	0.000	0.009	—	—
Pgm-1				
(N)	(57)	(57)		
A	0.000	0.018	—	—
B	0.026	0.044	—	—
C	0.974	0.939	—	—
Pgm-2				
(N)	(57)	(57)	(42)	(35)
A	0.070	0.000	0.000	0.014

Continued on p. 215

Table 2. contd.

Locus	*Inachus dorsettensis*		*Hyas coarctatus*	
	Bradda	Laxey	Bradda	Laxey
B	0.175	0.096	0.012	0.157
C	0.474	0.404	0.845	0.786
D	0.219	0.377	0.095	0.029
E	0.061	0.123	0.048	0.000
F	—	—	0.000	0.014
Sod-1				
(N)	(57)	(57)	(42)	(35)
A	0.026	0.044	0.000	0.014
B	0.974	0.956	0.238	0.214
C	—	—	0.762	0.771
Sod-2				
(N)	(57)	(57)	(42)	(35)
A	0.000	0.070	0.976	1.000
B	0.930	0.491	0.024	0.000
C	0.070	0.439	—	—
$P_{0.95}$	46.7	53.3	37.5	31.3
Ho (SE)	0.179 (0.057)	0.189 (0.059)	0.106 (0.035)	0.102 (0.032)
He (SE)	0.216 (0.063)	0.241 (0.069)	0.124 (0.042)	0.122 (0.040)

$P_{0.95}$= proportion of polymorphic loci when the frequency of the most allele does not exceed 0.95; H_o (SE)= observed (direct count) mean heterozygosity per locus with standard error in parentheses; H_e (SE)= unbiased mean expected heterozygosity (Nei, 1978) under Hardy-Weinberg equilibrium with standard error in parenthesis.

Table 3. Inachus dorsettensis and *Hyas coarctatus*. Summary of genetic differentiation between Bradda and Laxey populations. Data are for polymorphic loci only. Significant *F*-statistics, F_{ST} (Workman & Niswander (1970) test) are indicated by asterisks. Significant χ^2 values for the independence of allele frequencies (contingency χ^2 table) between Laxey and Bradda populations are also indicated by asterisks. Significance was determined after Bonferroni correction: ∗= $P<0.05$, ∗∗= $P<0.001$, ∗∗∗= $P<0.0001$

Locus	*Inachus dorsettensis*		*Hyas coarctatus*	
	F_{ST}	χ^2	F_{ST}	χ^2
Aat-1	–	–	0.007	1.88
Aat-2	0.004	2.30	–	–
Est-1	0.041∗∗∗	26.52∗∗∗	0.035∗	10.30
Gpi	0.001	6.79	0.000	0.02
Idhp-1	–	–	0.013	4.99
Ldh	0.031∗	5.61	–	–
Mdh-1	0.004	5.35	0.007	1.21
Mdh-2	–	–	0.047	7.02
Mep-1	0.116∗∗∗	23.87∗∗∗	–	–
Mep-2	–	–	0.007	1.06
Mpi	0.013	3.04	–	–
Pgdh	0.047∗∗	10.71∗	0.022	3.67
Pgm-1	0.005	2.57	–	–
Pgm-2	0.016∗	18.35∗	0.025∗	18.85∗
Sod-1	0.002	0.51	0.001	1.30
Sod-2	0.194∗∗∗	53.84∗∗∗	0.012	1.69
θ (over all loci)	0.086 ± 0.048		0.023 ± 0.017	

differences in allele frequencies between Bradda and Laxey.

The unbiased genetic identity (Nei, 1978) between Bradda and Laxey populations was lower in *I. dorsettensis* ($I = 0.972$) than in *H. coarctatus* ($I = 0.996$). The standardised variances across populations (F_{ST} and its estimator θ) were similarly higher in *I. dorsettensis* ($\theta = 0.086 \pm 0.048$) than in *H. coarctatus* ($\theta = 0.023 \pm 0.017$) (see Table 3), giving estimates of mean number of migrants per generation (N_em) of 2.6 (derived from θ) to 4.6 (derived from F_{ST}) between Laxey and Bradda populations of *I. dorsettensis*, while *H. coarctatus* gave estimates of 10.6–12.2 migrants per generation.

Discussion

There appear to be rather few published studies of the effects of pelagic larvae on gene flow in Crustacea. Several authors have provided data indicating that, as might be expected, pelagic larvae mediate widespread gene flow with little or no genetic differentiation over large distances (e.g. barnacles *Elminius*, Flowerdew, 1984; *Balanus*, Hedgecock, 1986; *Chthamalus*, Pannacciulli et al., 1997; decapods *Penaeus*, Mulley & Latter, 1981; *Artemesia*, Weber et al., 1993; *Panulirus*, Silberman & Walsh, 1994). Indeed, in some talitrids, little genetic differentiation with distance has been found even though there is no larval stage to aid dispersal (Bulnhein & Scholl, 1986; Conceição, 1995). However, in a few crustacean species, data indicate a restriction in gene flow despite an apparently dispersive pelagic larva. Knowlton & Keller (1986) concluded that very low dispersal in a Caribbean shrimp was a result of unexpectedly rapid larval settlement and in *Homarus americanus*, which showed genetic discontinuity despite a long-lived larva, differences were ascribed to an oceanic barrier (Tracey et al., 1975).

From the results of the present work, it is clear that between Bradda and Laxey, a distance by sea of only about 40 km, there is significant differentiation between populations of both *I. dorsettensis* and *H. coarctatus*. This differentiation indicates that in each species there is likely to be little or no gene flow between the two areas. Models of gene flow between populations vary greatly in their assumptions, but, from almost all models, a general conclusion is that (in the absence of abnormally strong selective forces), practically any level of migration, however small, will be sufficient to preclude genetic divergence (for a brief discussion of relevant models see e.g. Maynard Smith, 1989; detailed review of gene flow between populations by Slatkin, 1985, 1987). Thus, the differences at the locus *Pgm-2* in *H. coarctatus* and *Est-1*, *Mep-1*, *Pgdh*, *Pgm-2* and *Sod-2* in *I. dorsettensis* are only likely to have developed in the absence of effective gene flow. The overall levels of genetic differentiation are not great, but this is not surprising since, firstly, effective population sizes are probably large (both species are abundant over wide areas), which would have the effect of reducing rates of stochastic differentiation (genetic drift), and secondly, the geologically recent nature of the sea area concerned (since the last ice age, perhaps 10 000 years bp) has given little time for extensive differentiation.

Thus, in the two crab species, the interpopulational genetic variation indicates that differences in growth patterns described by Hartnoll et al. (1993) are likely to have a genetic basis, although some environmental effects are also possible. However, the high overall genetic identity values between populations mean that cryptic speciation (not uncommon in marine invertebrates, Knowlton, 1993; Thorpe & Solé-Cava, 1994) can be ruled out, since identity values between even closely related congeneric species are usually considerably lower (see Thorpe, 1982; Nei, 1987). The values obtained are well within the range (about 0.85–1.0) typical of allopatric conspecific populations.

The really notable feature of the results is that the observed levels of genetic differentiation between populations occur in species with well-studied larvae with a long and obligatory larval phase. As mentioned above, such results have implications for understanding the reasons why so many marine invertebrates have planktonic larvae. For many years, numerous authors have considered that the primary role of pelagic larvae is to provide a dispersive stage in the life cycle (reviews by e.g. Jablonski & Lutz, 1983; Grahame & Branch, 1985; Todd, 1985). This hypothesis has, however, been questioned (e.g. Strathmann, 1980, 1985; Todd et al., 1988). Based on mainly more recent genetic data it is clear that, whilst many species do apparently disperse widely, in others the intuitively apparent larval dispersal potential is not necessarily realised under natural conditions. One of the most striking examples is the work of Todd et al. (1988, 1993) who studied genetic differentiation between populations of the semelparous annual nudibranch *Adalaria proxima* in northern Britain. This species has a pelagic lecithotrophic larval stage, which (in laboratory studies) may remain metamorphically competent for several weeks. Contrary to expectations, *A. proxima* showed substantial temporally-stable differences in allele frequencies, indicating little or no gene flow, over distances as short as a few kilometres. Several other similar cases are known, but for many of these a plausible explanation for the limited dispersal is that, whereas in artificial conditions larvae may continue to swim around for extended periods, under natural conditions they mostly settle far more rapidly.

However, such explanations cannot hold for *Inachus dorsettensis* and *Hyas coarctatus* since both have long obligatory larval stages during which dispersal would be expected. Both species hatch from brooded eggs to form the first of two zoeal instars. These are followed by a megalopa, which is the settlement stage and the final larval instar before the larvae becomes a tiny crab. This then undergoes several prepubertal moults until it becomes an adult following the final ecdysis. As with all crabs, each larval instar changes into the next through moulting and concomitant growth. Both zoeal stages eat phytoplankton in the laboratory and plankton haul data indicate that the zoeae are confined to the top several meters of the water column, within which dispersal would be difficult to avoid. The megalopa, at least initially, is planktonic also (it is frequently found in plankton hauls), but presumably becomes benthic at or before metamorphosis (Crane, 1940; Hartnoll, 1992). The subsequent first crab stage is benthic and, indeed, is unable to swim. The necessity to go through the successive larval stages and to be able to grow as development proceeds results in a prolonged obligatory planktonic phase the duration of which is dictated largely by temperature and food supply. The main period for larval release in the seasonally breeding *H. coarctatus* is March–April, shortly after the water temperature has reached its annual low (Hartnoll, 1963). This low temperature slows larval development so that the duration from hatching to metamorphosis of the megalopa to the first crab is at least 80 days (Anger, 1984). In contrast, *I. dorsettensis* breeds throughout the year (Hartnoll, 1963), and larvae hatching in the summer and autumn, when water temperatures are close to the annual maximum, can have larval duration as short as 30 d (Lebour, 1927).

Explanations of genetic differentiation in spider crabs as being a result of small scale genetic patchiness are improbable since this would probably necessitate swarms of related larvae remaining together throughout the planktonic phase and settling together, with little or no subsequent mixing of adults. This is unlikely, particularly in species which are widely distributed over substantial areas.

The two areas from which samples were taken are on different sides of the Isle of Man and the residual surface currents (mainly north) are such that larvae may well not be transported directly from Bradda to Laxey or *vice versa*. However, the wide distributions of both species around the south of the Isle of Man (Bruce et al., 1963; Hartnoll, 1963) should ensure that at least the more southerly populations have gene flow to all of those further north on either side of the Island. Indeed, the generally strong tidal currents should enable larval transport from the south of the Isle of Man to Bradda or Laxey to occur possibly in hours rather than days. Any such transport should preclude genetic differentiation between populations off the south of the Isle of Man and either Bradda or Laxey and hence there should be no genetic differentiation between populations at Bradda or Laxey.

Although only two species have been studied, the species with the greater apparent potential for larval dispersal (*Hyas coarctatus*) is that with the lower genetic differentiation between populations. Thus, this aspect of the results is as would be predicted; it is only the geographical scale over which differentiation occurs which is unexpected. It should also be noted that the general distribution of populations of each species may favour wider gene dispersal in *H. coarctatus*. This species occurs generally on coarse grounds, as well as on the muddy sand areas sampled. The sea bed on the eastern side of the Isle of Man between its southern tip and Laxey is predominantly coarse (Jones, 1951), so there is likely to be more or less continuous distribution of *H. coarctatus* between the Bradda and Laxey grounds. In contrast, *I. dorsettensis* is not common on coarse grounds and is likely to be relatively scarce on the east side except for isolated pockets of abundance on areas of suitable substratum. A continuous (as in *H. coarctatus*) rather than disjunct (as in *I. dorsettensis*) distribution is more likely to favour gene flow.

Apparently, similar differentiation between populations on the east and west of the Isle of Man has been observed in a genetic study of queen scallop (*Aequipecten opercularis*) populations (Lewis & Thorpe, 1994a,b). This species also is thought to have a larval life of a few weeks. However, the genetic differences observed (mostly only significant over very large sample sizes) are much smaller than those in the spider crabs and, as with cases mentioned above, the duration of the larval life, in the absence of any fixed series of instars, is difficult to establish beyond doubt.

Whittington (1993), studying settlement of various invertebrate species on artificial substrata, has observed differences in recruitment between areas east and west of the Isle of Man. However, the data are based on a single relatively short period (about 3 months) of observation and interpretation is difficult.

As mentioned above, an oceanographic barrier has been hypothesised to cause the observed genetic discontinuity in population structure in *Homarus americanus* along the east coast of North America (Tracey et al., 1975). In most summers, an oceanographic front forms to the west of the Isle of Man (Simpson, 1971; Pingree & Griffiths, 1978), but this is not likely to preclude the movement of larvae because inshore tides move rapidly past the coastal areas affected. Furthermore, one species breeds in late winter (before the formation of the front) and the other breeds throughout the year.

Our results for *Inachus dorsettensis* and *Hyas coarctatus* are difficult to explain, but add to the growing number of studies showing genetic differentiation over distances which are comparatively small and which, therefore, should be readily spanned by pelagic larvae. A difference from most earlier work with similar results is that, because of the sequence of obligatory developmental stages (instars), the larval period is indisputably long and consequently the hypothesis of unexpected rapid settlement cannot be used to explain the data.

Acknowledgements

This work was supported by a grant from the Research Development Fund of the University of Liverpool (ref. 1629) to RGH.

References

Allen, J. A., 1967. The fauna of the Clyde Sea area Crustacea: Euphausiacea and Decapoda with an illustrated key to the British species. Scott. mar. biol. Ass., Millport, Scotland.

Anger, K., 1984. Development and growth in larval and juvenile *Hyas coarctatus* (Decapoda, Majidae) reared in the laboratory. Mar. Ecol. Progr. Ser. 19: 115–123.

Beaumont, W. I., 1900. The fauna and flora of Valencia Harbour on the west coast of Ireland. II. The benthos (dredging and shore collecting). VIII. Report on the results of dredging and shore collecting. Proc. r. ir. Acad. 5: 754–798.

Bruce, J. R., J. S. Colman & N. S. Jones, 1963. Marine fauna of the Isle of Man. University of Liverpool Press, Liverpool.

Bryant, A. D. & R. G. Hartnoll, 1995. Reproductive investment in two spider crabs with different breeding strategies. J. exp. mar. Biol. Ecol. 188: 261–275.

Bulnheim, H. P. & A. Scholl, 1986. Genetic differentiation between populations of *Talitrus saltator* and *Talorchestia deshayesii* (Crustacea: Amphipoda) from coastal areas of the north-western European continent. Mar. Biol. 92: 525–536.

Christiansen, M. E., 1969. Marine Invertebrates of Scandinavia, 2. Crustacea Decapoda Brachyura. Universitetsfarlaget, Oslo.

Conceição, M., 1995. Genetic investigations of talitrid amphipods. Unpublished Ph.D. thesis, University of Liverpool, Port Erin, Isle of Man.

Crane, J., 1940. Eastern Pacific expeditions of the New York Zoological Society. XVIII. On the post embryonic development of brachyuran crabs of the genus *Ocypode*. Zoologica 25: 65–82.

Flowerdew, M. W., 1984. Electrophoretic comparison of the antipodean cirripede *Elminius modestus* with immigrant European populations. J. mar. biol. Ass. U.K. 64: 625–635.

Grahame, J. & G. M. Branch, 1985. Reproductive patterns of marine invertebrates. Oceanogr. mar. Biol. ann. Rev. 23: 373–398.

Goudet, J., 1995. FSTAT(Version 1.2): a computer program to calculate F-Statistics. J. Hered. 86: 485–486.

Harris, H. & D. A. Hopkinson, 1976. Handbook of enzyme electrophoresis in human genetics. North Holland, Amsterdam.

Hartley, J. P. & J. D. D. Bishop, 1986. The macrobenthos of the Beatrice oilfield, Moray Firth, Scotland. Proc. r. Soc. Edinburgh 91B: 221–245.

Hartnoll, R. G., 1963. The biology of Manx spider crabs. Proc. zool. Soc. Lond. 14: 423–496.

Hartnoll, R. G., 1992. Megalopae and early postlarval stages of East African *Percnon* (Decapoda: Brachyura: Grapsidae). J. Zool. 228: 51–67.

Hartnoll, R. G. , A. D. Bryant & P. Gould, 1993. Size distribution in spider crab population spatial and temporal variation. J. crust. Biol. 13: 647–655.

Hedgecock, D., 1986. Is gene flow from pelagic larval dispersal important in the adaptation and evolution of marine invertebrates? Bull. mar. Sci. 39: 550–564.

Hedgecock, D., M. L. Tracey & K. Nelson, 1982. Genetics. In Abele, L. G. (ed.), Biology of Crustacea, 2. Embryology, Morphology and Genetics. Academic Press, New York: 283–403.

Ingle, R. W., 1983. Shallow water crabs: Keys and notes for the identification of the species. Cambridge University Press, Cambridge.

Jablonski, D. & R. A. Lutz, 1983. Larval ecology of marine benthic invertebrates: paleobiological implications. Biol. Rev. 58: 21–89.

Jones, N. S., 1951. The bottom fauna off the south of the Isle of Man. J. anim. Ecol. 20: 132–144.

Jones, N. S., 1956. The fauna and biomass of a muddy sand deposit off Port Erin. J. anim. Ecol. 25: 217–252.

Knowlton, N., 1993. Sibling species in the sea. Ann. Rev. Ecol. Syst. 24: 189–216.

Knowlton, N. & B. D. Keller, 1986. Larvae which fall far short of their potential: highly localized recruitment in an alpheid shrimp with extended larval development. Bull. mar. Sci. 39: 213–223.

Lebour, M. V., 1927. Studies of the Plymouth Brachyura. I. The rearing of crabs in captivity, with a description of the larval stages of *Inachus dorsettensis*, *Macropodia longirostris* and *Maia squinado*. J. mar. biol. Ass. U.K. 14: 795–280.

Lewis, R. I. & J. P. Thorpe, 1994a. Temporal stability of gene frequencies within genetically heterogeneous populations of the queen scallop *Aequipecten (Chlamys) opercularis*. Mar. Biol. 121: 117–126.

Lewis, R. I. & J. P. Thorpe, 1994b. Are queen scallops, *Aequipecten (Chlamys) opercularis* (L.), self recruiting? Can. tech. Rep. Fish. aquat. Sci. 214–221.

Macleod, J. A. A., J. P. Thorpe & N. A. Duggan, 1985. A biochemical genetic study of population structure in queen scallop (*Chlamys opercularis*) stocks in the Northern Irish Sea. Mar. Biol. 87: 77–82.

Maynard Smith, J., 1989. Evolutionary Genetics. Oxford University Press, Oxford.

Mulley, J. C. & B. D. H. Latter, 1981. Geographical differentiation of tropical Australian penaeid prawn populations. Aust. J. mar. Freshwat. Res. 32: 897–909.

Murphy, R. W., J. W. Sites, D. G. Buth & C. H. Haufler, 1990. Protein I: Isozyme Electrophoresis. In Hillis, D. M. & C. Moritz (eds), Molecular Systematics. Sinauer Associates, Inc. Publishers, Massachusetts: 45–126.

Nei, M., 1978. Estimation of average heterozygosity and genetic distance from a small number of individuals. Genetics 89: 583–590.

Nei, M., 1987. Molecular Evolutionary Genetics. Columbia University Press, New York.

Pannacciulli, F. G., J. D. D. Bishop & S. J. Hawkins, 1997. Genetic structure of populations of two species of *Chthamalus* (Crustacea: Cirripedia) in the NE Atlantic and Mediterranean. Mar. Biol. 128: 73–82.

Pingree, R. D. & D. K. Griffiths, 1978. Tidal fronts on the shelf seas around the British Isles. J. geophys. Res. 83: 4615–4622.

Poulik, M. D., 1957. Starch gel electrophoresis in a discontinuous system of buffers. Nature 180: 1477–1479.

Ramster, J. W. & H. W. Hill, 1969. Current systems in the northern Irish Sea. Nature 224: 59–61.

Redfield, J. A. & J. P. Salini, 1980. Techniques of starch gel electrophoresis of penaeid prawn enzymes (*Penaeus* spp. and *Metapenaeus* spp.). CSIRO Aust. Div. Fish. Oceanogr. Rep. 116: 1–19.

Rice, W. R., 1989. Analysing tables of statistical tests. Evolution 43: 223–225.

Siciliano, M. J. & C. R. Shaw, 1976. Separation and visualization of enzymes on gels. In Smith, I. (ed.), Chromatographic and Electrophoretic Techniques, 2. Heineman, London: 185–209.

Silberman, J. D. & P. J. Walsh, 1994. Population genetics of the spiny lobster *Panulirus argus*. Bull. mar. Sci. 54: 1084.

Simpson, J. H., 1971. Density stratification and microstructure in the western Irish Sea. Deep Sea Res. 18: 309–319.

Sinel, J., 1907. A contribution to our knowledge of the Crustacea of the Channel Islands. Rep. Trans. Guernsey Soc. nat. Sci. 5: 212–215.

Slatkin, M., 1985. Gene flow in natural populations. Ann. Rev. Ecol. Syst. 16: 393–430.

Slatkin, M., 1987. Gene flow and the geographic structure of natural populations. Science 236: 787–792.

Slatkin, M., 1993. Isolation by distance in equilibrium and nonequilibriium populations. Evolution 47: 264–279.

Strathmann, R. R., 1980. Why do larvae swim so long? Paleobiology 6: 373–376.

Strathmann, R. R., 1985. Feeding and nonfeeding larval development and life history evolution in marine invertebrates. Ann. Rev. Ecol. Syst. 16: 339–361.

Swofford, D. L. & R. K. Selander, 1981. BIOSYS-1 – A FORTRAN program for the comprehensive analysis of electrophoretic data in population genetics and systematics. J. Hered. 72: 281–283.

Thorpe, J. P., 1982. The molecular clock hypothesis: Biochemical evolution, genetic differentiation and systematics. Ann. Rev. Ecol. Syst. 13: 139–168.

Thorpe, J. P. & A. M. Solé-Cava, 1994. The use of enzyme electrophoresis in invertebrate systematics. Zool. Scr. 23: 3–18.

Thurston, M. H., 1970. The marine flora and fauna of the Isles of Scilly. Crustacea, Eucarida. J. nat. Hist. 4: 239–248.

Todd, C. D., 1985. Reproductive strategies of north temperate rocky shore invertebrates. In Moore, P. G. & R. Seed (eds), The Ecology of Rocky Coasts: Essays Presented to J. R. Lewis. Hodder & Stoughton, London: 203–219.

Todd, C. D., J. N. Havenhand & J. P. Thorpe, 1988. Genetic differentiation, pelagic larval transport and gene flow between populations of the intertidal marine mollusc *Adalaria proxima* (Alder and Hancock). Funct. Ecol. 2: 441–451.

Todd, C. D., W. J. Lambert & J. P. Thorpe, 1993. The genetic structure of intertidal population of two species of mollusc on the scottish west coast: some biogeographic considerations and an assessment of realized larval dispersal. In Baxter, J. M. & M. B. Usher (eds), The Islands of Scotland: A Living Marine Heritage. Scottish Natural Heritage, Edinburgh: 67–88.

Tracey, M. L., K. Nelson, D. Hedgecock, R. A. Schleser & M. L. Pressick, 1975. Biochemical genetics of lobsters: genetic variation and the structure of American lobster (*Homarus americanus*) populations. J. Fish. Res. Bd Can. 32: 2091–2101.

Weber, L. I., M. B. Conceição, M. L. Rufino & J. A. Levy, 1993. Population genetics of the shrimp *Artemesia longinaris* (Crustacea: Peneidae) on the southwest Atlantic coast. Comp. Biochem. Physiol. 106B: 1015–1020.

Weir, B. S. & C. C. Cockerham, 1984. Estimating F-statistics for the analysis of population structure. Evolution 38: 1358–1370.

Whittington, M. W., 1993. Scallop aquaculture in Manx waters: spat collection and the role of predation in seabed cultivation. Unpublished Ph.D. thesis, University of Liverpool.

Wright, S., 1978. Evolution and the genetics of populations, 4. Variability within and among natural populations. University of Chicago Press, Chicago.

Workman, P. L. & J. D. Niswander, 1970. Population studies on Southwestern Indian tribes. II. Local genetic differentiation in the Papago. Am. J. human Gen. 22: 24–49.

GPSR Compliance

The European Union's (EU) General Product Safety Regulation (GPSR) is a set of rules that requires consumer products to be safe and our obligations to ensure this.

If you have any concerns about our products, you can contact us on ProductSafety@springernature.com

In case Publisher is established outside the EU, the EU authorized representative is:

Springer Nature Customer Service Center GmbH
Europaplatz 3
69115 Heidelberg, Germany

Batch number: 10868240

Printed by Printforce, the Netherlands